AF705928

Catalysis in Chemistry and Biology

24th International Solvay Conference on Chemistry

Catalysis in Chemistry and Biology

Hotel Métropole, Belgium 18 – 22 October 2016

Scientific Editors

Kurt Wüthrich
The Scripps Research Institute, USA & ETH Zürich, Switzerland

R. H. Grubbs
California Institute of Technology, USA

International Solvay Institutes Editors

T. Visart de Bocarmé
Université libre de Bruxelles, Belgium

Anne De Wit
Université libre de Bruxelles, Belgium

World Scientific

NEW JERSEY • LONDON • SINGAPORE • BEIJING • SHANGHAI • HONG KONG • TAIPEI • CHENNAI

Published by

World Scientific Publishing Co. Pte. Ltd.
5 Toh Tuck Link, Singapore 596224
USA office: 27 Warren Street, Suite 401-402, Hackensack, NJ 07601
UK office: 57 Shelton Street, Covent Garden, London WC2H 9HE

Library of Congress Cataloging-in-Publication Data
Names: International Solvay Conference on Chemistry (24th : 2016 : Brussels, Belgium) |
Wüthrich, Kurt, editor. | Grubbs, Robert H., editor. | Visart de Bocarmé, Thierry, editor. | De Wit, Anne, editor.
Title: 24th International Solvay Conference on Chemistry : catalysis in chemistry and biology, Hotel Métropole, Brussels, 18–22 October 2016 / scientific editors, K. Wüthrich (The Scripps Research Institute, USA & ETH Zurich, Switzerland), R.H. Grubbs (California Institute of Technology, USA) ; International Solvay Institutes editors, T. Visart de Bocarmé (Université libre de Bruxelles, Belgium), A. De Wit (Université libre de Bruxelles, Belgium).
Description: New Jersey : World Scientific, 2018. | Includes bibliographical references and index.
Identifiers: LCCN 2017060718 | ISBN 9789813237162 (hardcover : alk. paper)
Subjects: LCSH: Chemical reactions--Congresses. | Catalysis--Congresses. | Biocatalysis--Congresses.
Classification: LCC QD505 .I5758 2016 | DDC 541/.395--dc23
LC record available at https://lccn.loc.gov/2017060718

British Library Cataloguing-in-Publication Data
A catalogue record for this book is available from the British Library.

Copyright © 2018 by World Scientific Publishing Co. Pte. Ltd.

All rights reserved. This book, or parts thereof, may not be reproduced in any form or by any means, electronic or mechanical, including photocopying, recording or any information storage and retrieval system now known or to be invented, without written permission from the publisher.

For photocopying of material in this volume, please pay a copying fee through the Copyright Clearance Center, Inc., 222 Rosewood Drive, Danvers, MA 01923, USA. In this case permission to photocopy is not required from the publisher.

For any available supplementary material, please visit
http://www.worldscientific.com/worldscibooks/10.1142/10907#t=suppl

Contents

The International Solvay Institutes
(October 2016)

Board of Directors

Members

Mr. Jean-Marie Solvay
President

Professor Paul Geerlings
Vice-President & Treasurer - Professor VUB

Professor Gino Baron
Secretary - Emeritus Professor VUB

Mr. Nicolas Boel
Chairman of the Board of Directors of Solvay Group

Mr. Eric Boyer de la Giroday
Chairman of the Board of Directors ING Belgium

Mr. Philippe Busquin
Minister of State

Professor Eric De Keuleneer
Chairman of the Board of Directors of the ULB

Baron Daniel Janssen
Former Chairman of the Board of Directors of Solvay S.A.

Mr. Eddy Van Gelder
Chairman of the Board of Directors of the VUB

Honorary Members

Professor Franz Bingen
Emeritus Professor VUB
Former Vice President and Treasurer of the Solvay Institutes

Baron André Jaumotte
Honorary Director
Honorary Rector and Honorary President of the ULB

Professor Jean-Louis Vanherweghem
Former Chairman of the Board of Directors of the ULB

Professor Irina Veretennicoff
Emeritus Professor VUB

Guest Members

Professor Anne De Wit
Professor ULB and Scientific Secretary of the International Committee for Chemistry

Mr. Freddy Dumortier
Secretary of the Royal Flemish Academy for Science and the Arts of Belgium

Professor Hervé Hasquin
Secretary of the Royal Academy for Science and the Arts of Belgium

Professor Franklin Lambert
Emeritus Professor VUB

Professor Alexander Sevrin
Professor VUB and Deputy Director for Physics and Scientific Secretary of the International Committee for Physics

Marina Solvay

Professor Lode Wyns
Deputy Director for Chemistry

Director

Professor Marc Henneaux
Professor ULB

Solvay Scientific Committee for Chemistry

Professor Kurt Wüthrich (Chair)
Scripps Research Institute, La Jolla, USA and ETH Zurich, Switzerland

Professor Gerhard Ertl
Fritz-Haber-Institut der Max-Planck-Gesellschaft Berlin, Germany

Professor Graham Fleming
University of Berkeley, USA

Professor Robert H. Grubbs
California Institute of Technology, Pasadena, USA

Professor Roger Kornberg
Stanford University, USA

Professor Harold W. Kroto
University of Sussex, Brighton, UK

Professor Henk N. W. Lekkerkerker
Utrecht Universiteit, The Netherlands

Professor K. C. Nicolaou
The Scripps Research Institute, USA

Professor JoAnne Stubbe
Massachusetts Institute of Technology, Cambridge, USA

Professor George M. Whitesides
Harvard University, Cambridge, USA

Professor Ahmed Zewail
California Institute of Technology, Pasadena, USA

Professor Anne De Wit
Université libre de Bruxelles, Belgium, Scientific Secretary

Acknowledgements

The organization of the 24th Solvay conference has been made possible thanks to the generous support of the Solvay Family, the Solvay Group, the Université libre de Bruxelles, the Vrije Universiteit Brussel, the Belgian National Lottery, the Foundation David and Alice Van Buuren, the Brussels-Capital Region, the Fédération Wallonie-Bruxelles, the Vlaamse Regering, the City of Brussels and the Hôtel Métropole.

Participants

Martin	Albrecht	University of Bern, Switzerland
Xinhe	Bao	Dalian Institute for Chemical Physics, China
Giuseppe	Bellussi	Eni SpA, Italy
Philip C.	Bevilacqua	Penn State University, USA
Steven G.	Boxer	Stanford University, USA
Ronald R.	Breaker	Yale University, USA
Stephen L.	Buchwald	Massachusetts Institute of Technology, USA
Erick M.	Carreira	ETH Zürich, Switzerland
John L.	Casci	Johnson Matthey Technology Centre, Billingham, UK
John	Christodoulou	Birkbeck College, University of London, UK
Christophe	Copéret	ETH Zürich, Switzerland
Avelino	Corma	University of Valencia, Spain
Mark E.	Davis	Caltech, USA
Brian	Dyer	Emory University, USA
Gerhard	Ertl	Fritz-Haber-Institut der Max-Planck-Gesellschaft, Berlin, Germany
Ben L.	Feringa	University of Groningen, The Netherlands
Graham	Fleming	University of California, Berkeley, USA
Hans-Joachim	Freund	Fritz-Haber-Institut der Max-Planck-Gesellschaft, Berlin, Germany
Karen I.	Goldberg	University of Washington, USA
Robert H.	Grubbs	Caltech, USA
Martina	Havenith-Newen	Ruhr University, Bochum, Germany
Stig	Helveg	Haldor Topsøe, Denmark
Daniel	Herschlag	Stanford University, USA
Donald	Hilvert	ETH Zürich, Switzerland
Graham J.	Hutchings	Cardiff University, UK
Enrique	Iglesia	UC Berkeley, USA
Christopher W.	Jones	Georgia Institute of Technology, USA
Karl Anker	Jørgensen	Aarhus University, Denmark
Judith P.	Klinman	University of California, Berkeley, USA
Marc T. M.	Koper	Leiden University, The Netherlands
Henk N. W.	Lekkerkerker	Utrecht University, The Netherlands
David M.	Lilley	University of Dundee, UK
Benjamin	List	MPI für Kohlenforschung, Mülheim an der Ruhr, Germany

Rudolph A.	Marcus	Caltech, USA
Ryuhei	Nakamura	RIKEN, Japan
Frank	Neese	MPI for Chemical Energy Conversion, Germany
Jens K.	Nørskov	Stanford University, USA
Kyoko	Nozaki	University of Tokyo, Japan
Marina V.	Rodnina	MPI für Biophysikalische Chemie, Göttingen, Germany
Melanie	Sanford	University of Michigan, USA
Joachim	Sauer	Humboldt University of Berlin, Germany
Gabor A.	Somorjai	University of California, Berkeley, USA
JoAnne	Stubbe	Massachusetts Institute of Technology, USA
Takashi	Tatsumi	Tokyo Institute of Technology, Japan
Rutger A.	van Santen	Eindhoven University of Technology, The Netherlands
Bert M.	Weckhuysen	Utrecht University, The Netherlands
Kurt	Wüthrich	ETH Zürich, Switzerland - The Scripps Research Institute, La Jolla, CA, USA
Darrin M.	York	Rutgers University, USA

Auditors

Kristin	Bartik	Université libre de Bruxelles, Belgium
Gilles	Bruylants	Université libre de Bruxelles, Belgium
Claudine	Buess-Herman	Université libre de Bruxelles, Belgium
Yannick	De Decker	Université libre de Bruxelles, Belgium
Frank	de Proft	Vrije Universiteit Brussel, Belgium
Dirk	de Vos	University of Leuven, Belgium
Anne	De Wit	Université libre de Bruxelles and Solvay Institutes, Belgium
Damien	Debecker	Université catholique de Louvain, Belgium
Joeri	Denayer	Vrije Universiteit Brussel, Belgium
Christophe	Detavernier	Universiteit Gent, Belgium
Gwilherm	Evano	Université libre de Bruxelles, Belgium
Eric	Gaigneaux	Université catholique de Louvain, Belgium
Abel	Garcia-Pino	Université libre de Bruxelles, Belgium
Pierre	Gaspard	Université libre de Bruxelles, Belgium
Paul	Geerlings	Vrije Universiteit Brussel, Belgium
Cédric	Gommes	Université de Liège, Belgium
Cédric	Govaerts	Université libre de Bruxelles, Belgium
Benoit	Heinrichs	Université de Liège, Belgium
Sophie	Hermans	Université catholique de Louvain, Belgium
Johan	Hofkens	University of Leuven, Belgium
Giulia	Licini	Università degli Studi di Padova, Italy
Remy	Loris	Vrije Universiteit Brussel, Belgium
Bert	Maes	University of Antwerp, Belgium
Johan	Martens	University of Leuven, Belgium
Olivier	Riant	Université catholique de Louvain, Belgium
Patrice	Soumillon	Université catholique de Louvain, Belgium
Bao-Lian	Su	Université of Namur, Belgium
Pascal	Van der Voort	Universiteit Gent, Belgium
Véronique	Van Speybroeck	Universiteit Gent, Belgium
Wim	Versées	Vrije Universiteit Brussel, Belgium
Thierry	Visart de Bocarmé	Université libre de Bruxelles, Belgium
Lode	Wyns	Vrije Universiteit Brussel, Belgium

Opening Address by Professor Marc Henneaux
Director of the International Solvay Institutes

Your Majesty,
Mrs. Solvay,
Members of the Solvay Family,
Ladies and Gentlemen,
Dear Colleagues, Dear Friends,

It is my great honour and pleasure to open the 24th Solvay Conference on Chemistry. Its theme is "Catalysis in Chemistry and Biology".

The International Solvay Institutes have the great privilege to benefit from the benevolent support of the Royal Family. This distinctive feature, which makes the Solvay Conferences unique, already held true for the very first Solvay Conference that took place in 1911. We are fortunate that this interest in the Solvay Conferences and fundamental science has been kept intact over the years.

Sire,

It is with a respectful gratitude that we acknowledge the continuation of this centenary tradition. Your Presence with us this morning is an enormous encouragement for basic scientific research.

Catalysis, the theme of the Conference that starts today, is fundamental to chemistry. There is probably no area of chemistry where catalysis is not essential. Traditionally, at the start of a Solvay Conference, I try to connect the Conference to those previous Solvay Conferences that had a related theme. In the current case, however, because catalysis pervades all areas of chemistry, I will simply give the list of all the previous Solvay Conferences on chemistry, where in one way or another, catalysis was indeed present.

Here it is*:

1. (1922) "Cinq Questions d'Actualité"
2. (1925) "Structure et Activité Chimique"

*The third meeting of the "Association Internationale des Sociétés Chimiques (AICS)" took place in 1913 in Brussels and was funded by Ernest Solvay. This meeting played an important role in Solvay's initiative to launch the Solvay Conferences on Chemistry.

3. (1928) "Questions d'Actualité"
4. (1931) "Constitution et Configuration des Molécules Organiques"
5. (1934) "L'Oxygène, ses réactions chimiques et biologiques"
6. (1937) "Les Vitamines et les Hormones"
7. (1947) "Les Isotopes"
8. (1950) "Le Mécanisme de l'Oxydation"
9. (1953) "Les Protéines"
10. (1956) "Quelques Problèmes de Chimie Minérale"
11. (1959) "Les Nucléoprotéines"
12. (1962) "Transfert d'Energie dans les Gaz"
13. (1965) "Reactivity of the Photoexcited Organic Molecule"
14. (1969) "Phase Transitions"
15. (1970) "Electrostatic Interactions and Structure of Water"
16. (1976) "Molecular Movements and Chemical Reactivity as Conditioned by Membranes, Enzymes and other Molecules"
17. (1980) "Aspects of Chemical Evolution"
18. (1983) "Design and Synthesis of Organic Molecules Based on Molecular Recognition"
19. (1987) "Surface Science"
20. (1995) "Chemical Reactions and their Control on the Femtosecond Time Scale"
21. (2007) "From Noncovalent Assemblies to Molecular Machines"
22. (2010) "Quantum Effects in Chemistry and Biology"
23. (2013) "New Chemistry and New Opportunities from the Expanding Protein Universe"

The scientific organization of the Solvay conferences is done by an international committee, the Solvay International Scientific Committee for Chemistry. I would like to express the deepest gratitude of the International Solvay Institutes to the Committee, and in particular to the committee chair, Kurt Wüthrich, as well as to Robert Grubbs, who accepted to co-chair the 24th Solvay Conference. Together with the session chairs, they put up a splendid program. I know that this meant a lot of work. Warmest thanks to all of them!

Thank you very much for your attention.

Marc Henneaux
Director of the International Solvay Institutes

Preface by Professor Kurt Wüthrich

Chair of the 24th Solvay Conference on Chemistry

The 23rd Solvay Conference, "New Chemistry from the Protein Universe", coincided with the celebration of the centennial anniversary of the first Chemistry Council in 1913. The 2013 Conference assembled scientists representing chemistry, computational biology, structural biology, biochemistry and cell biology, all of whom follow integrative multidisciplinary routes to investigate correlations between structures of biological macromolecules and their impact on the chemistry of physiological processes. One of the Sessions was devoted to catalysis of metabolic processes by protein enzymes. With the theme "Catalysis in Chemistry and Biology", the 24th Chemistry Conference now expands on this Session by addressing a broad range of areas concerned with catalysis. An illustrious group of scientists using a wide range of different experimental and computational techniques joined this venture, and the combination of different aspects of catalysis turned out to result in a most exciting Conference.

The format of the 24th Conference followed largely the format of the 2013 Conference. All participants contributed short articles presenting their current views on their Sessions themes. These papers were accessible to all the participants before the start of the Conference. At the Conference, each Session was opened with a general introduction by the Chairperson and short presentations by a group of panelists. This was followed first by a discussion among the panelists and then by a discussion involving all invited participants. The discussions were recorded and transcribed by "Auditors" recruited from Belgium Universities, who were in attendance at their Sessions. In the version presented in this volume, these transcripts have been edited by the discussion contributors for correct scientific content.

The theme of the Conference was presented in six Sessions, and the Proceedings are organized accordingly. The Session 1 on "Homogeneous Catalysis", chaired by Prof. Robert Grubbs, was devoted to reports on basic research on catalysis in homogenous solutions and applications thereof. The Session 2 on "Heterogeneous Catalysis and Characterization of Catalyst Surfaces", chaired by Prof. Gerhard Ertl, included extensive reference to industrial applications of catalysis on solid supports, and discussions on the experimental techniques used in this field presented special highlights. The Session 3 on "Catalysis by Microporous Materials", chaired by Prof. Mark E. Davis, was devoted to a detailed characterization of this particular class of solid support catalysts, with special emphasis on model analysis of the processes catalyzed by these materials. Session 4 on "Catalysis under Extreme

Conditions: Studies at High Pressure and High Temperatures – Relations with Processes in Nature", chaired by Prof. Henk N. W. Lekkerkerker, broadened the scope of the two preceding sessions with the presentation of exciting illustrations. The Sessions 5 and 6 on "Catalysis by Protein Enzymes", chaired by Prof. JoAnne Stubbe, and "Catalysis by Ribozymes in Molecular Machines", chaired by Prof. David Lilley, presented an exciting contrast to the initial four Sessions. The combination of the six Sessions provided an impressive overview of the field, and resulted in innovative insights into relations between catalysis in chemical processes and in biological systems. The meeting thus presented a unique opportunity for animated exchanges on anticipated developments during the coming years and in the more distant future.

It was a special privilege for me to chair this Solvay Conference on Chemistry, and I want to acknowledge the co-Chair of the Conference, Prof. Robert Grubbs, and the chairpersons of the individual Sessions for their support in generating the scientific program. To do proper justice to the widely different specialties of "catalysis" covered by the meeting, the panelists in the individual Sessions were selected by the respective chairpersons, so that I received a valuable education and could limit my involvement to solving organizational issues arising with the nominations received.

A special highlight of the meeting was the opening ceremony with the participation of His Majesty King Philippe of Belgium and the Solvay family represented by Mrs. Marie-Claude Solvay de la Hulpe, Mrs. Marina Solvay and Jean-Marie Solvay, the President of the "*Instituts Internationaux de Physique et de Chimie Solvay*". With short presentations by the Conference Chairs, the illustrious audience was provided with a survey of the science to be discussed at this meeting. We very much appreciated that our guests joined us also for a coffee break and entertained interested discussions with many of the Conference participants.

We greatly appreciate the generous support by the Solvay Institutes represented by Prof. Marc Henneaux. Special thanks go to Prof. Anne De Wit and Prof. Thierry Visart de Bocarmé, who supervised the recording and transcription of the discussions, and to the scientists who served as "Auditors" during the Conference and contributed to the discussion transcriptions, which are now a major part of these Proceedings. Dominique Bogaerts and Isabelle Van Geet from the Solvay Institutes made our lives easy and enjoyable with their commitment and dedicated assistance; we owe them our heartfelt thanks.

Kurt Wüthrich
Chair of the Conference

Session 1
Homogeneous Catalysis

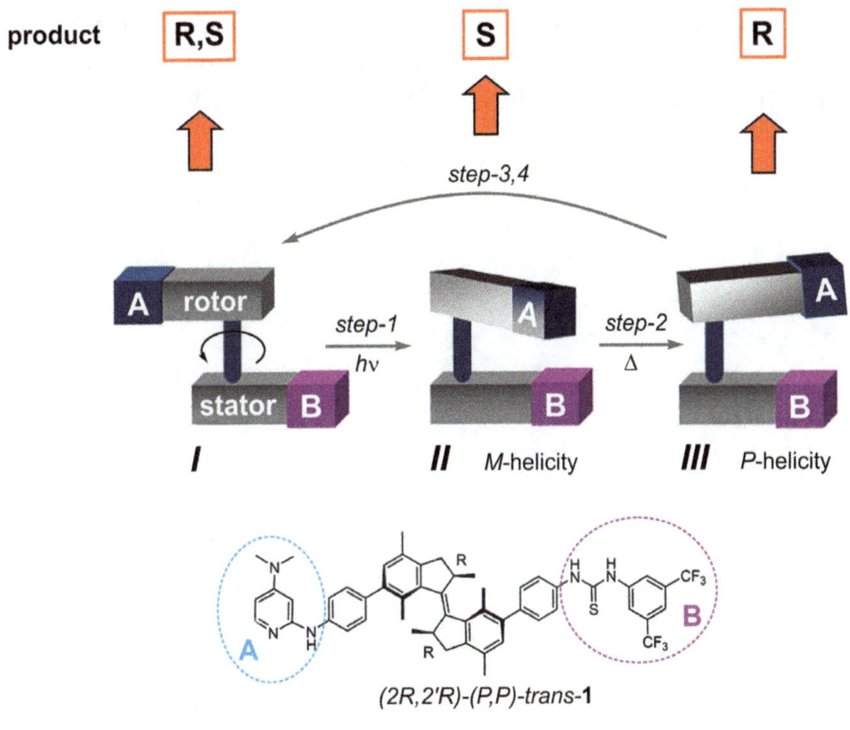

(2R,2'R)-(P,P)-trans-**1**

Molecular motor based multistage chiral organo-catalyst. The catalyst can be either used for the formation of a racemic mixture or preferentially for the formation of one enantiomer of a chiral product. Image by: Ben L. Feringa, University of Groningen.

HOMOGENEOUS CATALYSIS: ORGANOMETALLIC CATALYSIS AND ORGANOCATALYSIS

ROBERT H. GRUBBS

Division of Chemistry and Chemical Engineering, California Institute of Technology, Pasadena, CA 91125, USA

Although soluble catalysts have been used in organic chemistry for many years, until the 1960s they involved simple acids, bases, or metal salts. In most cases, the metals were ignored in any mechanistic considerations. For example, copper salts can catalyze the cyclopropanation of olefins using diazo compounds; however, the role of copper was not considered in early mechanisms. Simple soluble nickel and cobalt precursor complexes were used in some industrial processes but were not extensively studied. Heterogeneous catalysis especially for hydrogenation has a much longer history of use in organic synthesis.

The introduction of Wilkinson's hydrogenation [1] and hydroformylation [2] catalysts in the 1960s sparked an interest in understanding the reaction mechanisms of organometallic complexes. Early on, it was recognized that catalytic cycles could be broken down into fundamental reaction steps for individual study. With soluble systems, the methods developed in physical organic chemistry could be applied to analyze these elementary steps. Kinetics, the isolation of intermediates, and the study of model systems provide a firm basis for the establishment of catalytic cycles, which can lead to the discovery of new reactions and the improvement of known catalytic systems [3].

As in many new areas of chemistry, serendipity played a major role in the initial discovery of new catalysts and transformations that could be further improved through mechanistic understanding. The staggering effect that ligand selection can have on reaction rates and selectivities was quickly realized. As the limitations of simple ligands, such as halides and phosphines, were recognized, an array of new ligand classes were designed that resulted in homogeneous organometallic catalysis becoming a key component in the chemical enterprise. It is now difficult to find a modern academic or pharmaceutical synthesis that does not feature one reaction that involves a homogeneous catalyst. One of the appealing aspects of homogeneous catalysis has been the translation of fundamentals studies and discoveries into practical commercial applications. The following discussion provides a short discussion of the history of homogeneous catalysis as a basis for the presentations of the current state of the field.

Organometallic catalysis

Olefin hydrogenation was one of the first widely used heterogeneous reactions in organic chemistry and has played the same role in homogeneous systems. Use of chiral phosphine ligands allowed for the first asymmetric hydrogenation in the work of Knowles and the Monsanto group [4]. Halpern and his group developed a detailed mechanistic understanding of this reaction and provided the model for the study of subsequent asymmetric transformations [5]. These studies, which drew attention to the fact that not every complex that is seen is responsible for catalysis, have had a lasting effect on the way organometallic chemists analyze the kinetic competence of intermediates.

Noyori and his group expanded asymmetric hydrogenation to carbonyls and other functionalized systems and opened a new age of organic synthesis [6]. To complement these olefin reduction methods, Sharpless developed an asymmetric epoxidation system that provided new routes into highly functionalized asymmetric molecules [7]. Since the development of these early systems, the area of asymmetric homogeneous catalysis has exploded.

In an academic setting, considerable progress had been made in applying organometallic catalysis to more efficiently synthesize complex organic molecules. At the same time, new industrial processes were being developed that use homogeneous systems to prepare simple organic feedstocks on a staggering scale. The most widely used is the oxo process, also known as hydroformylation, in which an olefin, carbon monoxide, and hydrogen react to form aldehyde products [8]. This process has wide-ranging applications, from providing aldehyde intermediates in the synthesis of plasticizers to being a key component in the Shell higher olefin process (SHOP) for the synthesis of fatty alcohols from ethylene [9]. Early studies, aided by the diagnostic IR stretching frequencies of metal carbonyl complexes, provided a mechanistic basis for the selectivity of these reactions, and phosphine ligands were found to significantly improve the regioselectivity and rates of hydroformylation.

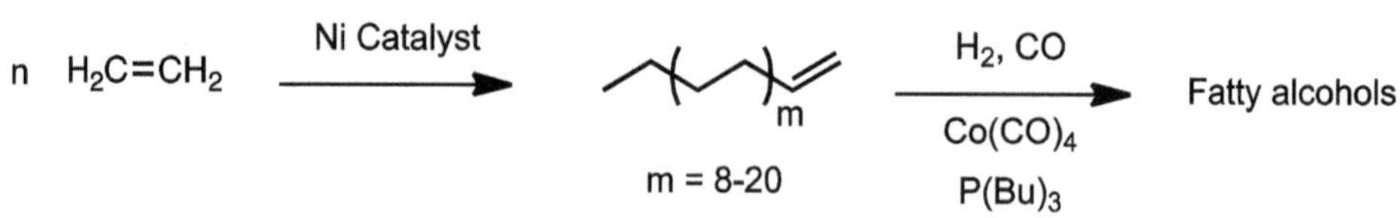

Another revolutionary commercial homogeneous reaction was the Monsanto acetic acid process, which efficiently converts methanol and carbon dioxide into acetic acid. Not only did this process establish that homogeneous reactions involving expensive metals could be commercially viable, it also was one of the first in which conclusions drawn from detailed mechanistic studies were used to improve the reaction efficiency. A similar process, which utilizes an iridium catalyst promoted by ruthenium, has process advantages over the original Monsanto system, and is now being practiced [10].

$$CH_3OH \xrightarrow[CO]{MI_2,\ HI} CH_3CO_2H$$

Monsanto process: M = Rh, 60 atm and 150°C

Cativa process: M = Ir (Ru) similar conditions to Monsanto process but uses less H_2O

Early ethylene and propylene polymerization catalysts developed by Ziegler and Natta were heterogeneous, and high efficiencies and selectivities were achieved through Edisonian discoveries (i.e. a trial-and-error approach) [11]. Introduction of homogeneous systems in the 1990s allowed for the rational design of new ligands to support catalysts that produced polymers with a wide array of useful properties [12]. In addition to the old high-density polyethylene that was prepared by Ziegler-type systems, there are now polymers accessible that produce exceeding strong films and provide an array of new consumer products [13]. The ligands and metal center govern the stereochemistry and allow the introduction of co-monomers to control the branching in the polymer backbone. Over the past several years, high throughput screening has been introduced for the discovery of catalysts that introduce very subtle changes in structures that result in greatly enhanced properties and processing ability of the resulting polymers [14]. Polyethylene is not just one material; through the precise design of ligands, polyethylenes can be produced that show an amazing array of structures and properties.

Carbon–carbon bond construction is the basis for all of organic chemistry and represents a research area that has exploded in popularity and utility in the past several decades. Heck [15] discovered a palladium-catalyzed method for coupling sp^2-hybridized centers that has been the starting point for a large range of other C–C coupling processes. There were a number of researchers involved in the development of methods for coupling aromatic and vinyl groups using palladium catalysts, including the groups of Negishi [16] and Suzuki [17]. These reactions, which couple aryl anion equivalents to a variety of other unsaturated fragments, have revolutionized how complex molecules are constructed. Recent breakthroughs that will greatly expand the synthetic possibilities involve the coupling of alkyl halides that contain sp^3 fragments.

Most pharmaceuticals contain carbon–nitrogen bonds. While traditional organic approaches to construct these bonds often require several synthetic steps, organometallic methods developed in the laboratories of Buchwald [18] and Hartwig [19] provide an array of catalytic methods for this critical bond construction. These reactions are playing a major role in complex molecule synthesis and have opened new structural motifs that may prove to be pharmaceutically active.

Another coupling reaction that was initially a heterogeneous polymerization reaction that was later optimized using well-defined homogeneous species is olefin metathesis [20]. Initial systems using ill-defined Ziegler-type catalysts were only useful in the catalysis of hydrocarbon reactions. New homogeneous catalysts based

Heck Reaction

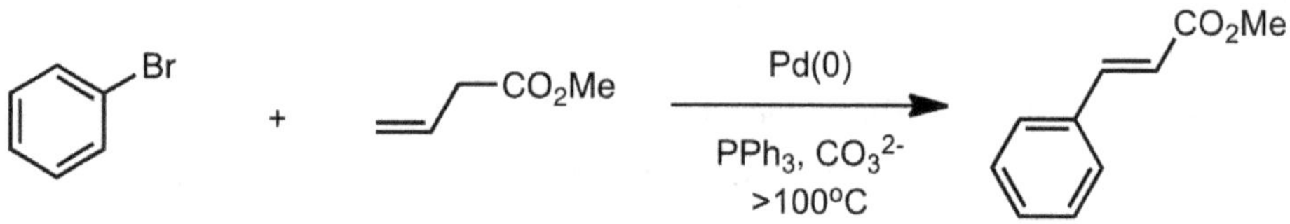

Suzuki Reaction

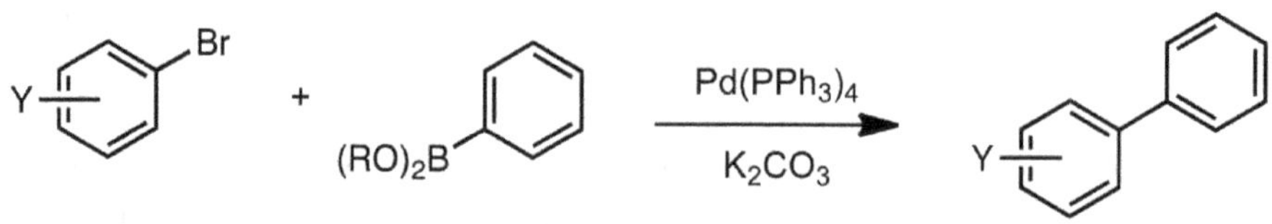

Negishi Coupling

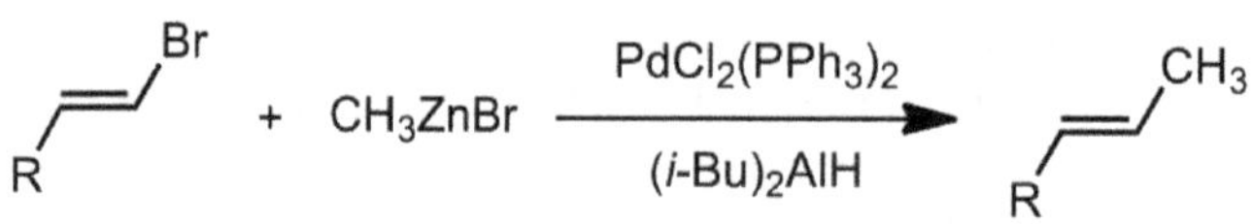

Buchwald-Hartwig Amination

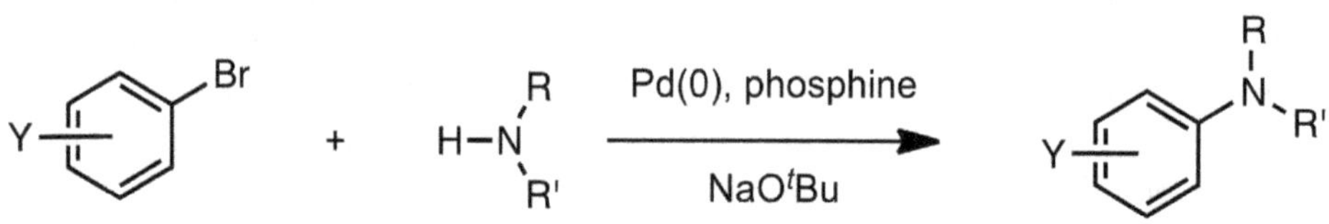

on Mo [21] and Ru [22] tolerated many organic functional groups and are now key to the synthesis of complex molecules. In the early 1990s, olefin metathesis became a standard organic synthetic reaction that has seen major use in the construction of ring structures. The first pharmaceutical synthesis using metathesis was part of the original treatment for Hepatitis C [23]. These catalysts are also being utilized for the synthesis of functional polymeric materials. For example, the ruthenium systems are now the basis for a family of new polymeric materials that are finding use in the production of oil and gas [24].

At the present time, major efforts are being focused on the direct activation of C–H bonds to introduce new functionality. Significant advances are being made in this area and are starting to impact the synthesis of complex molecules [25]. This topic will be a major theme of the discussions of the future of homogeneous catalysis.

Organocatalysis

Organocatalysis utilizes small organic molecules to catalyze a desired organic conversion. Although organic acid and base catalysts have been used for many years, the recent development of organocatalysts designed to carry out asymmetric reactions has revolutionized the field. In 2000, two papers appeared that opened the field of asymmetric synthesis utilizing chiral organic molecules as catalysts. The Scripps group of List, Lerner, and Barbas [26] introduced proline as a catalyst for asymmetric aldol reactions. The MacMillan group reported chiral amine catalysts that condensed with aldehydes to activate them for asymmetric Diels–Alder reactions and coined the term "organocatalysis" [27]. In the last 1.5 decades, many new organocatalyzed transformations have been introduced and these reactions have become a critical tool in the synthesis of complex molecules. Recently, process development has been achieved for several organocatalytic systems, demonstrating the utility of these methods for industrial-scale processes.

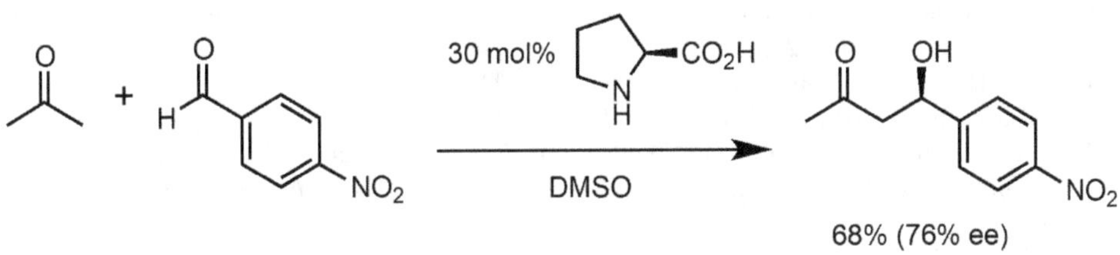

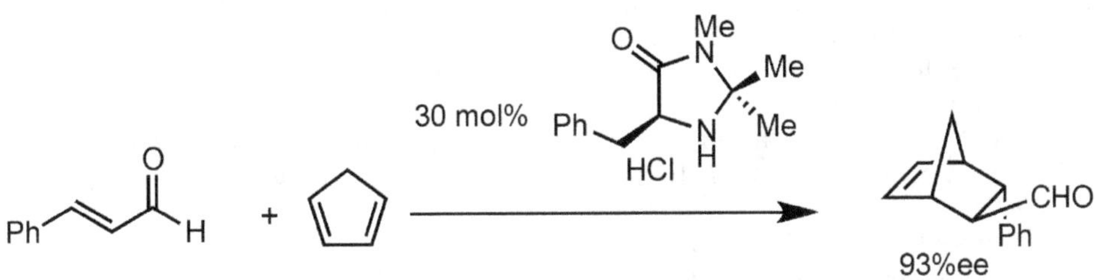

In the last 50 years, homogeneous catalysis has developed from a small area of inorganic chemistry to a major field in the synthesis of complex molecules and the production of commercially important molecules. Mechanistic studies have aided in the development of new reactions and have provided insight into the operation of heterogeneous systems as well. The following discussions will provide a basis to evaluate where we go from here.

References

[1] (a) F. H. Jardine, G. Wilkinson, *J. Chem. Soc. C*, 270 (1967); (b) S. Montelatici, A. van der Ent, J. A. Osborn, G. Wilkinson, *J. Chem. Soc. A*, 1054 (1968).

[2] C. K. Brown, G. Wilkinson, *J. Chem. Soc. A*, 2753 (1970).

[3] J. P. Collman, *Acc. Chem. Res.* **1**, 136 (1968).

[4] W. S. Knowles, *Angew. Chem. Int. Ed.* **41**, 1998 (2002).
[5] J. Halpern, *Science* **217**, 401 (1982).
[6] A. Miyashita, A. Yasuda, H. Takaya, K. Toriumi, T. Ito *et al.*, *J. Am. Chem. Soc.* **102**, 7932 (1980).
[7] T. Katsuki, K. B. Sharpless, *J. Am. Chem. Soc.* **102**, 5974 (1980).
[8] A. Stefani, G. Consiglio, C. Botteghi, P. Pino, *J. Am. Chem. Soc.* **99**, 1058 (1977).
[9] (a) E. F. Lutz, *J. Chem. Educ.* **63**, 202 (1986); (b) B. Cornils, W. A. Herrmann, C.-H. Wong, H. W. Zanthoff (eds.), *Catalysis from A to Z: A Concise Encyclopedia*, 2408 (Wiley-VCH, 2013).
[10] J. H. Jones, *Plat. Met. Rev.* **44**, 94 (2000).
[11] (a) P. Cossee, *Tetrahedron Lett.* **17**, 12 (1960); (b) R. Hoff, R. T. Mathers (eds.), *Handbook of Transition Metal Polymerization Catalysts* (John Wiley & Sons, Inc., 2010); (c) G. Natta, F. Danusso, (eds.), *Stereoregular Polymers and Stereospecific Polymerizations* (Pergamon Press, 1967).
[12] (a) J. A. Ewen, *J. Am. Chem. Soc.* **106**, 6355 (1984); (b) S. Costeux, P. Wood-Adams, D. Beigzadeh, *Macromolecules*, 2514 (2002).
[13] K. Kuwabara, H. Kaji, F. Horii, D. C. Bassett, R. H. Olley, *Macromolecules* **30**, 7516 (1997).
[14] T. R. Boussie, G. M. Diamond, C. Goh, K. A. Hall, A. M. LaPointe *et al.*, *Angew, Chem. Int. Ed.* **45**, 3278 (2006).
[15] R. F. Heck, J. P. Nolley, *J. Org. Chem.* **37**, 2320 (1972).
[16] S. Baba, E. Negishi, *J. Am. Chem. Soc.* **98**, 6729 (1976).
[17] N. Miyamura, K. Yamada, A. Suzuki, *Tetrahedron Lett.* **20**, 3437 (1979).
[18] S. Guram, R. A. Rennels, S. L. Buchwald, *Angew. Chem. Int. Ed. Engl.* **34**, 1348 (1995).
[19] J. Louie, J. F. Hartwig, *Tetrahedron Lett.* **21**, 3609 (1995).
[20] R. H. Grubbs, A. G. Wenzel, D. J. O'Leary, E. Khosravi (eds.), *Handbook of Metathesis*, 2nd Edition (Wiley-VHC, 2015).
[21] R. R. Schrock, J. H. Freudenberger, M. L. Listemann, L. G. McCullough, *J. Mol. Catal.* **28**, 1 (1985).
[22] S. T. Nguyen, R. H. Grubbs, J. W. Ziller, *J. Am. Chem. Soc.* **115**, 9858 (1993).
[23] C. Shu, X. Z. Zeng, M. H. Hao, X. D. Wei, N. K. Yee *et al.*, *Org. Lett.* **10**, 1303 (2008).
[24] Business Wire, April 11, 2016.
[25] J. J. Topczewski, M. S. Sanford, *Chem. Sci.* **6**, 70 (2015).
[26] B. List, R. A. Lerner, C. F. Barbas, *J. Am. Chem. Soc.* **122**, 2395 (2000).
[27] K. A. Ahrendt, C. J. Borths, D. W. C. MacMillan, *J. Am. Chem. Soc.*, **122**, 4243 (2000).

COPPER-CATALYZED HYDROFUNCTIONALIZATION REACTIONS

STEPHEN L. BUCHWALD

Department of Chemistry, Massachussets Institute of Technology, Cambridge, MA 02139, USA

Present view of homogeneous catalysis

The past three decades has seen an enormous increase in the use of homogeneous catalysis in the preparation of fine chemicals. In order for this to continue, workers in the field need to be increasingly concerned about the ability of the techniques that they are developing to be applicable to complex molecules that contain numerous functional groups. As always, a key challenge is achieving the right balance of reactivity and selectivity.

My recent research contributions to the field of homogeneous catalysis

During the last four years my research group has been increasingly interested in the development of new chemistry that converts readily accessible alkenes into more highly functionalized products. We have concentrated much of this effort on the use of chiral Copper Hydride Complexes (L*CuH). As an example, we have reported the selective asymmetric hydroamination of a variety of olefin (and alkyne) substrates [1]. These reactions proceed under mild conditions and with high levels of efficiency. Examples of these processes are shown in Fig. 1

Outlook to future developments of research on homogeneous catalysis

In the future, there will be an emphasis on the ability to functionalize increasingly complex molecules. This includes being able to modify and enhance the properties of proteins, antibodies and other biopolymers.

Acknowledgments

We thank the U.S. National Institutes of Health (GM046059 and GM058160) for support of our research program in this area. We are grateful to Dr. Christine Nguyen for help with the preparation of this manuscript.

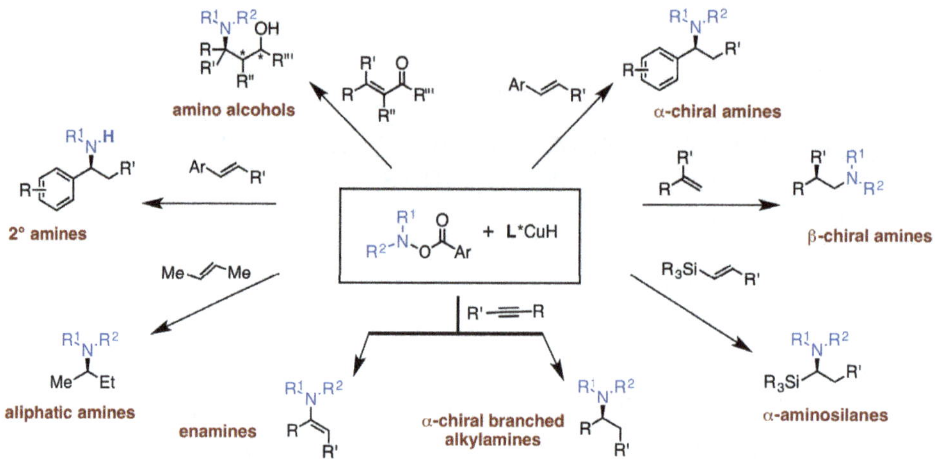

Fig. 1. The use of chiral copper hydride catalysts in carbon–nitrogen bond-forming processes.

References

[1] M. T. Pirnot, Y.-M. Wang, S. L. Buchwald, *Angew. Chem. Int. Ed.* **55**, 48 (2016).

CONTROLLING SELECTIVITY AND REACTIVITY IN CATALYTIC C–H FUNCTIONALIZATION REACTIONS

MELANIE SANFORD

Department of Chemistry, University of Michigan, Ann Arbor, MI 48104, USA

Present view of homogenous catalysis

Homogeneous catalysis is increasingly widely used in both academia and industry for the assembly and diversification of organic molecules. In particular, the past 20 years has seen tremendous research effort in the development of homogeneous catalysts for the selective functionalization of carbon–hydrogen bonds. Such transformations are finding particular application in medicinal and agrochemistry for the late-stage functionalization of bioactive molecules. However, despite the utility of these reactions, many challenges and opportunities remain with respect to catalyst design. In particular, the design of catalysts that modulate the site-selectivity of these processes is crucial for being able to transform a single starting material into a series of well-defined, chemically different products. Second, the development of strategies for controlling catalyst decomposition pathways is underdeveloped and will be crucial for larger scale applications of these systems. These two key challenges in the field of catalytic C–H functionalization (i.e., selectivity and catalyst longevity) mirror those in other areas of homogeneous catalysis.

My recent research contributions to field of homogeneous catalysis

Over the past 12 years, my research group has been highly active in the development of new approaches and catalysts for the selective C–H functionalization of complex organic substrates. Most recently, we have been particularly focused on the C–H functionalization of amines. Amines are particularly interesting substrates for two key reasons: (1) they are widely prevalent in bioactive molecules and (2) they present exciting (and largely unmet) challenges with respect to controlling catalyst reactivity, selectivity, and longevity in C–H functionalization reactions. In the course of these studies, we have developed catalytic C–H functionalization reactions in which the site selectivity is dictated by the metal catalyst (for examples, see Fig. 1) [1–3]. We have also made advances in developing ligands that can be used to modulate reactivity and reverse catalyst decomposition pathways.

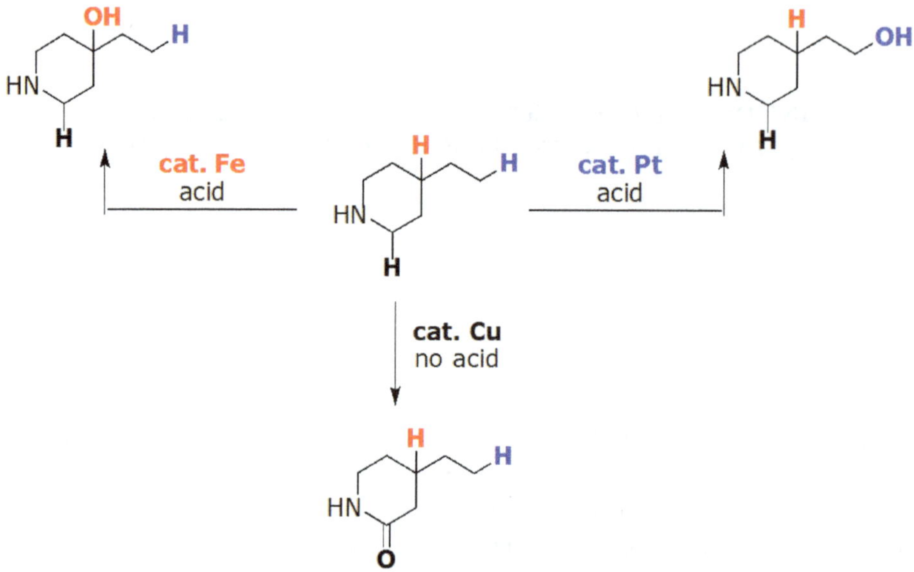

Fig. 1. Example of the use of catalyst to control selectivity in C–H functionalization of amines.

Outlook to future developments of research on homogeneous catalysis

Moving forward, there remain numerous opportunities for continuing to tune reactivity, selectivity, and stability as a function of catalyst and substrate structure. This will be particularly critical as these transformations find application in increasingly complex substrates, including active pharmaceuticals and agrochemicals, natural products, and biomolecules (proteins, sugars, nucleic acids).

Acknowledgments

This program is supported by the U.S. National Institutes of Health (1RO1 GM073836).

References

[1] J. J. Topczewski, P. J. Cabrera, N. I. Saper, M. S. Sanford, *Nature* **531**, 2202 (2016).
[2] M. Lee, M. S. Sanford, *J. Am. Chem. Soc.* **137**, 12796 (2015).
[3] C. T. Mbofana, E. Chong, J. Lawniczak, M. S. Sanford, *Org. Lett.* **18**, 4258 (2016).

DESIGNING NEW HOMOGENEOUS TRANSITION METAL CATALYSTS FOR AEROBIC OXIDATIONS

KAREN I. GOLDBERG

Department of Chemistry, University of Washington, Seattle, WA 98103, USA

Present state of research on homogeneous catalysis

Modern society is inextricably dependent on catalysis with more than 90% of all industrial chemical processes utilizing catalysts. Homogeneous transition metal catalysts comprise a vital and growing part of modern chemical production. They are used to promote reactions including hydrogenations, hydroformylations, carbonylations, oligomerizations, polymerizations, and metathesis, producing tens of millions of metric tons of chemicals each year. Mechanistic understanding of the individual reaction steps making up the catalytic cycles, and detailed studies of models for the intermediates in these cycles, have been essential in achieving major improvements in almost all of these homogeneously catalyzed commercial reactions. Using knowledge of the mechanisms of the individual steps, modifications to the ligands, metals and/or reaction conditions have been used to increase selectivity and catalytic activity. Insight into the mechanisms of these individual reactions also provides important clues that can be used in the design of new catalytic systems that can accomplish direct efficient transformations. Such new systems can replace current outdated production methods that require multiple reactions, significant energy input, isolation and purification of intermediates species, stoichiometric reagents, and/or generate undesirable waste streams. New catalytic methods are needed for both selective transformations used in fine chemical and pharmaceuticals production, as well as those to be used in large scale processes that can allow readily available carbon resources, including methane, alkanes, CO_2, and biomass materials, to be used directly as feedstocks for commodity chemicals and fuels.

My recent research contributions to homogeneous catalysis

My research group focuses on mechanistic studies of fundamental organometallic reactions with the goal of using this mechanistic understanding to design new homogeneous catalytic systems for the production of fuels and commodity chemicals. One area of particular interest is the development of catalysts for the selective aerobic oxidations of alkanes, including methane, to form alcohols or olefins. There are now numerous examples of transition metal complexes that can activate C–H bonds of saturated hydrocarbons, with many showing high selectivity for the desirable activation of primary C–H bonds [1]. The field now has a good understanding of the various types of ligands and metals that are needed for this first step in alkane oxidation.

A significant challenge is now the functionalization of the metal alkyl or alkyl hydride complex formed from the initial C–H bond activation step. For alkane oxidation to be effective in a large scale commercial process, a benign, widely available and inexpensive oxidant like molecular oxygen is necessary [2]. However, general and predictive understanding of how transition metal complexes react with oxygen is still limited, and this inhibits the development of new selective aerobic oxidation reactions. Thus, we have devoted considerable effort to exploring the reactivity of model transition metal complexes, including unsaturated metal centers, metal alkyl complexes, metal hydride complexes and metal alkyl hydride complexes, with molecular oxygen [3, 4]. We have uncovered a variety of promising oxygen reactions relevant to aerobic oxidative conversions of alkanes (Fig. 1). Our detailed mechanistic studies, combined with several recent reports by others, have allowed for a nascent understanding of the various pathways available for the reaction of oxygen with transition metal complexes [3, 4]. Furthermore, insight into the types of ligands and structures expected to be most effective in promoting productive reactions with oxygen is emerging, and this knowledge is being used to enable new selective aerobic oxidations of hydrocarbons.

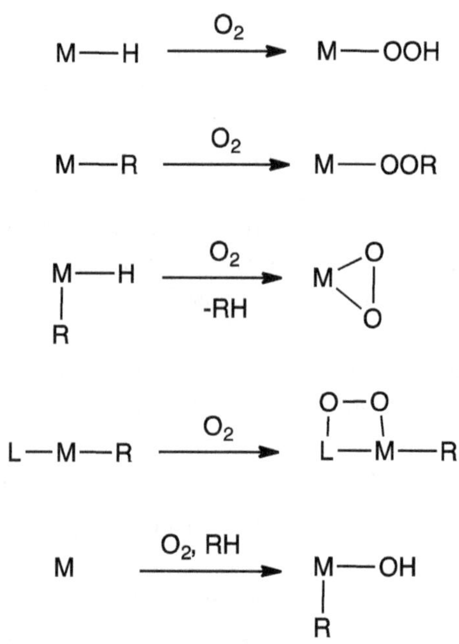

Fig. 1. A variety of observed reactions of O_2 with organotransition metal complexes.

Outlook to future developments of research on homogeneous catalysis

The last few decades have seen enormous advances in highly selective catalysis mediated by homogeneous transition metal complexes. Great progress has been made in reactions applicable to fine chemical and pharmaceutical production as well as large scale commercial processes. Recent demonstrations of the highly efficient abilities of transition metal complexes to selectively cleave and form bonds, particularly C–H bonds, bodes well for even more successes in the future. One of the most exciting such developments is the ability of transition metal complexes to selectively activate C–H bonds of saturated hydrocarbons. If we can translate this remarkable advance into viable large-scale commercial applications, for example efficient aerobic oxidations, we will have access to vast new resources of feedstocks for chemicals and fuels.

Acknowledgments

The United States (US) National Science Foundation (NSF, CHE-1464661), the US Department of Energy (DE-FG02-06ER15765), and the NSF Center for Chemical Innovation, the "Center for Enabling New Technologies through Catalysis (CENTC, CHE-1205189) are each acknowledged for support of different parts of this work.

References

[1] J. F. Hartwig, *Organotransition Metal Chemistry* (University Science Books, 2010).
[2] F. Cavani, J. H. Teles, *ChemSusChem* **2**, 508 (2009).
[3] L. Boisvert, K. I. Goldberg, *Acc. Chem. Res.* **45**, 899 (2012).
[4] M. L. Scheuermann, K. I. Goldberg, *Chem. Eur. J.* **20**, 14556 (2014).

ORGANOCATALYSIS — FROM LABORATORY SCALE TO INDUSTRIAL PROCESSES

KARL ANKER JØRGENSEN

Department of Chemistry, Aarhus University, DK-8000 Aarhus C, Denmark

Chirality is the basis for the origin of asymmetry in our body and culture [1]. Enzymes are the catalysts providing the asymmetry in our body and have also been applied as catalysts in academia and industry for over a century [2].

Chiral molecules formed by asymmetric catalysis are of great importance for the modern society and the catalytic asymmetric methods have expanded beyond the application of enzymes. In the last decades of the 20th century, transition-metal complexes dominated the field [3]; however, since the beginning of the new millennium organocatalysis has evolved to be the third pillar in asymmetric catalysis [4].

Organocatalysis has provided unique opportunities for the creation of novel reaction concepts to synthesize chiral molecular building blocks. These developments have been important for the success of organocatalysis, which has allowed the expansion of organocatalysis from academia to industrial processes, from simple functionalizations, to cascade, domino and tandem reactions, and the combination of organocatalysis with other fundamental reaction concepts, such as transition-metal catalysis.

Two different catalytic concepts have been introduced by organocatalysis depending on the functionality of the organic molecule to be activated: covalent and non-covalent catalysis. Our group has been involved in the development of new organocatalytic reactions based on both catalytic concepts. In the following presentation, the focus will be on one very general class of catalysts — the diarylprolinol silyl ethers — which act as a covalent catalyst. The diarylprolinol silyl ethers have expanded the activation modes of aldehydes beyond the classical enamine and iminium-ion activation modes, to novel activation modes which have allowed for the development of new enantioselective reactions and provided unprecedented molecular complexity [5].

The introduction of the diarylprolinol silyl ethers has played a key role in the development of new reactions and their applications. Figure 1 presents the enamine and iminium-ion activation modes and the novel activation modes, which became possible with the introduction of the diarylprolinol silyl ethers.

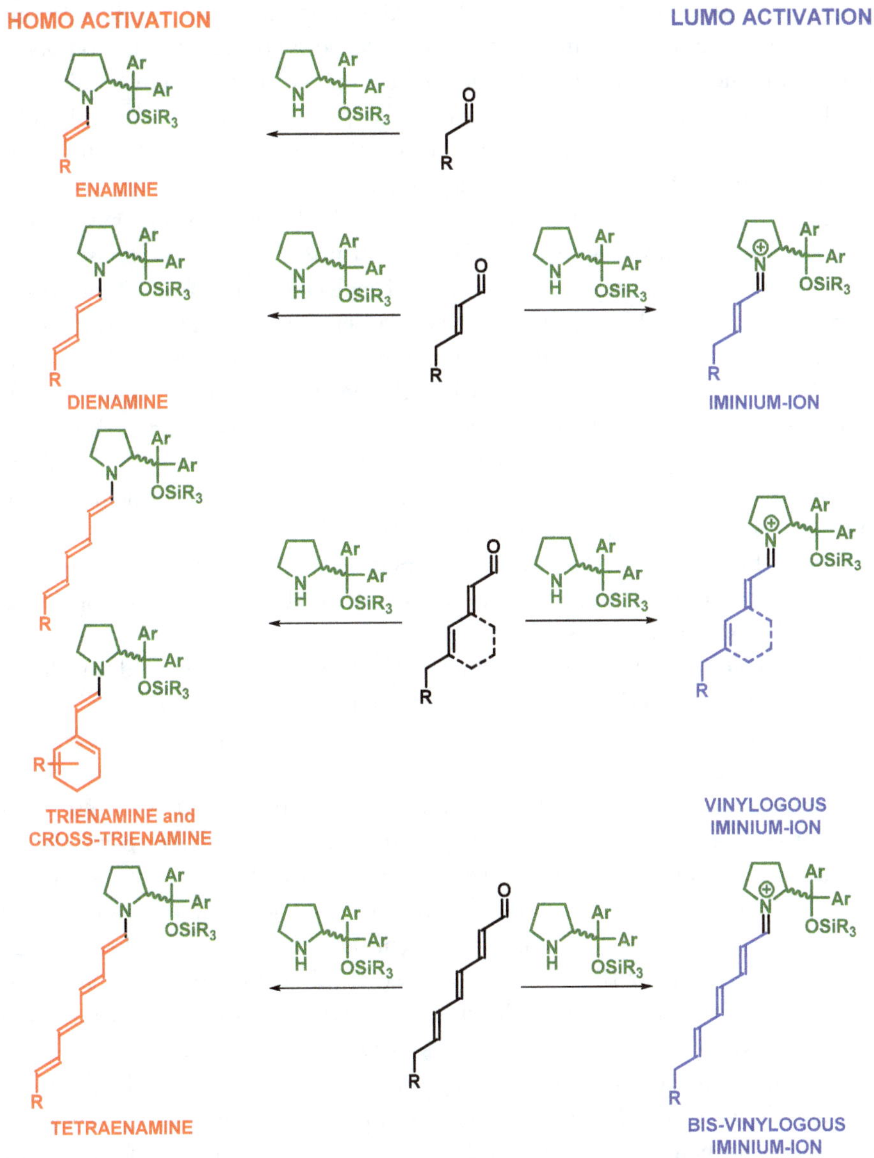

Fig. 1. Activation modes made possible by the diarylprolinol silyl ethers.

The two different activation modes in Fig. 1, the HOMO (highest occupied molecular orbital) and LUMO (lowest unoccupied molecular orbital) activation allow for a variety of different reactions. The HOMO activation has opened up for the stereoselective introduction of many different electrophilic reagents, including carbon and heteroatom moieties, reacting at the different nucleophilic positions in the red-molecule fragment to the left in Fig. 1. Furthermore, the different electron-rich π-systems in the HOMO activated systems can also undergo cycloaddition reactions, such as Diels–Alder, hetero-Diels–Alder and 1,3-dipolar cycloaddition reactions, generating a large number of different carbo- and heterocyclic compounds, which have found use in further chemical elaborations. The different addition and cycloaddition reactions can be controlled to take place both in the proximity of the stereocenter in the diarylprolinol silyl ether catalyst, or more remote, and in both cases with excellent regio- and stereoselectivity.

An important property of the diarylprolinol silyl ether catalysts is that these catalysts have acheive LUMO activation of polyunsaturated aldehydes. Thus, the same catalytic system can activate the same class of molecules, such as α, β-unsaturated aldehydes, to undergo both HOMO activation generating a dienamine intermediate or LUMO activation leading to an iminium-ion intermediate. As shown to the right in Fig. 1, the LUMO activation also leads to multiple different activation modes depending on the unsaturated aldehyde being activated. A variety of different nucleophilic compounds have been demonstrated to react with iminium-ion, vinylogous iminium-ion and bis-vinylogous iminium-ion intermediates. The electron-deficient π-systems in the LUMO activated intermediate can also react in Diels–Alder, hetero-Diels–Alder and 1,3-dipolar cycloaddition reactions. The reactions involving the diarylprolinol silyl ethers and LUMO activation are also found to be highly regio- and stereoselective.

The application of the diarylprolinol silyl ethers have expanded the field of organocatalysis from method development to applications in academia and industry for the formation of complex molecular architectures in total synthesis and compounds having important bioactivity. In Fig. 2 some selected examples are given to demonstrate the application of diarylprolinol silyl ethers in total synthesis and industrial processes. Nicolaou *et al.* have applied the epoxidation of α, β-unsaturated aldehydes via iminium-ion catalysis for the formation of the epoxide intermediate towards the synthesis of Hirsutellone B (Fig. 2a) [6]. The group of Toste reacted a ketoester with crotonaldehyde in the presence of the diarylprolinol silyl ether to generate an intermediate for the synthesis of Fawcettimine (Fig. 2b) [7]. A movement towards industrial application of the diarylprolinol silyl ethers has been demonstrated by Hayashi *et al.* for the synthesis of Oseltamivir (Tamiflu) (Fig. 2c) [8]. The application of the diarylprolinol silyl ethers in industry has also been documented by Merck for the synthesis of the anti-migraine compound Telecapant (Fig. 2d) [9] and Novartis has disclosed a patent for the formation of Aliskiren for the treatment of hypertension (Fig. 2e) [10].

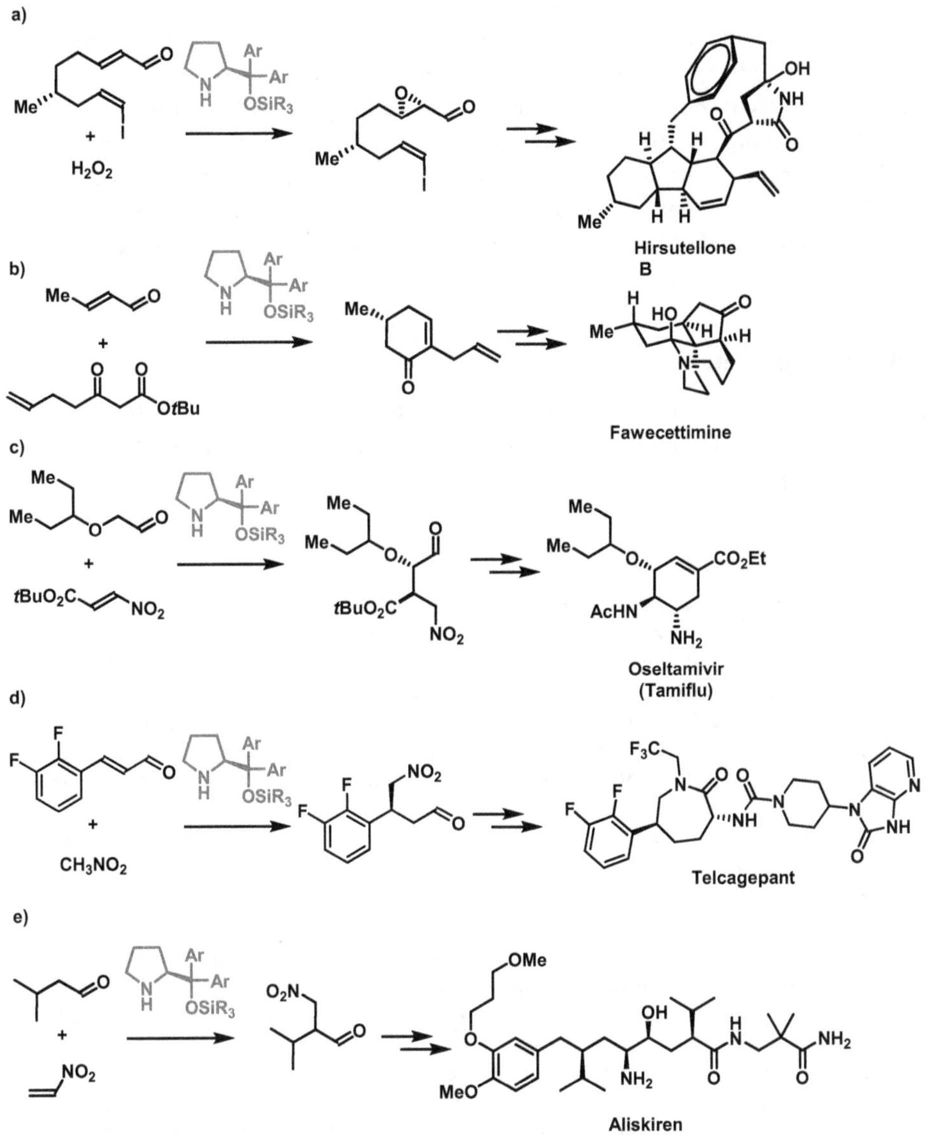

Fig. 2. Examples of the application of the diarylprolinol silyl ethers in total synthesis and industrial processes.

Outlook to future developments of research on organocatalysis

Organocatalysis has undergone a tremendous development since the turn of the millennium. A large number of novel catalytic asymmetric methodologies have been developed and applied for the synthesis of an enormous amount of chiral molecules. However, there are still many potentials and challenges in organocatalysis, which might expand the applications even further. Some of the challenges could be the development of catalysis that could give access to new and challenging transformations unattainable by a single catalytic system, by combining organocatalysis with other catalytic concepts. Since the catalytic loadings in organocatalysis are often much higher compared to *e.g.* transition-metal catalysis; strategies for moving the catalytic loadings from 5–20 mol% often applied in organocatalysis to less than 1 mol% is still an unmet challenge. The development of organocatalysis in such directions might hopefully provide *e.g.* the industry with even more environmentally and sustainable processes.

Acknowledgments

Thanks are expressed to Aarhus University, The Danish National Research Foundation, Carlsberg Foundation and Danish Council for Independent Research for generous support over the years.

References

[1] C. McManus, *Right Hand, Left Hand. The Origin of Asymmetry in Brains, Bodies, Atoms and Culture* (Harvard University Press, 2002).

[2] K. Drauz, H. Waldmann (eds.), *Enzyme Catalysis in Organic Synthesis* (Wiley-VCH Verlag GmbH, 2008).

[3] See *e.g.*: a) E. N. Jacobsen, A. Pfaltz, H. Yamamoto (eds.), *Comprehensive Asymmetric Catalysis* (Springer, 1999); b) L. S. Hegedus, *Transition Metals in the Synthesis of Complex Organic Molecules* (University Science Books, 1999); c) J. F. Hartwig, *Organotransition Metal Chemistry* (University Science Books, 2009).

[4] See *e.g.*: a) P. I. Dalko (ed.), *Comprehensive Enantioselective Organocatalysis: Catalysts, Reactions, and Applications* (Wiley-VCH, 2013); b) B. List (ed.) *Asymmetric Organocatalysis* (Springer, 2009).

[5] See *e.g.*: a) K. L. Jensen, G. Dickmeiss, H. Jiang, L. Albrecht, K. A. Jørgensen, *Acc. Chem. Res.* **45**, 228 (2012); b) B. S. Donslund, T. K. Johansen, P. H. Poulsen, K. S. Halskov, K. A. Jørgensen, *Angew. Chem. Int. Ed.* **54**, 13860 (2015); c) K. S. Halskov, B. S. Donslund, B. M. Paz, K. A. Jørgensen, *Acc. Chem. Res.* **49**, 974 (2016).

[6] K. C. Nicolaou, *Angew. Chem. Int. Ed.* **48**, 6870 (2009).

[7] X. Linghu, J. J. Kennedy-Smith, F. D. Toste, *Angew. Chem. Int. Ed.* **46**, 7671 (2007).

[8] H. Ishikawa, T. Suzuki, Y. Hayashi, *Angew. Chem. Int. Ed.* **48**, 1304 (2009).

[9] F. Xu, M. Zacuto, N. Yoshikawa, R. Desmond, S. Hoerrner *et al.*, *J. Org. Chem.* **75**, 7829 (2010).

[10] Novartis: WO2008119804.

HOMOGENEOUS CATALYSIS IN THE FUTURE

ERICK M. CARREIRA

Department of Chemistry and Advanced Biosciences, ETH-Zürich, Zürich, CH 8093, Switzerland;
Vladimir Prelog Weg 3, HCI H335, ETH Zürich, 8093 Zürich, Switzerland

My group has been involved in enantioselective homogeneous catalysis since the early days when the noteworthy milestones for the field included oxidative and reductive transformations, which were highlighted by the Nobel Prize in Chemistry to Sharpless (epoxidation and dihydroxylation) along with Noyori and Knowles (hydrogenation). In the intervening period, the discipline of homogeneous catalysis has made significant strides. Although, there are large swaths of reaction space that await exploration and discovery, in connection with the Solvay conference we focus on new directions for the discipline that involve its evolution beyond the objectives that have been traditionally pursued (Fig. 1). For the purposes of this discussion, these include: stereodivergent strategies, designing/managing multiple catalytic processes, use of reactive building blocks, and the interface with biocatalysis. The last of these items can be considered a subset of the second.

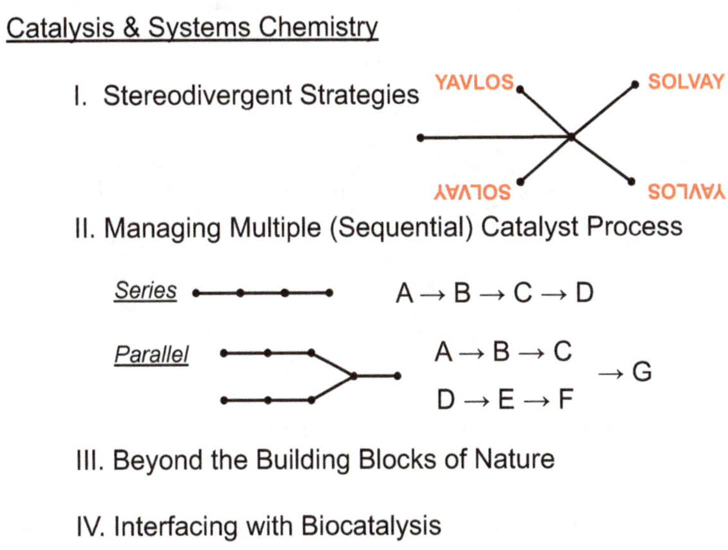

Fig. 1. New directions for catalysis.

A stereodivergent transformation is one in which access is made possible to all stereoisomers of a given product, including diastereomers and enantiomers, from the same set of starting materials under identical conditions [1–3]. The stereochemical features and associated permutations of an underlying structure can be considered a form of molecular diversity. Accordingly, exploration of the full capacity of a given optically active scaffold would necessitate synthesis of all possible stereoisomers. There are some key features of stereodivergent reactions which render them attractive for consideration in academia and industry. The implementation of these concepts in complex target oriented synthesis gives rise to stereodivergent synthesis strategies, whose implementation may be useful in exploration inter alia of polypharmacology and metabolic profiling [4]. A stereodivergent process, as we have defined it, results from the proper orchestration and inherent management of multiple catalytic cycles occurring concurrently or sequentially. Moreover, the problem is fundamentally interesting, as it demands deep insight into the reactive intermediates and transition states involved in a given reaction both individually and collectively. As it pushes the very boundaries of fundamental understanding of transition state ensembles, unanticipated emergent features can be expected. As such it provides fertile ground for interactions across traditional boundaries to include computational and theoretical research.

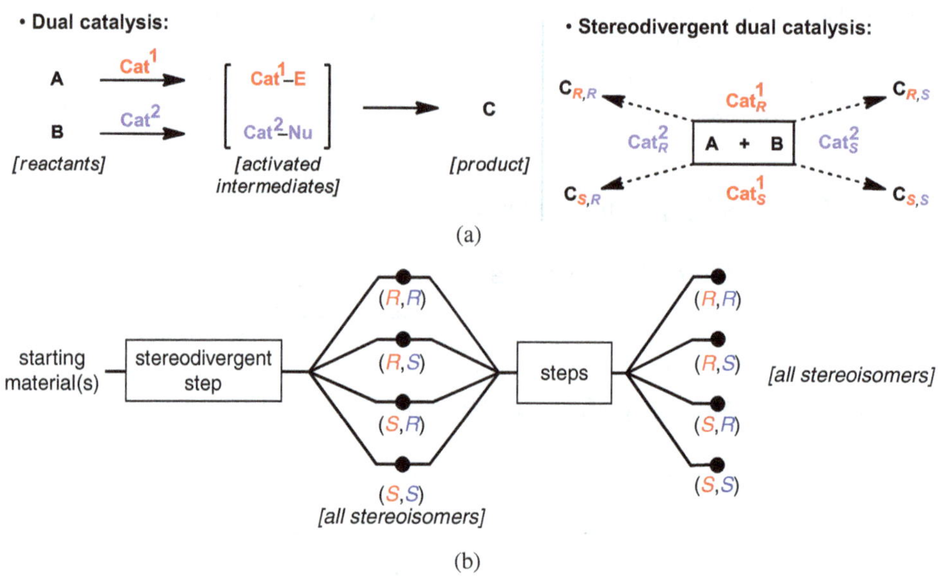

Fig. 2. (a) Stereodivergent dual catalysis concepts and definitions; (b) Stereodivergent synthesis strategies.

The typical research program in catalysis to date focuses on a specific reaction and catalyst family. Future endeavors can be envisioned to take an approach that can be termed "systems catalysis." Such an approach requires thinking well beyond isolated reactions involving conversion of a given starting material to product, A → B. An approach can be envisioned that focusses on end products and incorporates multiple catalytic processes coupled sequentially to meet the objective (A → B → C → ....Z). In effect, this would mimic biological ensembles in which numerous catalytic transformations operate in synergy, expeditiously, and in a timely manner. For the chemist, this leads to a level of complexity that compels the discovery of families of new catalysts, displaying mutual compatibility, wherein the product of one catalyst serves as the starting point for subsequent catalysts. The approach outlined can be envisioned to demand more from catalysis than current expectations. Such research endeavors offers wide ranging opportunities to think about complex phenomena in chemical systems, including entangled kinetic systems, novel media, compartamentalization, and how chemocatalysts can effectively interface with biocatalyst, in which (abiological) building blocks are linked with their biological counterparts.

We have initiated research efforts in two other areas outlined above. In the first of these, we are examining multiple catalytic processes that can be run in sequence to convert readily available starting materials to value added products. For example, we have been able to show that allylic substrates can be subjected to Ru-catalyzed alkene isomerization processes that result in the transient generation of a terminal olefin, whereupon hydroacylation produces esters with remote stereocenters [5, 6]. In this process the same catalyst is able to execute multiple transformations. One can envision a related construct that entails the use of copper(II) and ruthenium catalysts contemporaneously. The former furnish chiral propargyl amines while the latter effects alkyne isomerization and hydroacylations of the terminal alkyne, whereupon cyclization ensues.

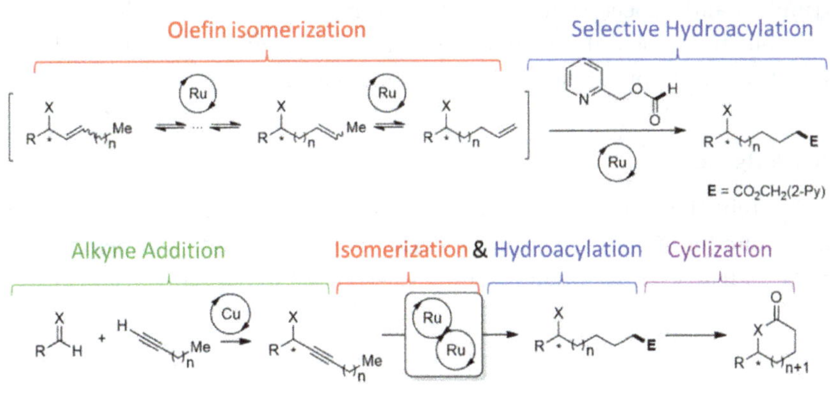

Fig. 3. Tandem serial catalysis.

A second direction we have undertaken seeks to employ reactive species that are uniquely "chemical" and offer opportunities for synthesis. Some of these are viewed as undesirable because of associated dangers of working with them, such as toxicity and violent decomposition. However, in general the risks are relevant upon attempts at their isolation or purification. Interestingly, these are made in aqueous media, where they are deemed stable and safely handled. In the past, catalysts and approaches have been employed in which these are isolated and used in organic media, where the typical catalysts have been designed to operate. The reagents may be generated under a variety of harsh conditions in water that include oxidants or reductants and strongly alkali or acids. The design and synthesis of catalytic active species that are able to employ these reagents under conditions in which they are generated, which are in a sense "extreme," presents noteworthy challenges for the discipline [7–12]. The successful developments offers great opportunities for exploration of uncharted structure space, where smart medicines and materials of the future may be discovered.

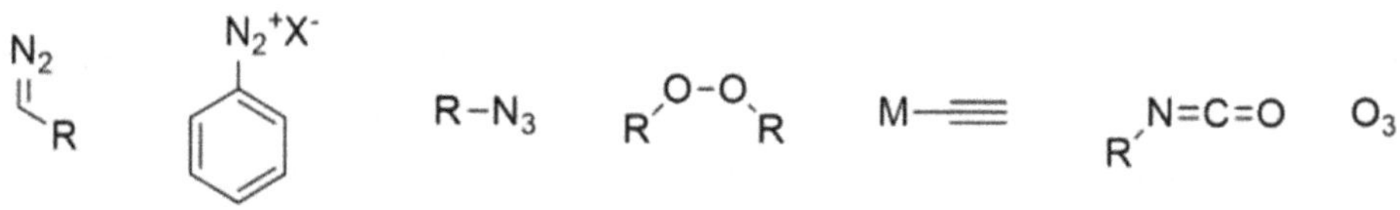

Fig. 4. Reactive species for use in new catalytic transformations.

In summary, the socioeconomic challenges of the new century demand energy efficient and environmentally benign catalytic processes under economically sustainable regimes. Although there have been great strides in homogeneous catalysis, great opportunities and expectations abound. Meeting these objectives will require advances in our fundamental understanding of reactions and mechanisms along with fundamentally new approaches to homogeneous catalysis.

Acknowledgments

We are grateful for generous funding and support over the years from the Swiss National Foundation, ETH-Zürich, and the Stipendienfonds der Schweizerischen chemische Industrie (SSCI) administered through ETH-Zürich.

References

[1] S. Krautwald, D. Sarlah, M. A. Schafroth, E. M. Carreira, *Science* **340**, 1065 (2013).
[2] S. Krautwald, M. A. Schafroth, D. Sarlah, E. M. Carreira, *J. Am. Chem. Soc.* **136**, 3020 (2014).
[3] S. Krautwald, H. F. Zipfel, E. M. Carreira, *Angew. Chem. Int. Ed.* **54**, 14363 (2015).
[4] A. Schafroth, G. Zuccarello, S. Krautwald, D. Sarlah, E. M. Carreira, *Angew. Chem. Int. Ed.* **53**, 13898 (2014).
[5] N. Armanino, E. M. Carreira, *J. Am. Chem. Soc.* **135**, 6814 (2013).
[6] N. Armanino, M. La France, E. M. Carreira, *Org. Lett.* **16**, 572 (2014).
[7] B. Morandi, E. M. Carreira, *Angew. Chem. Int. Ed.* **49**, 938 (2009).
[8] B. Morandi, E. M. Carreira, *Angew. Chem. Int. Ed.* **49**, 4294 (2010).
[9] B. Morandi, B. Mariampillai, E. M. Carreira, *Angew. Chem. Int. Ed.* **50**, 1101 (2011).
[10] B. Morandi, J. Cheang, E. M. Carreira, *Org. Lett.* **13**, 3080 (2011).
[11] B. Morandi, E. M. Carreira, *Angew. Chem. Int. Ed.* **50**, 9085 (2011).
[12] B. Morandi, E. M. Carreira, *Science* **335**, 1471 (2012).

EXPLORING CHEMICAL SPACE IN HOMOGENEOUS CATALYSIS

BEN L. FERINGA

Stratingh Institute for Chemistry, University of Groningen, 9747 AG Groningen, The Netherlands

My view of the present state of research on catalytic synthesis

The ability to synthesize new molecules and materials not being limited by Nature's design provides chemistry with unmatched creative and creating power. At the heart of many of these chemical transformations are homogeneous catalysts ranging from enzymes and organo-catalysts to a wide variety of metal-based systems that enable amazing reactivity and selectivity in organic synthesis and industrial processes. The versatility of *e.g.* enantioselective conversions, transition-metal based metathesis, cross coupling technology, gold catalysis, photo-redox catalysis or asymmetric organo-catalysis developed in recent decades is testimony of the potential of molecular design and precision in catalysis [1–4]. Being at a crossroad towards sustainable catalytic synthesis of the future the use of alternative feedstock *e.g.* renewables, waste products or small molecules, requires the development of entirely novel catalysts for example for bond breaking (deoxygenation), handling mixtures or compounds bearing multiple functionalities and activating rather inert molecules. The use of non-precious metals, functional group compatibility in total synthesis (avoiding extensive protection strategies), catalysis in aqueous media or novel catalysts materials (the interface of heterogeneous and homogeneous catalysis, supramolecular and biomimetic systems) for high selectivity low-energy conversions, all offer tremendous opportunities for discovery. Here the focus will be on some key questions i.e. how to achieve high atom economy, mild conditions, self-assembly or adaptive behavior in homogeneous catalysis?

My recent research contributions to sustainable catalytic synthesis

Transition metal catalysis, in particular Pd, Rh, Ir, Ru based systems, has dramatically changed the face of modern synthetic chemistry but equally impressive is Nature's use of non-precious metals like Fe in key transformations. In our approach towards C–N bond formation with low-environmental impact (low E-factor) the focus is on Fe-based catalysts. Taking advantage of hydrogen borrowing mechanisms, the goal was to achieve direct N-alkylation of amines with alcohols producing only water as waste product (Fig. 1) [5]. Amine alkylation is an essential chemical transformation typically requiring pre-activation of the alkyl component and frequently suffering from non-selective (multiple) alkylation and halide waste. Using a Knölker Fe-complex the direct N-mono-alkylation of a variety of amines is possible producing only water as side product. The transformation involves *in situ* alcohol dehydrogenation to the corresponding aldehyde, imine formation and reduction with the Fe-complex performing a dual function as alcohol dehydrogenation and imine hydrogenation catalyst. Taking advantage of diols as starting material a double alkylation procedure resulted in 5-, 6- and 7- membered N-heterocycles, valuable intermediates *e.g.* for pharmaceuticals. These findings have the potential to use bio-based alcohols in direct halide-free low-E-factor alkylation procedures for amines and N-heterocycles for materials, coatings or drug applications.

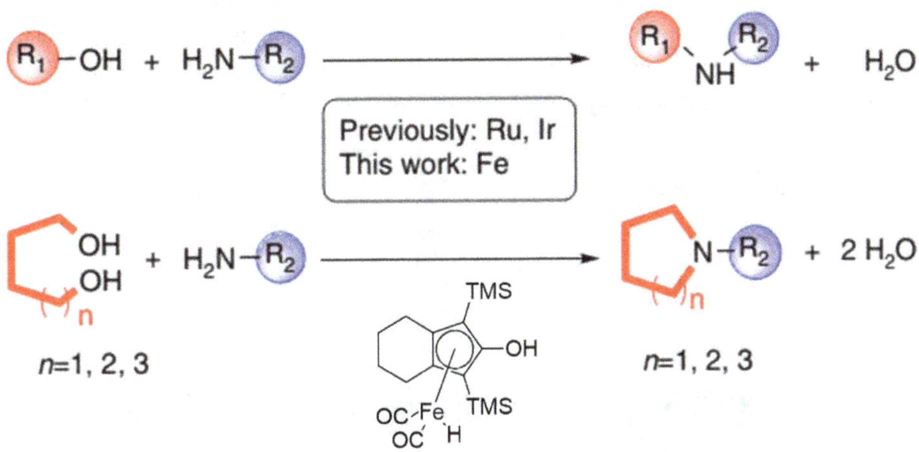

Fig. 1. Fe-based direct alkylation of amines with alcohols.

In the previous example water is the only waste product but using water as an "oxygen"-nucleophile or performing catalysis in water as a solvent (common to Nature's enzymatic transformations) offers equally challenging opportunities. Based on the Pd-catalyzed Wacker oxidation of olefins, typically showing Markovnikov selectivity in olefin conversion, Pd-nitro- catalysts were developed that show promising selectivity towards anti-Markovnikov oxidation for instance direct aldehydes formation by terminal oxidation of alpha-olefins [6]. An illustrative example of consecutive catalytic transformations, centered around this novel regioselective oxidation, is the use of Ir-catalyzed enantioselective allylic amination, Pd-catalyzed anti-Markovnikov oxygenation and Mn-TACN based oxidation providing valuable optically active beta-amino acids. In the emerging field of supramolecular catalysis advantage is taken of self-assembly and intermolecular interactions in order to enhance selectivity and achieve transformations under unconventional conditions. In contrast to the common use of DNA as an "information molecule" for instance for directed evolution approaches to novel enzymes, it provides a superb scaffold for supramolecular catalysis in water as has been demonstrated in *e.g.* copper-based highly enantioselective catalytic Michael additions, Diels–Alder and alkylation reactions. In a remarkable example of small molecule activation, stereocontrol and reactivity in water, a dynamic chiral supramolecular assembly of an achiral copper complex bound to oligonucleotides was shown to catalyze olefin hydration to provide the corresponding chiral alcohol with 82% enantiomeric excess [7]. Beyond the exclusive domain of hydratase enzymes, this catalytic asymmetric addition of water in water offers fascinating perspectives for catalysis in aqueous media and confined space exploiting the power of self-assembly.

Switchable and adaptive catalysis is another area with ample opportunities to achieve high control over chemo-, regio-, and stereo-selectivity with the possibility to perform multiple tasks by a single catalyst. (De)-activation in space and time and responsive (feedback) functions can readily be embedded in the catalyst offering adaptive behavior. In asymmetric catalysis dynamic control over chiral space was achieved with an organo-catalyst with an intrinsic molecular motor moiety (Fig. 2).

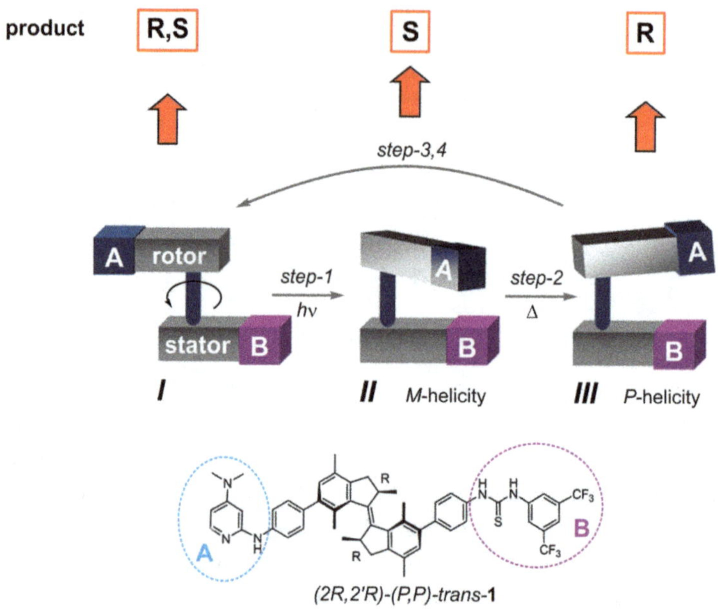

Fig. 2. Molecular motor based multistage chiral organo-catalyst for the formation of racemic or preferentially either enantiomer of a product.

The introduction of thiourea and DMAP moieties enabled effective asymmetric organo-catalysis in a thiol- or Michael addition while the presence of a unidirectional rotary motor acting as a multistage switch was used to control both geometry and stereochemistry in an non-invasive manner (with light) to modulate activity and enantioselectivity of the catalyst. Not only can a single chiral catalyst provide preferentially racemic or either enantiomer of a chiral product but also the sequence of formation of distinct catalytic species is governed by molecular design of the responsive system [8]. The recent demonstration of switchable chiral phosphine ligands for transition metal catalysis sets the stage for the broader exploration of responsive behavior in future catalysis allowing among others adaptive functions and multitasking in complex transformations.

While exploring chemical space with transition metal based homogeneous catalysts, the design of new ligands continues to be often the cornerstone to success. The introduction of phosphoramidites as monodentate chiral ligands allowed breakthrough discoveries in asymmetric conjugate addition and allylic alkylation with organometallic reagents (organozinc, Grignard and organolithium) [9]. Their versatile synthesis, compared to *e.g.* most phosphines, not only resulted in implementation in asymmetric hydrogenation including an industrial process, but also readily access to libraries of chiral ligands for rapid catalysts screening and optimization in enantioselective catalysis.

Finally the control of chemical reactivity being it rather inert molecules like carbon dioxide or extremely reactive ones like organolithium reagents poses additional challenges to catalyst design. Although catalytic cross-coupling has emerged as one of the privileged methodologies in chemistry in the past decades it is remarkable that the use of organo-lithium compounds, among the most common and widely used organometallic reagents, has hardly been explored in this context. Taming highly reactive organolithium reagents for Pd-catalyzed cross-coupling, in particular using N-heterocyclic carbene ligands, allowed amazingly selective C–C bond formation under extremely mild conditions (Fig. 3) [10, 11].

The fact that many common cross-coupling partners are made from organolithium reagents and the low environmental impact (E-factor, toxicity, mild conditions) associated with organo-lithium cross coupling mitigates the lower functional group compatibility compared to for instance organo-boron based methodologies.

Outlook to future developments of research on homogeneous catalysis

Catalytic function by molecular design has many facets ranging from novel ligands to photo- and redox-active systems and activation of rather inert molecules to control of selectivity in multifunctional substrates. Most likely in organic synthesis novel homogeneous catalysis methodology will enable direct (late stage) functionalization or conversions avoiding extensive functional group protection strategies enhancing atom and step economy and reducing E-factors. This includes diminished solvent use or in particular enabling transformation in water providing great prospects for further integration of organometallic and biocatalytic transformations. Small molecule activation is one of the most challenging areas in catalysis in the decades ahead and a great playground where homogeneous and heterogeneous catalysis will

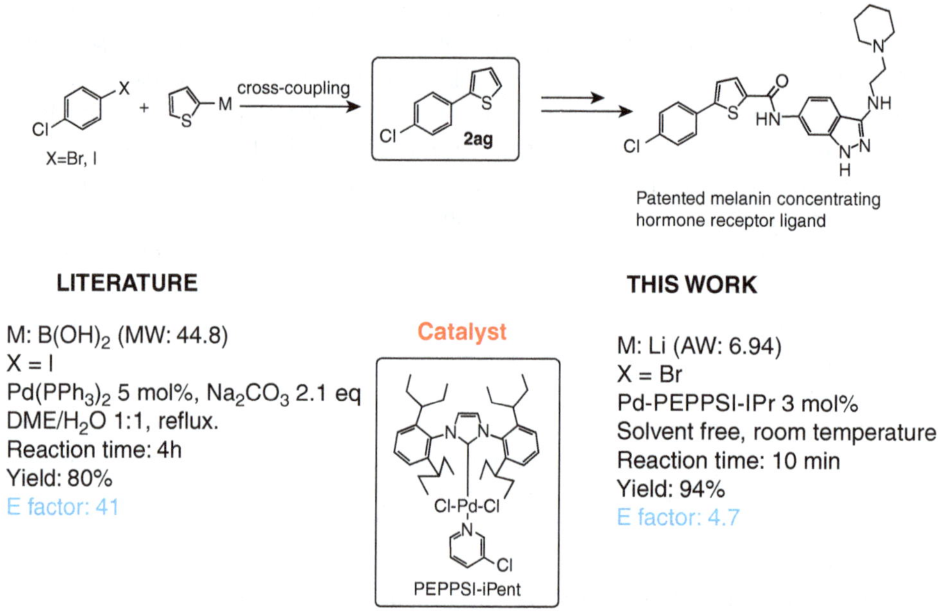

Fig. 3. Catalytic cross-coupling of organolithium reagents.

meet. Water, hydrogen, carbon dioxide, oxygen, methane activation under mild condition and achieving selectivity's that can match Nature's systems might need reconsidering some of our current concepts in catalysis. Supramolecular approaches, control of interface assembly and interaction and exploring novel materials architectures for catalysis are still largely in their infancy. Here precision catalysis by molecular programming, positional control and confined space as well as multimetal catalytic complexes by design offer tremendous opportunities. Photo-redox catalysis is rapidly emerging and the key question how to convert photons and electrons efficiently into chemical products and fuels will be a dominating factor in our field for years to come. Despite the great challenges ahead we can gain confidence from the intricate systems that evolved in biology, like the photosynthesis machinery and the rather simple set of components used to achieve such remarkable catalytic transformations under very mild conditions, as well as the accelerating capabilities of chemical synthesis and catalysis and the breath of catalysis research.

Acknowledgments

The Netherlands Organization for Scientific Research (NWO-CW), the Royal Netherland Academy of Arts and Sciences (KNAW) and the Ministry of Education Culture and Science (Gravitation program 024.601035) are acknowledged for financial support.

References

[1] C. C. C. Johansson Seechurn, M. O. Kitching, T. J. Colacot, V. Snieckus, *Angew. Chem. Int. Ed.* **51**, 5062 (2012).
[2] J. W. Tucker, C. R. J. Stephenson, *J. Org. Chem.* **77**, 1617 (2012).
[3] C. K. Prier, D. A. Rankic, D. W. C. MacMillan, *Chem. Rev.* **113**, 5322 (2013).
[4] A. Berkessel, H. Groeger, *Asymmetric Organocatalysis* (Wiley-VCH, 2005).
[5] T. Yan, B. L. Feringa, K. Barta, *Nature Comm.* **5**, 5602 (2014).
[6] J. J. Dong, W. R. Browne, B. L. Feringa, *Angew. Chem. Int. Ed.* **54**, 734 (2015).
[7] A. J. Boersma, D. Coquière, D. Geerdink, F. Rosati, B. L. Feringa *et al.*, *Nat. Chem.* **2**, 991 (2010).
[8] M. Vlatkovic, B. S. L. Collins, B. L. Feringa, *Chem. Eur. J.* **22**, 17080 (2016).
[9] J. F. Teichert, B. L. Feringa, *Angew. Chem. Int. Ed.* **49**, 2486 (2010).
[10] M. Giannerini, V. Hornillos, C. Vila, M. Fananas-Mastral, B. L. Feringa, *Angew. Chem. Int. Ed.* **52**, 13329 (2013).
[11] E. B. Pinxterhuis, M. Giannerini, V. Hornillos, B. L. Feringa, *Nature Comm.* **7**, 11698 (2016).

LIGANDS WITH INTRINSIC DONOR FLEXIBILITY FOR REDOX CATALYSIS

MARTIN ALBRECHT, MIQUEL NAVARRO and CANDELA SEGARRA
Department of Chemistry & Biochemistry, University of Bern, CH-3012 Bern, Switzerland

Current state of catalyst design

Efforts towards increasing the efficiency of homogeneous catalysts have focused to a substantial extent on gaining a profound mechanistic understanding, in particular through identification of rate-limiting steps and through characterization of supposed intermediates of a catalytic cycle. In metal catalyzed processes, ligand design has played a key role to gain such insights. Ligands were originally considered as true spectators of the metal center and hence serve as a crucial handle to tune and tailor the performance of the metal center as the catalytically active site. More recent work has demonstrated that ligands often actively participate in bond making and breaking processes [1, 2]. Proton transfer from a substrate to a ligand site and *vice versa* has been identified to be a key process, for example in various (transfer) hydrogenation catalysts [3, 4]. In contrast, reversible electron storage and release is a concept that has been used much rarer, even though it has long been known that acceptor orbitals in reduction processes are often ligand-centered, *e.g.* in $[M(bipyridine)_3]^{n+}$ complexes [5].

The assistance of the ligand in 'metal'-catalyzed processes has attractive benefits, as the active component to consider is extended from a purely metal-centered unit, as originally perceived, to a polyfunctional metal-ligand scaffold, which can for example transfer protons and electrons *simultaneously*, and thus lower the energy surface of a redox reaction [6, 7]. Intrigued by these considerations, we were considering that ligands with a dynamic rather than static donor ability would offer considerable benefits for redox transformations. Especially donor flexibility that imparts electron donating and electron-withdrawing features are expected to facilitate redox processes at the metal center, as one and the same ligand scaffold is able to stabilize high- and low-valent metal centers. Mesoionic ligands constitute a potent family of ligands for these purposes and for imparting flexible rather than static electronic stabilization of the metal center as they can adopt either a neutral or a formally anionic character when bound to a metal center.

My recent research contributions to developing efficient homogeneous catalysts

We have developed a set of C- and N-donor ligands, *viz.* triazolylidenes (trzs) and pyridylidene amides (PYAs), which are mesoionic and which can toggle, without chemical transformation, between neutral L-type bonding form and an anionic X-type ligation, represented by two limiting resonance structures (Fig. 1) [8–10]. Analyses in solution and in the solid state lend support to an electronically flexible bonding mode, which is influenced by numerous factors such as the polarity of the environment (solvent, non-coordinating anions) as well as the electronic situation at the metal fragment (different metals, different ancillary ligands). Various techniques allow for deducing the different contributions of the limiting resonance structures. In particular pyridylidene amide systems (PYA in Fig. 1) show distinct π basicity and acidity in their zwitterionic and neutral resonance forms, respectively [10]. Thus, apolar solvents and electron-rich spectator ligands X favor the neutral resonance structure. This structure is similar to pyridine and promotes metal-to-ligand charge transfer (π backbonding) and hence stabilizes electron-rich (low-valent) metal centers. In contrast, with electron-poor spectator ligands (*e.g.* solvento complexes) or in strongly polar solvents, the zwitterionic structure is predominant, which features a formal amide anion that acts as a π donor ligand and stabilizes high-valent electron-poor metal centers.

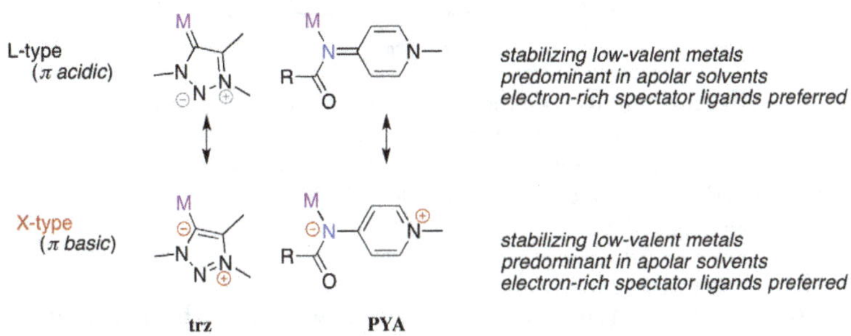

Fig. 1. Neutral and ionic metal bonding in donor flexible ligands illustrated with triazolylidene (trz) and pyridylidene amide (PYAs) complexes.

We have exploited these unique bonding properties in various homogeneously catalyzed redox reactions such as the transfer hydrogenation of ketones [11]. Solvent variation enables the catalytic activity to be tailored and optimized, which we attribute to the specific solvent-dependent properties of the PYA ligand. Moreover, the strongly donating π basic resonance form, predominant in polar media, has been utilized to generate highly efficient iridium water oxidation catalysts [12] in efforts to contribute to the development of efficient devices for artificial photosynthesis and the storage of transient energy. For example, PYA complexes achieve very high turnover numbers (TON > 90,000) under harsh catalytic conditions (pH 1, aqueous solution, highly oxidizing medium). The robustness of the catalytic cycle is assumed to originate from the donor flexibility of the ligand, which enables the stabilization of the various iridium oxidation states that have been proposed to be relevant for water oxidation. In agreement with this model, the PYA complexes perform better in water oxidation than analogous complexes containing a pyridine ligand, which lacks similar donor flexibility (Fig. 2).

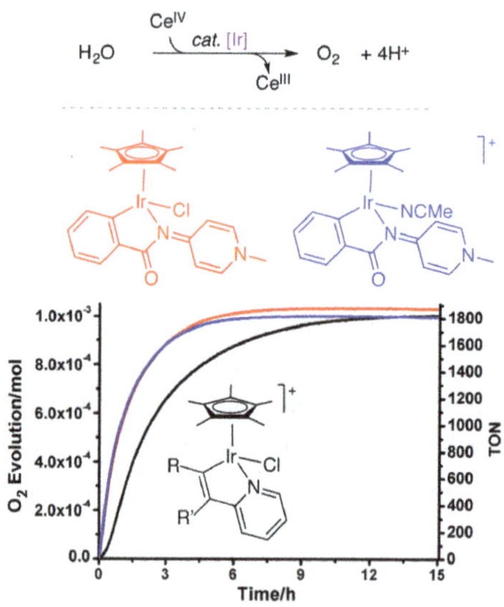

Fig. 2. Water oxidation catalyzed.by various iridium complexes; the rate enhancement entailed by the PYA ligand is demonstrated and it not related to the formal charge of the complex, nor the presence of a specific spectator ligand, but is dependent on the donor properties of the N-bound ligand; donor flexible PYA ligands induce substantially higher activity than pyridine ligands with relatively static donor properties.

Enhanced catalytic activity has also been observed in the periodate-mediated ruthenium-catalyzed oxidation of olefins to aldehydes. Substantial rate enhancement was noted when using PYA ligands as compared to N-heterocyclic carbenes or pyridine ligands.

Outlook to future developments of research on homogeneous catalyst design

While the utilization of functional ligands as redox active units or as transient proton reservoirs has ample precedence, the exploitation of donor flexibility through enhanced or decreased mesoionic contributions has been underdeveloped. Yet the fact that no proton or electron transfer is required to change the properties of the ligand makes this concept highly appealing for stabilizing different metal oxidation states and different degrees of nucleo- or electrophilicity of the metal center, with obvious implications for catalytic cycles that involve formal reduction or oxidation of the metal center. We anticipate that this donor flexibility will have a marked impact on various catalytic processes. Obviously, the concept is not constrained to the two ligand classes presented here, and may thus provide new generations of highly active catalytic systems.

Acknowledgments

Our work in this area has been supported financially by the European Research Council, the Swiss National Science Foundation, and Science Foundation Ireland. We thank Mo Li and Stefan Bernhard (Carnegie Mellon University) for continuous support with water oxidation catalysis.

References

[1] J. R. Khusnutdinova, D. Milstein, *Angew. Chem. Int. Ed.* **54**, 12236 (2015).
[2] K. Hindson, B. de Bruin, *Eur. J. Inorg. Chem.* **2012**, 340 (2012).
[3] Y. Shvo, D. Czarkie, Y. Rahamim, *J. Am. Chem. Soc.* **108**, 7400 (1986).
[4] R. Noyori, T. Ohkuma, *Angew. Chem. Int. Ed.* **40**, 40 (2001).
[5] P. J. Chirik, *Inorg. Chem.* **50**, 9737 (2011).
[6] D. Milstein, *Top. Catal.* **53**, 915 (2010).
[7] M. H. V. Huynh, T. J. Meyer, *Chem. Rev.* **107**, 5004 (2007).
[8] For trz ligands, see: K. F. Donnelly, A. Petronilho, M. Albrecht, *Chem. Commun.* **49**, 1145 (2013).
[9] For PYA ligands, see: J. Slattery, R. J. Thatcher, Q. Shi, R. E. Douthwaite, *Pure Appl. Chem.* **82**, 1663 (2010).
[10] V. Leigh, D. J. Carleton, J. Olguin, H. Müller-Bunz, L. J. Wright *et al.*, *Inorg. Chem.* **53**, 8054 (2014).
[11] K. F. Donnelly, C. Segarra, L. Shao, R. Suen, H. Müller-Bunz *et al.*, *Organometallics* **34**, 4076 (2015).
[12] M. Navarro, M. Li, H. Müller-Bunz, S. Bernhard, M. Albrecht, *Chem. Eur. J.* **22**, 6740 (2016).

DEVELOPMENT OF NEW CATALYSTS TOWARD UTILIZATION OF RENEWABLE RESOURCES

KYOKO NOZAKI

Department of Chemistry and Biotechnology, The University of Tokyo, Bunkyo-ku, Tokyo 113-8656, Japan

Present view of homogeneous catalysis

Chemical industry made progress utilizing various raw materials, especially fossil resources such as coal, petroleum oil, and natural gas. If we view the chemical transformations from a different viewpoint, usage of renewable resources, we find there are many reactions to be discovered. The fossil resources are made of highly reduced form of carbon atom, such as elemental carbon or hydrocarbons. Contrastively, renewable resources such as carbon dioxide and biomass consist of more oxidized form of carbon atoms. In addition, the carbon resources in biomass are highly complex structures. Accordingly, reduction and degradation would be the key words we are facing at present.

My recent research contribution to the field of homogeneous catalysis

Carbon dioxide is one of the most attractive renewable C1 resources, which has many practical advantages such as abundance, economic efficiency, and lack of toxicity. We recently reported copolymerizations of carbon dioxide and olefins by using a meta-stable lactone intermediate which is formed by the palladium-catalyzed condensation of carbon dioxide and 1,3-butadiene [1].

For more than half a century, various selective oxidation processes have been developed to convert fossil resources to value-added chemicals. Unlike petroleum, renewable resources, plant dry matter for example, is mostly composed of highly oxidized carbon compounds. Thus, development of efficient technology involving the chemical reduction becomes a great challenge. Direct and selective hydrogenolysis of sp^2 C-OH and sp^3 C-O bonds recently developed in our laboratory will be presented [2].

While the hydroformylation, a reaction to add carbon monoxide and dihydrogen to an unsaturated carbon–carbon multiple bond, has been widely employed in the chemical industry since its discovery in 1938, the reverse reaction, retro-hydroformylation, has seldom been studied. Here we report the retro-hydroformylation reaction to convert an aldehyde into an alkene and synthesis gas (a mixture of carbon monoxide and dihydrogen) catalyzed by cyclopentadienyl iridium complexes [3].

Outlook to future developments of research on homogeneous catalysis

The more complex the substrates are, the more selective the catalysts need to be. Continuous development of catalyst focusing on ligand design will open unprecedented transformations to provide useful chemicals out of complex renewable resources. Less costly separation technology will be also needed.

Acknowledgments

The program is supported by JSPS KAKENHI (Grant Number JP15H05796) in "Precisely Designed Catalysts with Customized Scaffolding".

References

[1] R. Nakano, S. Ito, K. Nozaki, *Nature Chem.* **6**, 325–331 (2014).
[2] S. Kusumoto, K. Nozaki, *Nature Commun.* **6**, 6296 (2015).
[3] S. Kusumoto, T. Tatsuki, K. Nozaki, *Angew. Chem. Int. Ed.* **54**, 8458–8461 (2015).

CHALLENGES FOR ORGANOCATALYSIS

BENJAMIN LIST

Department of homogeneous catalysis, Max-Planck-Institut für Kohlenforschung, 45470 Mülheim an der Ruhr, Germany

My view of the present state of research on organocatalysis

The last 17 years have witnessed an astonishing development in the field of asymmetric synthesis. *Organocatalysis*, the catalysis with small organic molecules, rose from being entirely neglected to what is now probably the single most actively researched area within the field [1, 2]. Who would have thought back in the late 1990s that in 2016, the majority of publications dealing with enantioselective synthesis focus on organocatalysis? However, as remarkable as this revolution in asymmetric synthesis has been, organocatalytic processes for the industrial production of enantiomers, while also significantly on the rise, are still relatively rare. Industrial asymmetric catalysis is still largely dominated by transition metal-catalyzed hydrogenation technologies and biocatalytic processes. Partly, this situation may be explained with the relative youth of the field of organocatalysis. But it is also clear that there are several remaining challenges for organocatalysis, that will need to be addressed before its current academic dominance will translate into general industrial utility. Here, I will discuss three challenges that I consider relevant in this regard: (1) developing heterogeneous organocatalysis, (2) solving problems of chemical synthesis through organocatalysis, and (3) developing high-performance organocatalysis.

My recent research contributions to organocatalysis

Developing heterogeneous organocatalysis

The one inherent potential advantage of organic catalysts over transition metal complexes and biocatalysts is that they can be easily, reliably, and permanently immobilized onto a solid support via covalent bond formation. In principle, such immobilized organocatalysts should be infinitely recyclable and are therefore highly attractive research targets. However, while the immobilizability of organic compounds is well-established since the pioneering contributions by Bruce Merrifield, making efficient, selective, and recyclable supported organocatalysts has not been a trivial undertaking and has not led to widely used catalyst systems — despite intense research efforts.

We have recently proposed a new approach to the heterogenization of organocatalysts that is based on their covalent fixation onto *textile materials*. Upon irradiation of simple Nylon fibers with UV-light, surface radicals are presumably generated, which readily react with organic molecules that contain carbon-carbon double bonds, forming very stable covalent bonds. The resulting *organotextile catalysts* have shown to be highly active and enantioselective. Moreover, one such catalyst, which is made by immobilizing a commercially available cinchona alkaloid-based organocatalyst that naturally comes with an olefin appendant, can be recycled several hundred times, without loss of activity or selectivity (Fig. 1) [3].

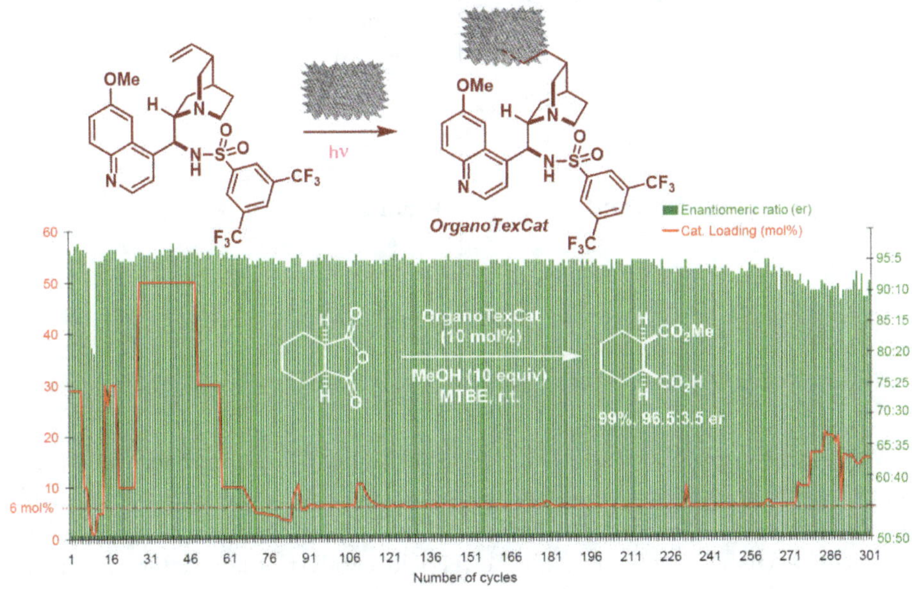

Fig. 1. Immobilization of an organocatalyst onto nylon, and use of the resulting heterogenous catalyst in recyclable asymmetric catalysis.

Solving problems of chemical synthesis through organocatalysis

We recently became interested in designing and developing organic Lewis acid catalysts. As a fundamental activation mode, Lewis acid catalysis enables key reactions in chemical synthesis, such as the Diels–Alder and Friedel–Crafts reactions, and various aldol, Mannich, and Michael reactions. Consequently, substantial efforts have been directed towards the development of enantiopure Lewis acids, which have enabled important asymmetric variations of such reactions. Despite the plethora of elegant catalysts and methodologies developed in this context, a key limitation of enantioselective Lewis acid catalysis is the frequent need for relatively high catalyst loadings, which result from issues such as insufficient Lewis acidity, product inhibition, hydrolytic instability, and background catalysis.

We have recently developed *in situ* silylated disulfonimide-based organocatalysts, which address some of these problems in various highly enantioselective Mukaiyama-type reactions involving silicon-containing nucleophiles with unprecedentedly low catalyst loadings. As an example of asymmetric counteranion-directed catalysis (ACDC) [4], these reactions proceed via silylation of an electrophile, generating a cationic reactive species that ion-pairs with an enantiopure counteranion and reacts with a silylated nucleophile (Fig. 2).

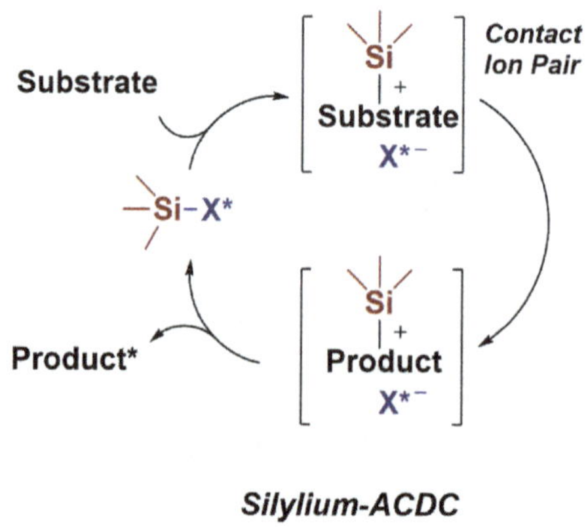

Fig. 2. "Silylium-ACDC" as a new approach to asymmetric Lewis acid organocatalysis.

We became interested in advancing this "silylium-ACDC" approach to, in principle, all types of Lewis acid catalyzed reactions, including those that do not involve silylated reagents. One first realization of this concept has been with the development of extremely active organic Lewis acid catalysts that enable asymmetric versions of highly challenging Diels–Alder reactions [5].

Developing high-performance organocatalysis

Organocatalytic methodologies have frequently required relatively high catalyst loadings. Especially in the early phase of the field, catalyst amounts of 20 mol% or even higher were not uncommon [6]. While high loadings may not be problematic with catalysts such as (*S*)-proline, which is edible, the more recently developed organocatalysts are often more precious and possibly also not non-toxic. Nonetheless, the relatively frequent use of large amounts of the organocatalyst has occasionally led to the opinion that organocatalysis *always* requires high catalyst loadings while transition metal catalysis and biocatalysis, in contrast, *always* function at very low catalyst loadings. Recent work in my group and elsewhere however, has convinced us that this opinion is not entirely accurate. We have found that with newly developed catalyst motifs we were able to reach catalyst loadings approaching the sub-*ppm* level in reactions of high relevance to chemical synthesis [7].

Acknowledgments

I thank all previous and current members of my group for their contributions to the results discussed here and the MPG, the ERC and the DFG for funding.

References

[1] B. List, J. W. Yang, *Science*, **313**, 1584 (2006).
[2] D. W. C. MacMillan, *Nature*, **455**, 304 (2008).
[3] J.-W. Lee, T. Mayer-Gall, K. Opwis, C. E. Song, J. S. Gutmann *et al.*, *Science*, **341**, 1225 (2013).
[4] M. Mahlau, B. List, *Angew. Chem. Int. Ed.* **52**, 518 (2013).
[5] T. Gatzenmeier, M. van Gemmeren, Y. Xie, D. Höfler, M. Leutzsch *et al.*, *Science*, **351**, 949 (2016).
[6] J. W. Yang, C. Chandler, M. Stadler, D. Kampen, B. List, *Nature*, **452**, 453 (2008).
[7] Unpublished results.

SESSION 1: HOMOGENEOUS CATALYSIS

CHAIR: ROBERT GRUBBS
AUDITORS: S. HERMANS[1], G. EVANO[2]

[1]*Université Catholique de Louvain, Institut de la Matière Condensée et des Nanosciences (IMCN), Place Louis Pasteur 1, B-1348 Louvain-la-Neuve, Belgium*
[2]*Laboratoire de Chimie Organique, Service de Chimie et PhysicoChimie Organiques, Université Libre de Bruxelles, Avenue F. D. Roosevelt 50, CP160/06, 1050 Brussels, Belgium*

Discussion among the panel members

Robert Grubbs: I'd like to thank all the people who talked this morning. I think we have a very good overview now of the area of homogeneous catalysis. Basically, we've covered from general organometallic chemistry to organocatalysis, we've seen very complex molecules being prepared and also very subtle reactions that people are looking at to do transformations. So we're now supposed to spend a half hour having a general discussion. Does the panel have any general questions they would like to ask or discuss?

Stephen Buchwald: I think one of the things that I have seen is the need for long-term funding. So, if I look for example at your metathesis research and, similarly, to our work in ligand development and carbon–nitrogen bond formation — which I didn't talk about today — I think one of the things is that one needs to have long-term funding if you're going to try to address problems that are significant. It takes a while, longer that we'd like, to actually not only discover the basic chemistry and understand the mechanisms, but also to work it through past the low-hanging fruit, which is the simplest, the first one or two papers, and then get it to a point where you can do transfer to industry or to researchers in a variety of fields. And I think, at least in the United States, this is something that is sorely lacking. I would also make a pitch here for basic research, even though maybe the end result is applied research. I think the applied research is all, completely, hundred percent dependent on having an increased level of funding for basic research and I would also reject the idea that everybody should be necessarily focused on doing something that has, in the end, an application.

Benjamin List: I'm always marveling about the fact that only we, Germans, have this amazing Max Planck Society, which exactly does that. Where you have long-time view on your research, and you can work on really tough problems, and you can take all the time you need. I mean, hopefully you get it done before you retire, but several decades is absolutely no problem. I don't understand why no other country has this kind of model that you give at least to your very best people. They made a mistake with me but let's say Steve and you, why don't the US give you, or Melanie, funding until you retire, for the next 40 years? We know you guys are driven by yourselves, you're creative, you're original and you're hard-working, so why not? What's the problem about this? I really agree with you Steve.

Robert Grubbs: I was lucky. During the generation I grew up, we had pretty long-term funding but I think that's gone in the US. I think I've been very lucky that I've been able to work on one reaction for a very long time but that's probably not going to happen again. I mean even with our National Science Foundation, which supports fundamental work, for every proposal, you have to have a statement about what can be used and what can be done with it. So it's getting much more difficult to do this long-term. And I agree, it would be great to have a Max Planck Society, but we don't. Anyone else want to comment on that? Melanie, you're in the business, young, getting started... I think longer terms are low again. I think in many places the emphasis on short term stuff is even more difficult. And so I think that's going to be a big problem.

Ben Feringa: I fully agree that this is worrying, because if you think about fundamental problems — it was mentioned today, methane activation, or CO_2 conversion, or whatever — if you really want to make a catalytic process and industry wants to use it, we need long-term commitment. But it's very difficult, at least in Holland, in the National Science Foundation and so to get any money for this kind of long-term research. Now we started to convince some industries — Bert Weckhuysen is sitting there — to build up a consortium with industries for a 10-year program, which is called an advanced research center, exactly to try to do this, to build a long-term 10-year program with both more fundamental research, and more applied research, which is then more governed by the industry. We have to see: we will start this year, or early next year, with the program, we will see how it will work out. We have some funding from the government but we have to find out if that will be continued in the future. It is a tough struggle. We have been fighting for it for three years — Bert that's correct? — to get the noses pointing in the same direction. But we should indeed pursue that method, and I also don't know exactly how to do that. Because you would like to do that in a much broader context, like for instance an European initiative, or even a US-Japanese-European initiative, because we all know it is extremely difficult to bring CO_2 to a fuel in a couple of years and we cannot wait 50 years for a solution. So yes, any mechanism that can help or where we can join forces or we can advocate "look we need fundamental research", and we need persistence over many years and we need, maybe — and this is another aspect that I would mention — also the input of various disciplines. I mentioned it briefly, we are good in homogeneous and heterogeneous catalysis, or biocatalysis, but we should learn from each other but also maybe, from the chemical engineers. How to make a real process out of that?

Kyoko Nozaki: Could I take this opportunity to ask a question to the heterogeneous players and the biocatalysis players that we, homogeneous catalysis researchers working on methane utilization or the utilization of renewable resources, have? Heterogeneous catalysis might be much more superior to work on hydrocarbon oxidation or might biocatalysis superior to work on renewables. So I just want to have comments from those players on the subjects we presented today.

Robert Grubbs: The key is that we have to identify, I think, as you say, which reactions are best to do by heterogeneous catalysts and which are best to do by homogeneous catalysts, and how to combine the two. And also the whole issue, as you raised in your last topic, is: how do you turn a homogeneous catalyst into a heterogeneous system? And under what circumstances do you do that? And is that advisable and what are the features?

Karen Goldberg: Maybe I'm going to make a comment that kind of combines both of those areas in terms of funding and combining heterogeneous and homogeneous engineers. One funding mechanism that I do think has become very powerful in the United States are these large centers. So we have, in the chemistry division of NSF, Centers for Chemical Innovation, which are large centers bringing together faculty from different disciplines, all chemistry-focused though. So, for example, in our catalysis center we have heterogeneous chemists, homogeneous chemists, those interested in theory and mechanism, and all working together on a lot of these issues. And I think that that mechanism of funding — and that has been now for 10 years, ours will end in 2017 — has been very powerful in getting people together to make progress on a lot of these problems.

Robert Grubbs: I agree this is important. My problem with most centers is that I found that you end up getting enough money for one person which makes it very difficult to go very long term. So the idea is great, but the level of funding I think is really to be increased.

Erick Carreira: Just before the conferences started, I had the pleasure of seating next to Judith Klinman and we were talking together... Well, she brought up the issue that one aspect is that we are really very fragmented as a community. There are probably pairwise combinations of people that could seat next to each other that, then, wouldn't understand each other if they talked about what they do in catalysis. So I think one important issue, and we keep bringing up just homogeneous and how it interfaces with heterogeneous, but as Professor Goldberg brought up, there are also interfaces with the theoreticians and the chemical engineers and how do we break down those barriers? And this is not necessarily about money. Although you would think that as you breakdown those barriers you would generate funding. So I think that's an important topic that we need to consider.

Karen Goldberg: I would agree, and language is really a significant barrier. So when we have discussions between chemists and chemical engineers, we often go for 15 minutes before realizing that we're using different terms to talk about the exact same thing. The more those conversations happen, the easier it would be to get beyond those issues.

Stephen Buchwald: One thing that I think is helpful, (and again money issues often prevent people from doing this), and one of the things that I like to do, and has been very beneficial, is to have as many postdocs as possible from areas outside of the mainstream of what goes on in my group. So that I can bring on a theoretician, or somebody in the material science, somebody more traditional inorganic chemist, chemical engineering and then they can teach me and the other people in the group. And then they can also facilitate the interactions with people, at MIT for example, particularly with chemical engineering or other areas. I have the luxury of being able to afford that and many times people don't. So that's another reason that funding can help facilitate that sort of interaction.

Robert Grubbs: Some of my most important things that have happened to me have been postdocs coming in from all kinds of different areas. And so, I think again about collaborations, and if those collaborations can involve exchange of personnel, that would be very helpful. Is that happening in your NSF center?

Karen Goldberg: Yes, it does. Another issue that I wanted to bring up along those lines: You know we've separated these sessions into homogeneous, heterogeneous, biocatalysis. And the idea of tandem reactions was brought up and cascade reactions, and doing some of those where you're either combining heterogeneous and homogeneous or an enzyme and a homogeneous catalyst. All of those can be really powerful if we can get the community working together on it.

Christophe Copéret: It actually takes time and humility. And we are back to the money point. We don't need so much more money to work. All of us around the table, we have no real money problem, even though we have problem with time. So the problem is to have actually enough money over time. If you have a long program, it takes a lot of humility and time to talk to our colleagues, to understand what they are talking about... And today, time is very precious because the problem we have in our society is that we have no more time. I mean if you look at how many papers we would have to read. How many papers we would have to write? Then, when do you have the time to actually meet your colleagues in the corridor? At the end, it's all coffee break/table-business... That's why, anyway, as Professor Buchwald said, you want to foster in your group this time by bringing all these people together within a group but, at the end, the big problem is time. And I think this is something that we should pass on to the government: we need more time to discuss and less time to write proposals to get money.

JoAnne Stubbe: I would say on the interface between biology and chemistry, there is no way you can survive unless you collaborate, and you absolutely have to have collaborations across all kinds of disciplines: inorganic, organic, biophysical and biological. And that's the only way you can tackle big problems rapidly. And so establishing those collaborations, I think it's key to your success.

Graham Hutchings: In the UK, we've sort of started to address this issue of collaborating. Ben was saying that what they've been doing in the Netherlands has been a superb example for the last twenty years to try to organize catalysis in particular ways. What we've done is to set up what we call the UK catalysis hub, but it took us about seven years to get the funding in place for this and a lot of dogmatic perspiration went into this. So Richard Catlow and I put this idea together and in the end, the EU funding agencies put five-year money on the table. But we've got 45 universities working together now and they have to work together in terms of bringing engineers, bioscientists... So it's across the whole spectrum that we're bringing to this, but you end up with a different way of working. You end up with lots of PI's and one postdoc. So it comes back to the same problem, if you're only getting seed money, you're not going to tackle the really big problems. This is not possible within the way we fund things at the present time. But we are making a start on it. And maybe that's a model that could be used in other societies. Of course, concerning the UK funding agencies, whether we get another five years afterwards, I have no idea. It seems to be very successful, so I hope we will, because then after another five years, as it seems to run its course... This is the problem we have that we never get sustained funding when there's a good thing going.

Ben Feringa: Maybe I can add to this briefly. With this new initiative we started in Holland, we built on an initiative which was chaired by Rutger van Santen for many years in the Dutch school of catalysis. And there, we had a bit of a similar experience because we set the targets high to bring together the heterogeneous, homogeneous catalysis, etc... But it was maybe a little bit too distributed so, in the new program that we are setting forward with the industry we will focus on specific challenges: what kind of expertise would we need from heterogeneous, from homogeneous, from theory, from engineering? And at least that's our dream: to bring these things together with different expertise to work for instance on small molecule activation. Maybe Bert wants to add to this. And I'm really excited about this opportunity, and also the people from industry, because then you learn more about what are their questions as that's often the problem also, at least for me. What is your real question? What do you want us to do? And I must say, what I mostly learnt there is to appreciate the problems from heterogeneous catalysis and try to translate a catalyst to an immobilized one. Why do we have so much difficulties to immobilize catalysts and have them still work? And why are the heterogeneous then so successful? And why do the homogeneous ones, once immobilized, go down in activity or selectivity? To learn from each other's expertise, I think that it was an eye-opener, at least to me.

Bert Weckhuysen: Three years ago, we started in December 2013 if I recall well, we got from the government five and a half million euros, and that was actually our starting money, and they stated: build an advanced research center! Can you make something where industry and academia can meet? We, Ben and myself, thought: can we learn from the past in the Netherlands — the model which has already been eluded to, Rutger van Santen and the top research school catalysis, etc. — and we stated: well, maybe we have to be more selective: less academic groups. They have to apply and we call that "the best what the Netherlands has to offer." So, what we want to do is to bring limited number of scientists, who have to apply, and these people would then work together on topics which are also of interest for the longer term, for industry. So it's CO_2 activation, methane activation, but also you will work on the new materials, what we call coatings, etc... And what is so funny was that we found that we had allies in industry. When we depicted our wish, we thought: ok, they will only go for short term. No, they all said: we want a ten year horizon. So I felt that that was something rewarding to see.

Avelino Corma: I think that, when we try to put together the two communities, we have to take into account that the origins we come from are different. So, people working on heterogeneous catalysis, their education comes mostly from, say, inorganic chemistry, surface science, chemical engineering, and so on... While the people coming from homogeneous catalysis, besides enzymatic of course, they have a very big background on organic chemistry, and inorganic chemistry also. So, in order to put the two of them together, what I believe, and what we are trying to do, is to make teams. And even, making a team in which you have people coming from different backgrounds, I mean from, say, inorganic chemistry, material science, chemical engineering, organic chemistry... that takes time! Because you have to find the right subject and, little by little, transfer ideas to common projects, and get happy working together. That is the reality. And the second point is that, myself as a catalytic man, I don't care about if it's homogeneous or heterogeneous, provided that the thermodynamics works, whichever works better.

Rutger van Santen: I'd like to comment about the previous discussion. What our experience has been over the past 20 years in these processes is that education helps a lot. So what we started to do, 25 years ago, was to bring together people from different communities to create one course. And it turns out that it's much easier to get together if you try to make that course than to try to start research projects. So that you know, maybe, that we have these books of NIOK (Nederlands Insituut voor Onderzoek in de Katalyse) that deal with this. So that part, that's one thing I wanted to say. Second point is, it's much easier now that it was in the past. And that relates also to this development of insights in molecular catalysis. Because heterogeneous catalysis, 25 years, 30 years ago, was engineering. But now of course, part of heterogeneous catalysis is molecular catalysis. So the use of theory and molecular insights that we have now, also in heterogeneous catalysis, help of course immensely to bridge the gaps and also to see the differences, because there are such differences and, out of this, that's actually how we will get it more clear where are the differences and the correspondences. And number three, related to the engineering community, we should have a discussion on reactors. Because as soon as we talk about reactors, we talk about engineering. And that relates to thermodynamics, process conditions, but also of course, the lifetime of the process that we need. And that's also in the program that we have with Bert and Ben now in the Netherlands, of course, a very important point.

Kurt Wüthrich: Let's have now the coffee break and continue the discussion after the coffee break.

General discussion

Kurt Wüthrich: From the discussions so far, I am tempted to conclude that the only important catalyst is money! And I have been asked by some participants that we might perhaps change from money talk to other catalysts. And here I have a question: what about the solvents used? I have looked at some of the structures of these homogeneous catalysts, and they don't look as if they would be very well soluble in water. So, how do the expenses for solvents compare when you use homogeneous catalysis or heterogeneous catalysis? And what are the most widely used solvents when you apply homogeneous catalysts?

Robert Grubbs: It depends on the reaction you're doing. In the pharmaceutical industry, I think Steve would probably be best to answer this. For other kinds of reactions, it's best not to use solvents. So for example, I know of a process using our catalyst, which is converting seed oils into products, and no solvent is used. So you just put the catalyst into the substrate and no solvent, and off you go. The other question is about catalysis and water and I think that's a topic that we should have a broader discussion on. So, do you want to start Steve?

Stephen Buchwald: Partly it's a case by case issue. So, in academia we might use THF and "pharmawise" they might use 2-methyl-THF because separation from water is a lot easier and then they can recycle the solvent. So there is a lot of recycling of the solvent to get done. One of the issues of using water, if you're just using water, is that to get rid of water, you still have to pay for the waste disposal of water if it has a certain degree of contamination. So it's not the panacea that it is sometimes meant out to be. So I think trying to do reactions with high volumetric productivity by running reactions very concentrated, with the smallest amount of solvent, particularly ones that could be easily separable, is one way to do that. With many ligands, if you do want to use water, there are often sulfonated versions that allow the reactions to take place in water. But my experience has been, in consulting for several different pharma companies, there are very few reactions that they're doing in water. And that's one of the things that looks great in theory, but in practice, it seems to be either too expensive or it takes too long to do. So it doesn't actually get used.

Erick Carreira: You have to remember that the algorithms that have been developed to determine the efficiency of a process teach us that it's actually a multivariable equation. So it almost never suffices to focus on one solvent or ligand. And you do have to weight all of the various components that are going to be determining in the efficiency and the success of a process towards manufacturing.

Joachim Sauer: I think this question of the solvent brings us also to differences between the different types of catalysts. When you introduced the subject, the solvent didn't seem to exist. When you said that you know what the catalyst is and you know the mechanism, for you and for all the other presenters here, the solvent was absent. And if I look at most of the simulations, I also see that the solvent is absent. Still, there seems to be complete understanding, although, in the few cases where modeling has been made, the methods are often inadequate to treat the solvent properly. When you bind your substrate, something has to happen with the solvation shell. I would like to understand how important this is, and this could be a point where we see some differences to other types of catalysis.

Melanie Sanford: I think that's a great question. And this is one of the things I was thinking about as the discussion got started on solvent because one of the things that you also didn't see in any of the talks is that many of these reactions are extremely sensitive to the solvent. So you change the solvent: it doesn't work. You change the solvent: the yield goes down, the selectivity changes. In the homogeneous catalysis community in general, we tend to think about solvent as just a dielectric, right. That's just some sort of constant dielectric, and we don't think about it in a more sophisticated way, but yet, it has such a profound effect on these reactions. One of the big challenges in the field is really understanding what the actual molecular basis of some of the solvent effects that we see in these reactions are. So I think it's a really great point. And to add something that we were just discussing briefly during the break: the effect of water on biocatalysis. I think that we are much less sophisticated in terms of understanding solvent.

JoAnne Stubbe: What I was surprised at, in terms of the many many slides we saw, is that there were really no turnover numbers, I mean the number of turnovers we should get before the enzyme became inactive. And then I focused on something that I'm interested in, which is oxygen-dependent reactions with metals. There was a slide in Melanie's presentation where she was trying to do hydroxylation reactions that had an iron hydroperoxide or whatever. But I was wondering what we really knew about those systems, and how they are controlled because in biological systems, I think that a major focus is to try to identify the reactive intermediates in heme/non heme-iron/dinuclear non heme-iron systems.

Melanie Sanford: I would say, just to the turnover numbers that, at least in my systems, it's very dependent on the catalyst, it's very dependent on the reaction. Generally in most of these cases, we're talking about tens to hundreds, maybe up to a thousand or something like that and I would say that it's probably true for most of the examples we heard about today. That is a situation where the application is important. One of the powerful things about homogeneous catalysis is that there are applications. I was just talking to Steve at the break about cases where people in pharma and medicinal chemistry might be happy with a reaction that was actually stoichiometric in palladium because they could get a product they really wanted and they could use it. So I think, sort of back to what Erick said as well, that when you're thinking about the applications of these sorts of things and the usefulness, you don't want to focus on one metric, not just on turnover numbers, catalyst lifetime or selectivity because sometimes you might want different things out of your catalyst. That is one of the nice things about the field actually, that there are different applications and different possible things you can do. In terms of the specific hydroxylation chemistry, I tried to get across, but very ineffectively because I was going too fast — maybe now too — I think we can talk in more detail about mechanisms. In that particular case it's probably hydroxy radical chemistry but you can change the iron catalyst to something like a porphyrin and change the selectivity because you're changing the mechanism and we could talk about some specific details. One of the powerful things about that chemistry is that in fact you can access different mechanisms, and when you do access different mechanisms, you can get different mixtures of products and that can be really powerful and useful. We can talk about details of mechanism later but that's what's going on there.

Robert Grubbs: So again the whole question of turnover numbers and efficiencies, etc., is one of the real growing areas in homogeneous catalysis because, as I pointed out, in heterogeneous catalysis, you can reactivate the catalyst, easily separate it and it's easy to do it. In homogeneous catalysis that's much tougher to do. So if you're going to scale up reactions and do reactions on a big scale, you have to go to very low catalyst loadings. That is one of the directions that we, and I think a lot of other people, are working on. How efficient can a catalyst be? Can you get it down to parts per million or less? And I'm really glad to see in organocatalysis that the loadings are now getting down quite low.

Karen Goldberg: Part of it I think has to do again with the language issue and so synthetic chemists will just talk about catalyst loading in a percent yield on a reaction rather than reporting a turnover number. I also wanted to go back to the issue of solvents. I think that's a critical issue, particularly in C–H bond activation. If you want to do something like methane and you want something to react with methane, what are you going to use as the solvent that your catalyst is not going to react with? And those are huge issues. Going back to cost issue, another one I would like to bring up is the cost of the catalyst. There's been a huge move in recent years to go to non-precious metal catalysts — and we get that from the funding agencies — but the feedback that we are getting from industries is that they want efficient catalysts and using a precious metal catalyst, if it has many turnover numbers and lives for a long time, they can recycle it. They would rather use that, that goes into capital cost, and I think we have to be mindful of this whole multi-variable issue of maximizing what we are trying to get out of a catalytic system and so not all precious metal catalysts are bad. Many are used, as we saw, in large commercial processes. As an example, in the Cativa process that Bob brought up, the catalyst is iridium. They moved on to iridium from rhodium. Rhodium is used a lot in hydroformylation reactions. These are all huge scale homogeneous systems. So there isn't as much emphasis from the companies, as there are from funding agencies, that we need to move to non-precious metals.

<u>Gerhard Ertl:</u> I have a more fundamental question: somebody told before there is a difference between homogeneous and heterogeneous catalysis because the members of the communities have a different background, and I think that's true. I heard a lot of organic chemistry from your group, but very little about kinetics, and you have well-defined molecules reacting with each other and well-defined catalysts in contrast to heterogeneous catalysis, but I was wondering if you really know how a catalyst is operating, I missed that point.

<u>Stephen Buchwald:</u> We have 10 minutes to speak so we cherry pick up what we are going to talk. I think probably almost everybody here at this table who has done the work, has done kinetic analysis of their catalyst systems. If you try to talk about kinetics you're gonna be on slide one when we get the waving. Can I just say one thing about the turnover number? I think turnover numbers can be important but let's just take pharma for example. We have a lot of catalysts that give 100 000 turnovers. If you pick simple systems you can get 10 billion turnovers. If you want to do a Heck reaction, there's almost no limit to the number that you can get. Not in anything that anybody wants but if you think about pharma, 99.5% of the chemists in pharma are medicinal chemists and they are not process or manufacturing chemists and they don't care what the turnover number is. What they care about is generality and ease of use. They don't want to use a glovebox, they want everything to be commercially available and frankly they don't want to weight anything out, so everything is pre-weighted, they're happy and they would do it in molten palladium if that would work better.

<u>Robert Grubbs:</u> I think generally, in the area of homogeneous catalysis, there's been a tremendous amount of kinetic and mechanistic analysis done. In fact Melanie, when she was a student in my group did a lot of kinetic analysis of our systems,

understanding all the details there. So I think it's there, and as Steve said, we didn't talk too much about it, we were talking more about the products. But I think there is a very strong background in this field because many of the people in the field came from physical organic chemistry, which is the field of trying to understand kinetics and reaction mechanisms. So it's been very well covered I think.

Ben Feringa: I would completely agree with that. In the mechanistic studies on homogeneous catalysts, etc., it helps a lot also in designing your next step, your next ligand and so on, to go from random ligand screening to really designing your catalyst. It helped us tremendously and I think also many of our colleagues. On the other hand, we have also to realize our level of understanding. We have now performed asymmetric catalysis for 50–60 years. The whole community of homogeneous catalysis almost jumped on asymmetric catalysis, and help me if I'm exaggerating or if I'm wrong you can correct me, but I would love to know how I can predict a chiral catalyst. I have no idea, there is no Woodward–Hoffman rules for asymmetric catalysis, even after 60 years, we have no idea. We screen ligands and we are pretty good at saying "Oh, this ligand of Steve will work for my thing and this one will work maybe and give some selectivity" but it's a lot of hand waving still. So the level of understanding, especially when you go to, for instance, highly selective catalysts is still rather modest, let me say it this way.

Erick Carreira: I think that's JoAnne's point. You can never do enough mechanism in essence and that's how the field moves forward. We are all the beneficiaries of physical organic chemistry in its golden age, in the 1960–70s perhaps, and we need to continue to do that mechanistic investigation as most of us do, to propel forward in the next generation of catalysts and processes.

Benjamin List: I found your question very inspiring Professor Ertl and I should say this upfront, we are doing a lot of kinetics, we have computational chemists, we have spectroscopists in the lab. I think this is actually the best way to do these collaborations within the group and not sort of externally, ideally, in an ideal world. I still would like to make this point that chemistry has always been a twofold science, those who are more the understanding community people, like maybe you are, and then there is the other side which is more on the creative side [laughs]. My name is Steve Buchwald from MIT [laughs]. Don't get me wrong, you know what I mean [laughs]. In the groups you have the students, they grab the project and they work on the topics they're interested in and it's true, it's just the reality. There's a lot of trial and error, playfulness intuition in catalysis design. And I think this was also true for Haber and Bosch perhaps, right, if you think about it.

Martin Albrecht: I'd like to differentiate a bit from what we have talked about — mechanistic understanding — I think we do have very sophisticated understanding of hydrogenations, of hydroformylations, of a good number of reactions like metathesis. But then there are other types of reactions that are homogeneous, for which we have plenty of kinetic data but we don't understand the mechanism at all. I'd say many of these reactions are related to oxidation so that's going back to what JoAnne brought up. About water oxidation, I think we have 17 mechanisms out there, we have plenty of kinetic studies, and we still have no clue what are relevant intermediates. So there're still lots to do and I think as well we should start mixing heterogeneous and homogeneous concepts more and more. So that's where symposiums like this are useful.

Gerhard Ertl: This touches the core of my question. In fact, I wonder if you can really predict and design a homogeneous catalyst for a desired reaction.

Stephen Buchwald: From scratch? I would say I have more fingers than the important reactions in metal catalysis that have been designed from scratch. It is an iterative thing and once you sort of know a certain amount, then you can make more and more logical changes to affect, and hopefully perfect, the reactions that you're doing. But just starting 'back of an envelope' and doing calculation and *ab initio* starting with new reactions, there — to my knowledge — aren't very many.

Robert Grubbs: As you know in catalysis, it's catalytic cycles, and you have many intermediates. But we're still lacking the intuition about relative stabilities and relative rates between those. And you may be able to do one but it's difficult to do 5 or 6 or 7. And computationally it's still quite hard to do, trying to get all the different energy levels properly so you can predict this whole cycle. You can do individual steps but doing a total cycle, it's quite difficult.

Christophe Copéret: I think one of the big problems in catalysis is that catalysis is a 1 kCal problem. And so, discovery is one thing, and we saw a lot of discoveries this morning where you can find your reaction, and then from that you can design, and then computational chemistry can help a lot to design, and even to predict in some ways. But from scratch today, humility is important for us because I don't think, as a molecular chemist or heterogeneous catalysis people or chemical engineer, you can predict something which is 1 kCal. We have to be clear on that. That doesn't mean we shouldn't aim at trying to predict, and I think our colleagues from computational chemistry can probably reply. They are trying obviously to approach this. I personally believe it would be impossible because it is intrinsically very complex, we'll come back to the solvents — the solvent has an importance in everything we do — unless we have not a super computer but a mega mega super computer. At the end you are better off doing yourself and discovering yourself, and then we can go from that. Coming back to a question of Professor Ertl, I think — and that's a comment I had actually with Rutger von Santen during the coffee break — physical organic chemistry and physical inorganic chemistry are the core of the matter, and probably we don't do it enough. That's actually one of the questions that you have, maybe we don't do it enough. There are always several stages in research in catalysis: there's discovery, which is basically unpredictable, or close to being unpredictable, and then, if you find something good, it's maybe time to spend some time trying to understand how it works. And then physical tools are essential to go forward. And again that's why you start to bridge and to talk to your colleagues because the big problem we have today is that science is so big that we cannot know everything, and that's why you need to have all these postdocs which are top-level people of different fields, your colleague next-door might know more than you do on one field and that's how you can progress again.

Hans-Joachim Freund: I'm still a little puzzled with the discussion because I'm in heterogeneous processes and we don't know much about what's happening and so forth. And you claim you know everything but remember the first question that Joachim Sauer asked: "what does the solvent do?" How can you claim, even with the most sophisticated kinetic analysis as you do, that you have a detailed understanding of the mechanism if you don't understand what the solvent does?

Stephen Buchwald: Most of us will never claim that we know everything. All I said was that we had done some kinetic analysis and we have a working model, which is why we don't put in the solvent. Because if we did, again we would be on slide 1 before we'd be on slide 0.1. So there's a lot that needs to be done in terms of that but I want to change the focus just a tiny bit. The mechanistic stuff is great — I like doing it — but I want to argue against trying to force everybody to do things the same way. I think if you don't want to do mechanism, that's perfectly fine. Empiricism is good (Edison was pretty effective) to steal a line from my friend David MacMillan. He hadn't figured out everything, and he probably never did. So I think what we want is diversity in approaches and that's going to be the fastest way to success. And to respect others' approach, don't say "mine is more intellectual, mine is more effective or more practical." People are doing what they feel comfortable with, what they're good at, and the fact that they are different is going to get us to the right answer faster than if everybody thinks the same way.

Robert Grubbs: We don't claim to know everything and so we try to study mechanisms to understand to a certain level, and then you operate from that level. I've been working on a system for 25 years and I still get surprised very often because we don't understand everything. But you try to understand what you can, and sort of try to know what you don't know, and you operate pretty well. If you want to spend an amazing amount of time, Jack Halpern for example really analyzed the system in great detail, understood the factor of solvent, etc., but he spent probably 10–15 years studying one system, and we don't have time for that anymore. So you try to reach the level you can and then you go from there. You just don't have the time.

Ben Feringa: I really like this and I'm looking forward to the discussions later on heterogeneous catalysis and I tell you why: because maybe we have to think this week about redefining what homogeneous and heterogeneous means. Because I'm not so sure anymore that the classic definition that we use holds because I don't know exactly how homogeneous is my dissolved catalyst. What does it mean at the nanoscale with the solvent molecules around it? Is it one species? The conformational flexibility, the different states, what does it mean on a surface? Is this really heterogeneous or is the active metal coming off the surface a little bit? And then you are much closer to what we see for instance in organometallic complexes. I would love to see that discussion during this week.

Giuseppe Bellusi: It can be even simpler than that. By considering your previous comment that when trying to immobilize a homogeneous catalyst, one very often sees a loss of activity and selectivity. I had a limited experience in the heterogenization of homogeneous catalyst and it has been my experience that I always had such kind of losses. My perception was that it was not only a matter of chemical interaction with the support but it could have been very often that by immobilizing a catalyst, the reactive site was diluted in the reaction volume, and this generates a problem that has been tackled with the instruments of the chemical engineering and the reactor design. This makes important also the knowledge of the kinetics for a proper design of the reactor. A proper reactor design in many instances could solve the problem of the homogeneous catalyst immobilization.

Christophe Copéret: Our experience is actually if you immobilize properly homogeneous catalysts, they are always much better than the homogeneous ones, and by large. So the big problem is again, how much time we spend in trying to understand. The problem we are facing today is a problem of surface chemistry, surface science. The beauty of molecular chemistry is that you have so many tools to investigate the system very fast. When you go to surfaces, you dilute, as Bellusi said. You dilute your system on a surface and then it gets much more complicated at what we are looking at. And so sometimes it doesn't work, just because actually you never made what was described as a ChemDraw structure. This is one of the big problems. How do we know that we actually make what we intend to make? This is one of the biggest challenge in surface chemistry. Often it degrades, maybe it's incompatible or maybe it was actually never made. This is a big problem, even before discussing surface interactions, can they decrease activity? There's a big challenge and I think 21st century surface science has evolved so much that we have now tools to address these questions but again it takes time. It will take a lot of time to be able to bring these tools. And if you see what they are able to do now at the atomic scale, you should be able to do that at some point on powder catalyst and if we can bring that together, maybe we can actually move forward. But again, it will take time to take the same tools, to think molecularly. So at the end, we are back to "can we think as molecules, whether you are on a surface, whether you are in solution, what is this entity doing to my substrate?" That's what we should be basically questioning.

Graham Hutchings: There are lots of similarities between the homogeneous catalysis approach and the heterogeneous catalysis approach. After this afternoon, let's see if that's true. But I think one of the key things we share is the role of the solvent, and however we are using the catalyst, there is a fluid phase. That's something I think as a heterogeneous community, we may take as given in some circumstances, because we have to think about the things we can use as most available. But it's something for which I think theory, to go back to Joachim's point, is what we are really lacking at the present time. If we want to get bottom-up design, the role of the solvent, the fluid phase and what we are doing, it's going to be critical. So the methane oxidation or the C–H activation we grab hold with this and water was the only thing we could trust because we could do NMR in water, we could do the analysis and methane was not going to be very reactive, we could look at very small amounts of product. And then we thought it would be clever to start using methanol as the solvent because that's what we are going to make and we just make more methanol and the catalyst didn't work at all, didn't do anything. So, it was a great idea, and we got somewhere, and we couldn't use it. Somewhere along the line the solvent needs to be very early in in our discussions, and it isn't at the moment I suspect.

Martina Havenith: I'm heading a center of excellence on solvation with the focus on this. I just want to add to my colleagues that I'm not optimistic on that. It's now time that they can go along with theory and experiment and address this question, so upscaling up so I think it's time to do this. So as experimentalists, we think we have come to the point that we can "see" a single solvent molecule and what it does and have methods to address this. Meanwhile, the theoreticians have proceeded and have developed methods for upscaling to the bulk. So, for us, we are more optimistic on that, this is the time to approach now from both sides and address exactly this question: addressing the role of the solvent.

Steven Boxer: This is a problem that we've worried about a lot recently and it turns out that there are actually very good force fields now for many many organic solvents. We all think about water, having the most advanced force field and that's certainly true, but there are lots of good force fields that have been carefully parametrized and we've actually tested them pretty systematically and they are very very reliable. So there's no reason at all why we can't include solvent in simulations of these things. The best solvent parameters we know of were published in 2012, and I think there was a big European consortium. That's how recent that has been in terms of being widely available, but people should use it.

Joachim Sauer: The force field may be good if there is a non-specific role of the solvent, but if the solvent molecule binds in a more interesting way to your catalyst before the substrate is approaching the site, or if you would like to describe what happens along the reaction path, then I'm afraid it's not enough to look at force fields because you cannot make them feeling these chemical and structural changes, unless you have a reactive force field but they are rarely available and difficult to parametrize.

Steven Boxer: If you look at the solvent from the right perspective, you can do exactly what you described. What you want to do is not being hung up on things like solvent polarity and sort of classical ways of thinking but you want to have a molecular view and a molecular view of these non-covalent interactions requires a different view. We now have ways of measuring those things, not just calculating them, and that is to focus on the electric field that's created by the solvent, and that's a specific and molecular characteristic. I think then you can bring it into consideration of reactions.

Karl Anker Jørgensen: Moving from Lewis acid catalysis to organocatalysis 16–17 years ago was a relief because we don't worry about solvents. It works in water, it works with oxygen, it works in a variety of different solvents and what we have been told by industry it's so easy to scale up because we don't need to worry too much about solvents. So that's one of the changes moving from metal catalysis to also catalysis in water and other solvents. It just has one disadvantage and that's with the students: I can tell you that it has cost me thousands of euros because the students want to try to run organic catalytic reactions in beers, wines, various types of gins. We know what the difference in enantiomeric excess is in a variety of different Carlsberg products. But they did it, Friday and Saturday. Thank you.

Martin Albrecht: We would probably understand quite a bit when it comes to the principal role of solvents in transition metal reactions, in complexation and ligation and substitutions. I think where it matters then really is the very weak interactions at remote sites of the catalyst that changes the structure only a slight tiny bit, and then we go back to this 1 kCal question that Christophe Copéret addressed. So very very small changes make those modifications, but I think that's because a catalytic cycle is *per se* dynamic: it's flexible, it's changing all over the place. So I think we're having an awful hard time to model each intermediate properly without the solvent interactions. The remote ones, not the ones close to the reactive site.

Frank Neese: From a theoretical point of view, even if you convince me that you can do the molecular dynamics efficient enough to be conversion, even if you convince me that your force field is good to a 1 kCal/mol, I still don't think that you solved the catalysis problem with it because many of these catalytic reactions that we have seen here proceed *via* openshell transition metal intermediates and here, it is really very electronic structure, not geometric structure. The electronic structure is absolutely crucial and you will not have shred of a chance to incorporate that in a force field. There are many, many reactions of that kind, basically almost all metallo-protein reactions outside the field of, say, zinc enzymes. I'm sure force fields can solve and productively contribute to great many problems, but there is a very large number of problems especially in catalysis where this will not cut it, I'm afraid.

Kurt Wüthrich: Well, I think that we should actually get to a close of the session. I think that you have really well-demonstrated that homogeneous catalysis is an immature field [laughs] which needs much more funding for basic research and which is a great area for young scientists to get into because so many problems are still open. I don't know what your conclusions are but maybe you'll close the session.

Robert Grubbs: Thanks, and I totally agree with you, you've understood things properly. So I think it's time to end, thank you for all your contributions, we've brought up some very interesting points and I think the ideas of the sort of magnitude of differences, the kilocalorie problem, the calorie problem when you're talking about asymmetric induction, is a very serious problem both from a qualitative viewpoint and also, as you indicate, from a quantitative viewpoint of trying to calculate and do things. So I think we still have a long way to go and I'm looking forward to learning what's happening in the heterogeneous field and the bio field, so we can move forward. Thanks everyone, thanks to the Panel, thank you for all your questions.

Session 2

Heterogeneous Catalysis and Characterization of Catalyst Surfaces

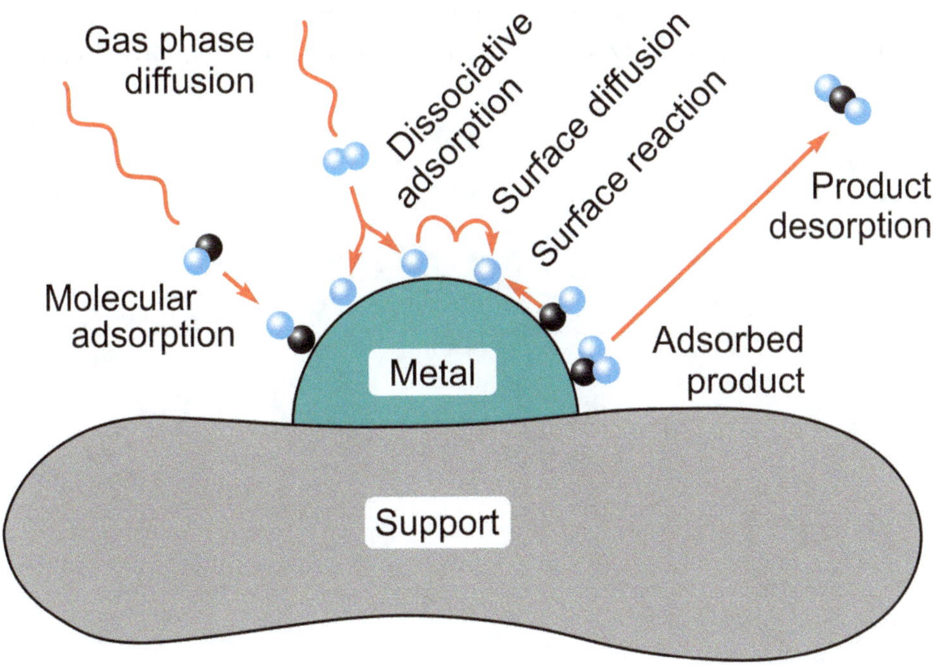

Scheme of the key steps in a typical process of heterogeneous catalysis on a supported metal nanoparticle.
Image by: G. Ertl, Fritz Haber Institut der Max Planck-Gesellschaft.

HETEROGENEOUS CATALYSIS: WHERE ARE WE?

GERHARD ERTL
Department of Physical Chemistry, Fritz Haber Institut der Max Planck-Gesellschaft, 14195 Berlin, Germany

The organisation of the first Solvay Conference in 1911 was suggested by Walther Nernst who also played a decisive role in the development of one of the most important industrial processes based on heterogeneous catalysis: the synthesis of ammonia from the elements. In 1905 he had established his heat theorem, later known as the Third Law of Thermodynamics, which allows evaluating the equilibrium constant of a chemical reaction, and a controversy with Fritz Haber prompted the latter to work again on the catalysis of ammonia synthesis. This eventually lead to the Haber–Bosch process which started to be implemented in 1913. The successful catalyst (a *doubly promoted* iron catalyst) had been found by A. Mittasch after checking several thousand samples. This strategy, nowadays denoted as *high throughput screening*, is still widely in use for optimizing existing and finding novel catalysts.

However, much progress has been made in the meantime from a more fundamental level: For example, the mechanism of the ammonia synthesis reaction is now well-understood after almost seven decades of intensive research on this topic. This success was achieved with a *bottom up* strategy denoted as *surface science approach*. The complex catalyst surface is reduced to the model of a clean single crystal surface where the elementary processes can be investigated on atomic scale by a whole arsenal of physical techniques which were developed during the past decades. In parallel, strong progress of theory took place, so that for simple systems, bonding to the surface and even reaction rates can be evaluated *ab initio* with fair accuracy on the basis of quantum chemistry and statistical mechanics.

These model systems are of course usually quite different from the conditions of *real* catalysis. Several of the physical techniques are based on the interaction of electrons with the surface and can hence only be operated under high vacuum conditions. This gave rise to the so-called *pressure gap*. Since the state of the surface is affected by the ongoing catalytic reaction, methods which can be applied under working conditions *(operando)* are needed. Even more severe, however, is the *materials gap*. Real catalysts often consist of small particles on a support. Their controlled preparation relies among others on the methods of colloid chemistry. Hence, heterogeneous catalysis was a nanotechnology long before this term was invented. Particles consisting of less than about 100 atoms will exhibit altered electronic properties and hence also reactivities. A striking example is offered by gold which as bulk sample is practically inert while very small particles may have very interesting catalytic activities. Additional effects can come into play by structural defects and by interaction with the support. Controlled preparation and characterization of the *active sites* of such samples will hence be of crucial importance.

Another possibility for achieving high specific surface areas consists in the use of porous materials with large inner surfaces. In this connection, zeolites are for example widely in use in the oil industry. Related materials with well defined inner structure can serve as *single site catalysts* where the *active sites* are well separated from each other and well characterized. Such systems can be considered as forming a bridge between *heterogeneous* and *homogeneous* catalysis.

Research in heterogeneous catalysis has still to solve the following main problems:

- Characterization of the catalyst surface and identification of the nature of the *active sites* under working conditions.
- Determination of the sequence of reaction steps (*i.e.* the reaction mechanism) leading eventually to formulation of the reaction rate (*microkinetics*).
- Since catalysis is a kinetic phenomenon, a clear definition of the catalytic activity is required.

Further progress of the interplay between experiment and theory in order to design *ab initio* a catalyst with desired activity and selectivity under optimum working conditions as the ultimate goal.

INTEGRATION OF THE THREE FIELDS OF CATALYSIS: HETEROGENEOUS, HOMOGENEOUS, AND ENZYME

GABOR A. SOMORJAI, RONG YE, TYLER J. HURLBURT and KAIRAT SABYROV

Department of Chemistry University of California, Berkeley, CA 94720, USA
Chemical Science Division, Lawrence Berkeley National Laboratory, Berkeley, CA 94720, USA

View of the present state of research on molecular catalysis

In the past 25 years, catalysis has become a molecular science. This was, in part, due to the rise of nanomaterial synthesis since most catalysts are nanomaterials that participate in chemical processes to produce molecules and chemical changes in various technologies and the human body. An example of a heterogeneous, homogeneous, and enzyme catalyst are each shown in Fig. 1.

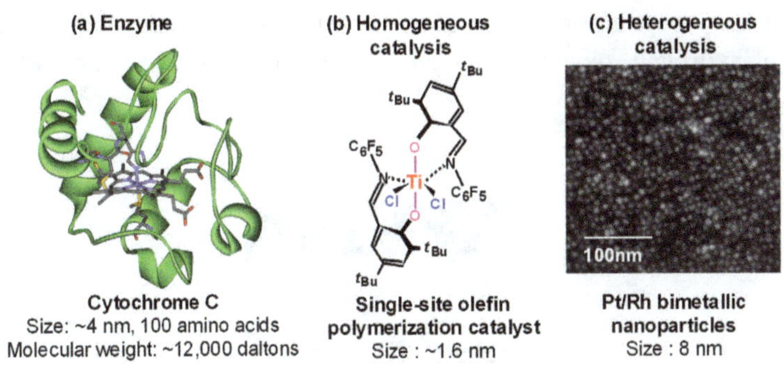

Fig. 1. Examples of an enzyme, homogeneous, and heterogeneous catalyst in nanometer size.

In addition to nanomaterials, instruments that can study the catalysts' composition, atomic and molecular structure, and oxidation states with high spatial and time resolution have become available. These instruments can monitor the dynamic changes that occur during catalytic reactions, not just measuring the state of the catalyst before and after reaction. Many of these instruments are listed in Table 1 [1], which includes photon and electron scattering instruments that were made available by the use of a synchrotron.

My approach to molecular catalysis was from the direction of surface science. The most important types of sites in catalysis science are covalent, forming covalent bonds between metals and molecules, and acid-base, where a charged species such as an electron or proton is transferred to the reactant molecule. We can readily prepare heterogeneous metal catalysts in the 1–10 nm size range in various shapes, thus varying the covalent type catalysts. These nanoparticles can be supported on mesoporous, high surface area oxides which have different acidic or basic properties. Our aim is to produce high selectivity systems that have high turnovers and slow deactivations, while understanding why the system behaves the way it does.

Table 1. In situ surface techniques.

- Sum Frequency Generation Vibrational Spectroscopy
- Synchrotron
  - Ambient Pressure XPS
  - X-ray Absorption Near-Edge Structure
  - Extended X-ray Absorption Fine Structure
  - IR and X-ray Micro-Spectroscopy
- High Pressure Scanning Tunneling Microscopy
- Nanodiode

Starting with single crystals, it was discovered that surface defects, steps and kinks, with low coordination sites have the chemical influence to dissociate the molecules of hydrogen, oxygen, nitrogen, and even break C–H and C–C bonds as well (Fig. 2) [1].

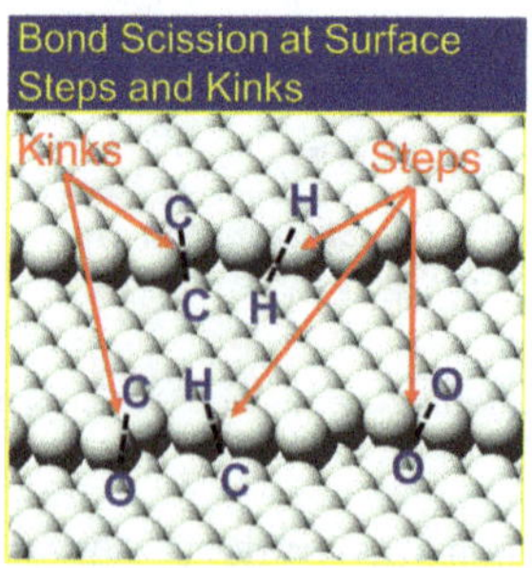

Fig. 2. An illustration of surface steps and kinks on the surface of a single crystal. Panels adapted with permission from Ref. [1], Wiley.

With the onset of nanoparticle synthesis and the control over size and shape, dependence of changes in bonding was amplified further as compared to the single crystal defect behavior. Since catalysts are mostly nanoparticles, the reaction selectivities are heavily influenced by the nanoparticle structure and size. Metal nanoparticle synthesis and examples of size and shape control is shown in Fig. 3.

Metal nanoparticles for size-dependent covalent bond catalysis

The major technique for the synthesis of nanoparticle catalysts is colloid chemistry (Fig. 3A). These nanoparticle-based catalysts are produced with precisely controlled sizes in the 1–10 nm range, and can be placed in two dimensions using the Langmuir–Blodgett technique [2] or placed within porous three-dimensional supports. The nanoparticles, mostly metals, are placed in microporous and mesoporous supports made out of silica or transition metal oxides having between 5–25 nm pores. Templating methods exist whereby a template is coated by an oxide precursor. The template is removed by chemical leaching to leave behind porous oxide which is then loaded with nanoparticles for catalytic studies. The oxide itself, as it will be shown, is often a very important ingredient for catalytic reactions [1].

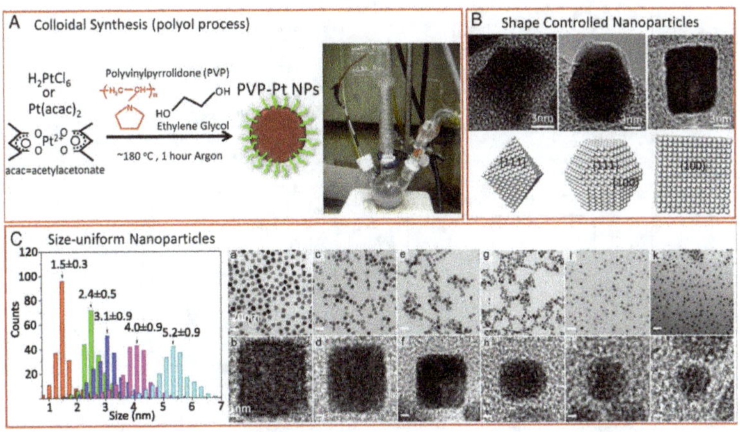

Fig. 3. Examples of size and shape control of nanoparticles. (A) Schematic of Pt nanoparticles synthesis by the polyol reduction method. (B) Transmission electron microscopy (TEM) images and ball models of Pt nanoparticles with different shapes. (C, Left) Particle size distribution histograms of the Pt/SBA-15 series catalysts obtained from TEM images. The number inserts indicate the mean particle diameter and SD for each sample. (Right) TEM and high-resolution TEM (HRTEM) images of Pt nanocrystals with different shapes and sizes. TEM images of (a) 9-nm nanocubes, (c) 7-nm nanocubes, (e) 6-nm nanocubes, (g) 5-nm nanocubes, (i) 5-nm nanopolyhedra, and (k) 3.5-nm nuclei. HRTEM images of a single (b) 9-nm nanocube, (d) 7-nm nanocube, (f) 6-nm nanocube, and (h) 5-nm nanocube along the [100] zone axis. HRTEM images of a single (j) 5-nm nanopolyhedron, and (l) 3.5-nm nucleus along the [111] zone axis. (Scale bars: Tem images, 20 nm; HRTEM images, 1 nm.) Panels adapted with permission as follows: B, Ref. [2], American Chemical Society; C, Left, Ref. [4], Elsevier; C, Right, Ref. [5], American Chemical Society.

In our work, we found that the size and shape of metal nanoparticles control both catalytic reaction rates and selectivities. We also learned that all of these multipath reactions show size dependence in their turnover rates and selectivity [3]. It is possible to achieve different shapes of platinum nanoparticles (Fig. 3B) [2] and the size regime of platinum nanoparticles can be sharply focused in the 1.5 to 8.0 nm range (Fig. 3C). Such a well-defined particle size and distribution are essential to detecting the differences in turnover rates of the hydrogenation of either benzene or toluene, as they are structure sensitive. We observed a fourfold change in turnover rates between nanoparticles of platinum in the 2–4 nm range and in the 4–6 nm range. The size and shape dependence of nanoparticles can readily be controlled by colloid science technology (Fig. 3C) [2, 4, 5]. The isomerization of methylcyclopentane is much more shape dependent on the platinum nanoparticles than size dependent [6] (Fig. 4A). In the case of the Fischer–Tropsch CO hydrogenation reaction over cobalt nanoparticles, the product distribution is size dependent [7] and the turnover rate also increases fivefold with increasing size (Fig. 4B).

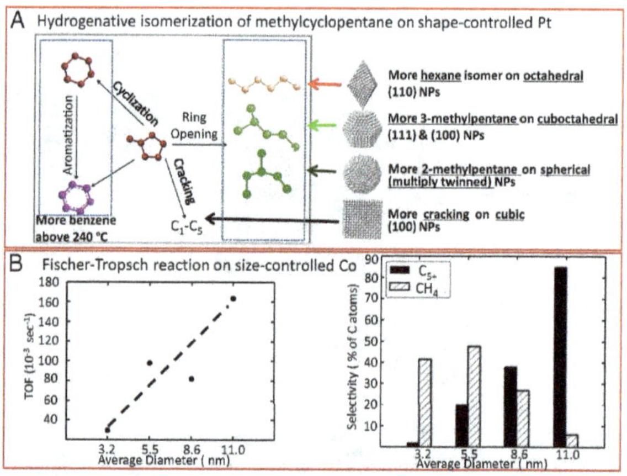

Fig. 4. Examples of shape and size dependence of nanoparticle catalysts. (A) Reaction pathways and possible products of methylcyclopentane hydrogenation reaction catalyzed by Pt nanoparticles with different shapes. (B, Left) CO consumption turnover frequency for the CO hydrogenation at 5 bar (H_2:CO = 2:1) for various sizes of cobalt nanoparticle catalysts supported on MCF-17. The turnover frequency (TOF) corresponds to the number of CO molecules converted in time divided by the number of cobalt atoms at the catalyst surface. (Right) Selectivities toward hydrocarbons with a carbon number of 5 and higher (C_{5+}) and methane selectivities (SCH_4) as a function of cobalt crystallite sizes for hydrogenation of carbon monoxide (H_2:CO = 2:1) at 5 bar and 250°C. Both selectivities are expressed relative to the total number of carbon atoms converted. Panels adapted with permission as follows: A, Ref. [6], Springer; B, Ref. [7], Springer.

Synthesis of catalyst nanoparticles has been rapidly developed in all three fields (heterogeneous, homogeneous and enzyme) but they were developed independently by investigators with expertise in only one of the three fields of catalysis. The outcome of our studies with nanoparticles synthesized to carry out heterogeneous catalysis is that the size and shape of the nanoparticles determine the reaction rates and selectivity. Also, a dynamic restructuring of surfaces was observed as either the reaction conditions changed from oxidizing to reducing or the temperature changed, producing a different catalyst surface composition. These surface changes were very apparent in bimetallic systems and oxide-metal interfaces [1].

The large mean free path of electrons in metals lead to the discovery of high kinetic energy hot electrons, which provided charges at oxide-metal interfaces (Schottky barrier) that generated charged molecular ions at the interfaces leading to strong metal surface interaction (SMSI) with unique acid-base interactions at these sites and unique catalytic chemistry. The structural rearrangements of catalytic surfaces during catalytic turnover detected by scanning tunneling microscope (STM) and extended X-ray absorption fine structure (EXAFS), and other techniques revealed the dynamic nature of the catalytically active interfaces.

With all these molecular changes made possible through the synthesis of nanoparticles, and the instrumentation developments that could monitor surfaces during reaction turnovers, the molecular dynamics of active catalysts was uncovered mostly in heterogeneous catalysis at solid-gas and solid-liquid interfaces. However, the three fields of catalysis (hetero, homo and enzyme) continued and pursued catalysis studies as separate fields. This happened in spite of the fact that all three fields of catalysis were frequently undertaken through similar or identical reaction chemistry. Examples of these are shown in Table 2 [8–16]. Since catalysis is an interface science that is performed at solid-liquid or solid-gas interfaces, it appears that the separation can be, and should be, overcome. It is likely that hybrid systems that would be created this way will produce new chemistry yielding desirable product molecules.

Table 2. Example papers of the same type of reactions catalyzed by all three fields of catalysis. See Refs. [8–16].

Type	Alcohol Oxidation	Isomerization	Esterification
Heterogeneous	Solvent-Free Oxidation of Primary Alcohols to Aldehydes Using Au-Pd/TiO_2 Catalysts. (2006)	A Porous Coordination Network Catalyzes an Olefin Isomerization Reaction in the Pore. (2010)	Esterification reaction using solid heterogeneous acid catalysts under solvent-less condition. (2005)
Homogeneous	Copper Catalyzed Oxidation of Alcohols to Aldehydes and Ketones: An Efficient, Aerobic Alternative. (1996)	In Situ Generated Bulky Palladium Hydride Complexes as Catalysts for the Efficient Isomerization of Olefins. (2010)	Bulky Diarylammonium Arenesulfonates as Selective Esterification Catalysts. (2005)
Enzyme	Alternative pathways and reactions of benzyl alcohol and benzaldehyde with horse liver alcohol dehydrogenase. (1993)	Determination of kinetic constants for peptidyl prolyl cis-trans isomerases by an improved spectrophotometric assay. (1991)	A highly selective enzyme-catalysed esterification of simple glucosides. (2005)

Recent research contributions to molecular catalysis

The tenets that direct our catalysis research involve nanoparticle synthesis, characterization under reaction conditions, and reaction studies using these nanoparticles to determine kinetics, selectivity, deactivation and other kinetic parameters. These variables are studied in the same research group because they are the underpinning of molecular catalysis. The hypothesis that we are striving to support is that the three fields of catalysis (heterogeneous, homogeneous and enzyme) behave similarly on the molecular level. These ongoing studies are the subjects of this paper.

Structure sensitivity of catalytic reactions on transition metal surfaces

Figure 5 shows a number of catalytic reactions that were studied in my laboratory over the years. Their turnover rates and selectivities were studied as a function of metal nanoparticle size and all found to be dependent structure sensitive. Their kinetics are dependent on the size of the metal nanoparticle catalysts in the 1–10 nm range [17].

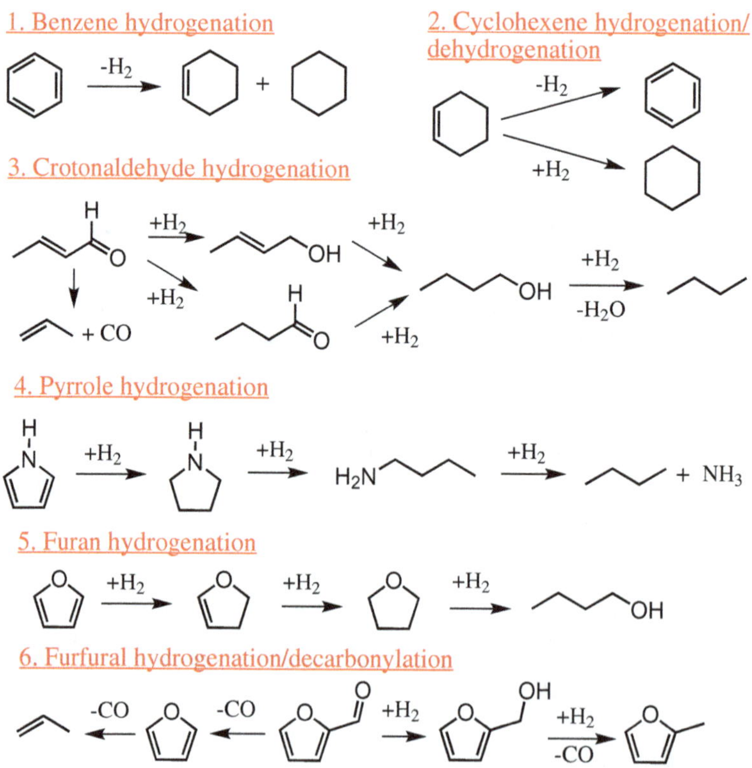

Fig. 5. A list of structure sensitive reactions.

Oxide-metal interfaces as active sites for acid-base catalysis

Platinum is an excellent hydrogenation catalyst of many organic molecules, such as crotonaldehyde. When platinum nanoparticles of the same size are placed on two different oxides — silica or titania — one can see that the turnover rates and the selectivities are much higher when titania rather than silica is used as a support. The importance of the oxide support for metal catalysts to change selectivities and product distributions is well known. This phenomenon of the oxide support effect on catalytic reaction rates where the oxide alone does not carry out the same or any reaction is called positive strong metal support interaction (SMSI) in the literature [18]. It is the charge transfer ability of reducible oxide supports that acts on the performance of metal catalysts, but how does charge regulate catalytic processes?

An explanation was found by surface physics studies of hot electron emission under light illumination that was carried out by exothermic catalytic reactions on metal surfaces [19], such as CO oxidation or hydrogen oxidation (Fig. 6A). The deposited energy produces high kinetic energy electrons that have a mean free path within the metal in the range of 5 to 10 nm. The chemical energy deposition in metals to produce hot electrons has been well demonstrated by Wodtke *et al.* [20] using highly vibrationally excited NO molecules impinging on gold as compared to lithium fluoride surfaces. On gold, the NO molecules in the 15th vibrationally excited state lose 1.5 eV energy to produce molecules in the 8th vibrational state. On the other hand, the vibrationally excited NO molecules lost no energy when they scattered from lithium fluoride, which has no free electrons. The hot electron generation can be observed by using exothermic catalytic reactions on a Schottky diode on platinum and titanium oxide where the platinum is less than about 5 nm in thickness. The charge flow between the platinum and the titanium oxide allows one to determine the current flow in the battery configuration shown in Fig. 6A. One can detect a so-called chemicurrent, which is correlated with the turnover rate of exothermic CO oxidation or hydrogen oxidation reactions [21]. Theoretical calculations showed that the transition state in these processes involves CO_2^- or H_2O^- which yields the chemicurrent that is proportional to the turnover rate. In this study, acid-base catalysis is correlated with charge concentration, and not with surface area. However, covalent bond catalysis is known to be surface area dependent. These two modes of catalysis are the major ways chemistry occurs in most catalytic processes.

When one places metal nanoparticles into a mesoporous oxide support, many oxide-metal interfaces are produced within the mesopores between the metal and the oxide. Studies have found that these oxide-metal interfaces have major effects on catalytic reactions. The isomerization of n-hexane on naked oxide alone results in the cracking of the n-hexane molecules, while mesoporous oxides give rise to 100% selectivity of n-hexane isomers of high octane numbers [22] in the presence of platinum (Fig. 6B). The next figures (Figs. 6C–6E) show similar effects when

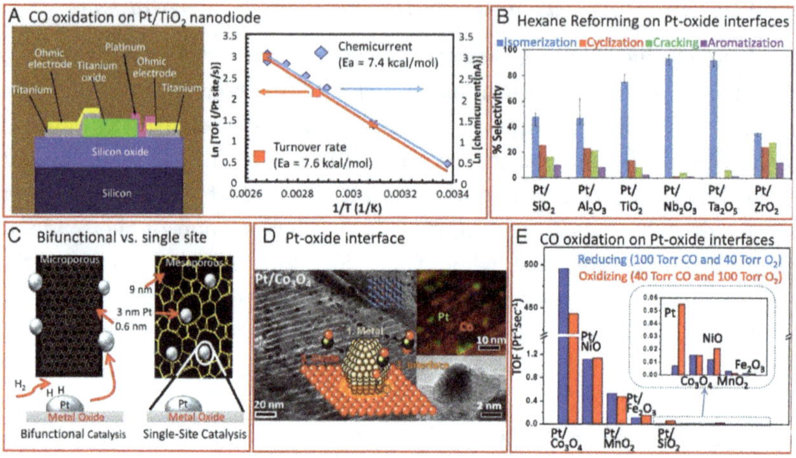

Fig. 6. Evidence for the importance of the metal oxide support on catalysis. (A, Left) Schematic of Pt/TiO_2 device. (Right) Arrhenius plots obtained from chemicurrent and turnover measurements on a Pt/TiO_2 diode with pressure of 6 torr of H_2 and 760 torr O_2. Both give similar activation energies, which implies that hot electron generation under hydrogen oxidation is proportional to the catalytic turnover rate. (B) Product distributions of n-hexane isomerization over 2.7 nm Pt nanoparticle catalysts supported on different kinds of oxide supports at 360°C. (C) Schematics of the differences between bifunctional and single-site catalysis. (D, Left) TEM image of Pt/Co_3O_4 catalysts. (Top Right) Energy-dispersive spectroscopy (EDS) phase mapping of Pt/Co_3O_4 catalysts, showing the merged image of the Co K (red) and Pt L (green) lines. (Bottom Right) High-resolution TEM image of Pt/Co_3O_4 catalysts. (Inset Top) Illustration of mesoporous-oxide-supported Pt nanoparticle catalysts. (Inset Bottom) An illustration showing the potential reaction sites of Pt-nanoparticle-loaded oxide catalysts during CO oxidation. (E) Comparison of TOFs at 473 K of CO oxidation over Pt-nanoparticle-loaded oxide and pure mesoporous oxide catalysts. Panels adapted with permission as follows: A, Left, Ref. [19], American Vacuum Society; A, Right, Ref. [21], American Chemical Society; B, Ref. [22], American Chemical Society; D and E, Ref. [23], American Chemical Society.

platinum is placed on various mesoporous oxides. Platinum nanoparticles produce very little CO oxidation within mesoporous silica, but when they are placed on mesoporous cobalt oxide, more than a thousand-fold increase in catalytic turnover for CO oxidation kinetics is found [23].

Figure 6C (right) shows mesoporous transition metal oxide supported platinum nanoparticles. This is the oxide-metal interface, which produces large, strong metal support effects. There is a charge transfer between the metal and the oxide under reaction conditions. If the oxide is alone, as in the case of n-hexane conversion with pure oxides of niobium oxide, titanium oxide, and other oxides, only the cracking of the n-hexane molecules is observed. However, if the platinum is in the mesopores, a 100% selectivity to isomerization is produced, which is very high and an important factor in making high-octane gasoline.

Figure 6C (left) shows platinum nanoparticles in contact with microporous oxides when the platinum nanoparticle is much larger and cannot fit into the micropores. In this case, the chemistry that occurs — known as bifunctional catalysis —

is the sum total of the chemistry of platinum and the microporous oxide, which act in parallel or consecutively. In the previous case, when the mesoporous transition metal oxide can accommodate the 3 nm platinum inside its mesopores, there are oxide-metal interfaces, where charge transfer and acid-base catalysis occur, which are uniquely selective in many circumstances. Similar results are seen when CO oxidation is carried out on platinum supported by silica or another transition metal oxide [23]. The turnover rate on silica is small, equal to pure platinum turnover; however, when cobalt oxide is the mesoporous support, the turnover rate is amplified by a thousand fold (Fig. 6E). This is indeed a major increase in catalytic activity.

Oxidation state of nanoparticles change with decreasing size: conversion of heterogeneous to homogeneous catalysis

When CO oxidation was studied on rhodium nanoparticle surfaces as a function of size, below 2 nm the CO oxidation rates increased by thirtyfold. Ambient pressure XPS studies [24] indicated that the higher turnover rates are due to the oxidation state of rhodium changing from metallic rhodium to rhodium^{3+}. Similar studies on platinum indicated that platinum nanoparticles above 1.5 nm are metallic; however, the studies also found that platinum below 1.5 nm and as low as 0.8 nm are in the 2^+ and 4^+ oxidation states [25, 26]. Because very few bulk atoms are available for these nanoparticles, they become dominated by low coordination surface atoms, and as a result, their electronic structure changes. Nørskov *et al.* [27] have studied this process and found that the adsorption energy of oxygen on gold nanoparticles changes as a function of the decrease in size. Because there is a decrease in the gold coordination number at the adsorption sites, the gold becomes oxidized to gold 1^+ and 3^+ instead of metallic gold.

Homogeneous catalysts are usually single transition metal ions, surrounded by ligands. As a result, we tried to use these small nanoclusters, which have controlled, high oxidation states to carry out homogeneous catalysis. We adsorbed the small nanoclusters on dendrimers, treelike polymers that hold these nanoclusters throughout its branches (Fig. 7). We found that these are excellent homogeneous catalysts, so we managed to heterogenize homogeneous catalysts by using nanoparticles composed of 40 atoms of rhodium, palladium, gold, or platinum for reactions including hydroformylation, decarbonylation, and other commonly known homogeneous catalytic reactions [25, 28–34]. We demonstrated that catalytic reactivity and selectivity could be tuned by changing the dendrimer properties, in a similar fashion to ligand modification in an organometallic homogeneous catalyst [32, 34]. X-ray absorption spectroscopy studies (XANES and EXAFS) showed [26, 35, 36] that the nanoparticles dispersed to small, low coordination clusters under oxidizing conditions, but reassembled to the original 1 nm particles when under reduction by reactants or products on the dendrimers. This process is also reversible.

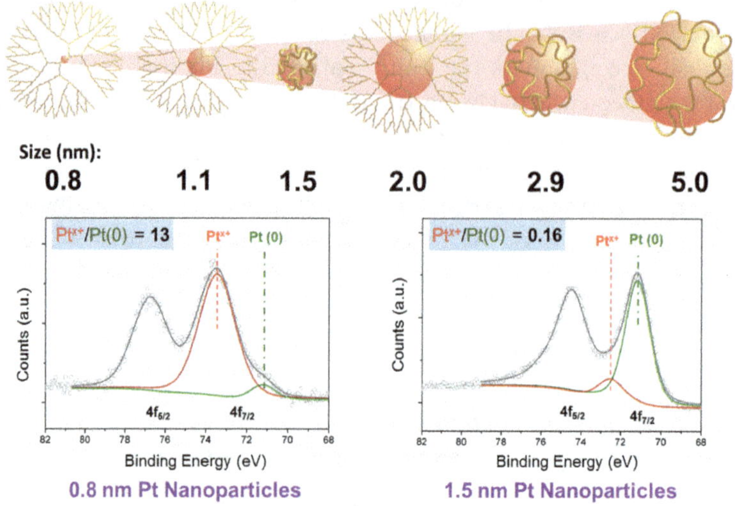

Fig. 7. The effect of oxidation state changes with the size of Pt nanoparticles. For clarity, the deconvoluted peaks for Pt $4f_{5/2}$ are not shown in the XPS spectra.

Likewise, single site homogeneous catalysts are low coordination systems comprised of ligands that control electronic structure and chemistry at the molecular level [26, 37]. By controlling types and binding of ligands, a high level of selectivity (regio- and enantio-selectivity) is obtained in homogenized catalytic reactions. Capping agents used in colloidal synthesis, similar to ligands in single site organometallic complexes, can be utilized to control selectivity in heterogenized homogeneous reactions, which, as of today, remains a great challenge.

Synthesis of heterogenized homogeneous catalyst

The synthesis protocols were reported in our previous works [30, 33, 38], see Fig. 8. A diluted solution of polyamidoamine (PAMAM) dendrimers with –OH terminal groups (G4OH) is mixed with 15–50 mole equivalents of an aqueous solution of 0.01 M metal precursor, *e.g.* K_2PdCl_4. After an appropriate time for complexation, the metal precursor is reduced by a solution containing excess $NaBH_4$ with vigorous magnetic stirring. The mixture was stirred and then purified by dialysis. Purified NPs were loaded onto the mesoporous SBA-15 silica. SBA-15 was added to a colloidal solution of NPs and the resulting slurry was sonicated. The NP supported SBA-15 was separated from the solution by centrifugation. The catalyst was dried at 100°C overnight. The resultant catalyst can be used as a heterogenized homogeneous catalyst upon addition of an oxidizer in a solvent.

We developed the synthesis of the catalyst Pt_{40}/G4OH/SBA-15: dendrimer G4OH encapsulated 40-atom clusters loaded into mesoporous silica SBA-15. The catalyst was highly active over hydroalkoxylation reactions, a type of π-bond

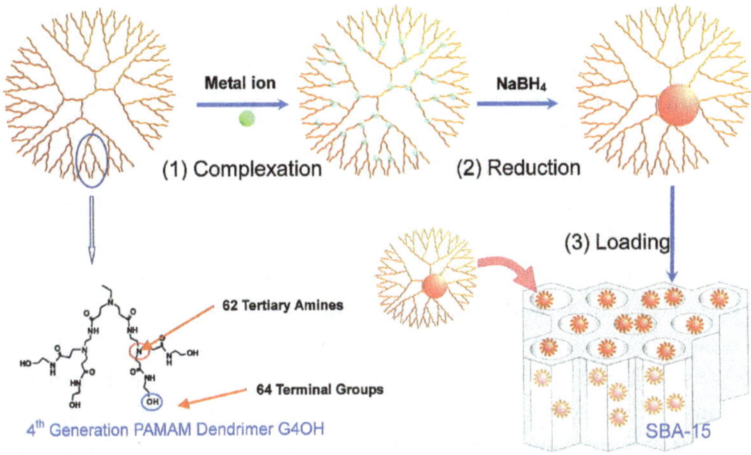

Fig. 8. An illustration of the synthesis of dendrimer encapsulated nanoparticles supported in a mesoporous silica SBA-15. Panels adapted with permission from Ref. [30], American Chemical Society.

activation reaction in the presence of $PhICl_2$, an organic oxidizer. No Pt leaching occurs during reactions with nonpolar solvents: confirmed by three methods [31]. We achieved this reaction in a flow reactor, and obtained kinetic parameters of the reaction [29]. We confirmed that the active catalyst is Pt ions by X-ray absorption spectroscopy [26]. We then used gold clusters in a similar way to catalyze cyclopropanation reactions [32]. We modified the dendrimer properties to tune catalytic reactivity and selectivity (even asymmetric catalysis) like ligand modification in a homogeneous catalyst [34]. We achieved the controlled product selectivity by changing the reactant flow rate in a cascade organic reaction, monitored by *in situ* IR and X-ray high spatial-resolution microspectroscopy [29]. Recently, we studied selective carbon-carbon bond activation catalyzed by Rh clusters at room temperature [39].

Comparing catalytic alcohol oxidation at solid-liquid and solid-gas interfaces

If we are to correlate the three fields of catalysis, heterogeneous, homogeneous and enzyme, we should understand how the catalytic chemistry changes when reactions occur at the much higher molecular density liquid (homogeneous or enzyme) as compared to the gas (heterogeneous) interfaces. We studied alcohol oxidation over platinum at both interfaces and the effects of water coadsorption; from reactants methanol to butanol, we found major variations in reaction kinetics. For example, the turnover rates are two orders of magnitude slower than at the gas interface and certain oxide supports, MnO_2 and CeO_2 of platinum nanoparticles are orders of magnitude more active than silica, SiO_2 [40].

DNA directed immobilization of enzymes to glass

Enzymes are able to maintain remarkably high selectivity towards their substrates while still retaining high catalytic rates. By immobilizing enzymes onto surfaces we can heterogenize these biological catalysts, making it practical to study, use, and combine them in an easily controlled system. In this work, we develop a platform that allows for the simple and oriented immobilization of proteins through DNA directed immobilization (DDI).

We use DDI to immobilize the aldolase enzyme onto glass slides [41]. This is done through the bottom up assembly of two separate components. The first is coupling of aminophenol DNA to aniline functionalized glass slides. In tandem, aldolase is modified at the N-terminus with a complementary DNA strand, also substituted with the aminophenol coupling partner. The subsequent hybridization of the surface oligomer with the complementary oligomer-protein conjugate results in the oriented display of aldolase on the glass surface (Fig. 9A). Fluorescence and atomic force microscopy studies confirmed the chemistries occurring at each step. Additionally, an activity assay of the aldolase displaying surface showed enzymatic activity only in the case where there was sequence complementarity between the two DNA strands, with only minimal background activity (Fig. 9B). This method of DDI also allows for several features such as: the reusability of surface immobilized aldolase, the ability to modulate surface coverage levels by varying annealing temperatures, and recyclability of the single stranded DNA modified surfaces.

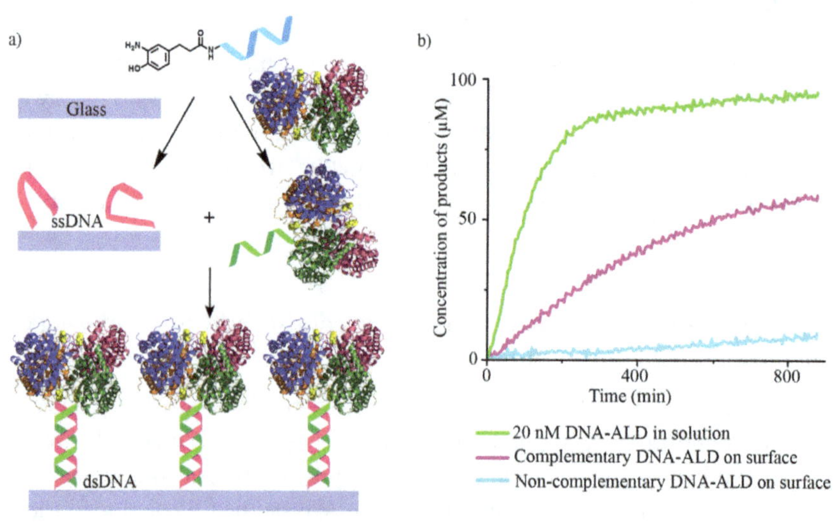

Fig. 9. (A) DNA directed immobilization of a site-selectively modified DNA-protein conjugate onto a glass surface displaying complementary single stranded DNA. (B) Activity assay of DNA-aldolase exposed to a glass surface displaying the complementary DNA strand (pink), the non-complementary DNA strand (blue), and free in solution at a concentration of 20 nM (green). The activity assay is based on the conversion of fructose 1,6-bisphosphate to glyceraldehyde 3-phosphate and dihydroxyacetone phosphate by the aldolase enzyme.

Taking advantage of the transparent nature of the glass surfaces used in these studies, we are also seeking to characterize these surfaces using alternative spectroscopic techniques, such as Sum Frequency Generation and X-Ray Photoelectron Spectroscopy, to gain information about the orientation and coverage of the protein. Similar studies are carried out by adsorbing dehydrogenase enzymes on glass surfaces that oxidize alcohols to aldehydes and aldehydes to acids, respectively.

Outlook for future developments of research on molecular catalysis

Hybrid systems

It is clear that we can heterogenize homogeneous catalysts. However, enzymes [42–46] are also very important catalytic systems, and recent studies have focused on how to synthesize pure enzymes as well as on how to look for similarities between enzymes and all three catalytic systems on the molecular scale. For example, it is known that when enzymes are immobilized on a surface they gain the reusability and ease of separation seen in heterogeneous catalysts, as well as in some cases becoming more stable to wider pH and temperature ranges [47–49]. We plan to advance previous work by immobilizing enzymes on a solid surface and study the enzyme-solid interface at the molecular level, and to develop a molecular understanding of all three types of catalysts under similar conditions of reactions and chemical environments.

In our attempt to focus on the chemical correlations between the three catalysis groups — heterogeneous, homogeneous, and enzymatic — the future looks very promising for molecular catalysis science studies. Catalysis of homogeneous, heterogeneous, and enzymatic origin alike involve nano-sized materials. These nanocatalysts are comprised of inorganic and/or organic components. Charge, coordination, interatomic distance, bonding, and orientation of catalytically active atoms are molecular factors shared by all three field of catalysis. By controlling the governing catalytic components and molecular factors, catalytic processes of a multichannel and multiproduct nature could be run in all three catalytic platforms to create unique end-products. This is the promise of a molecularly unified catalytic scheme of the future.

Road to catalytic complexity: new instrumentation

New cells for X-ray spectroscopy studies (XPS and XAS) of solid-liquid and solid-gas interfaces

We have developed two new cells for x-ray spectroscopy studies under gases and liquids [50]. The first cell is dedicated to X-ray absorption (XAS) measurements, with total electron yield collected *at the sample electrode* (Fig. 10A). The short mean free path of electrons in the liquid phase makes the technique surface sensitive. To make possible measurements of electrode-electrolyte interfaces in the presence of electric fields the X-ray intensity is modulated via piezo-driven chopper and lock-in detection. With this design, the structure of interfacial liquid layers can be studied *in situ* and under electric fields. The second cell is dedicated to X-ray Photoelectron Spectroscopy (XPS) under atmospheric gas pressures and liquid environments. In this new flow cell we use graphene membranes to separate the ambient pressure gas or liquid from the vacuum of the synchrotron beamline (Fig. 10B). Because of its atomic thickness, photoelectrons with $E_{kin} \sim 10^2 - 10^3$ eV can pass through the membrane with low attenuation for XPS analysis. While perfect graphene is strong enough to support the large pressure differential across the membrane, because of the unavoidable presence of defects and grain boundaries in CVD grown graphene, a special support is required. In our design the membrane covers an array of micrometer holes in a gold (or other metal) coated Si_3N_4 membrane. This structure has proved strong enough to withstand the pressure of gases at atmospheric pressures (tested to 2.5 atm), or liquids.

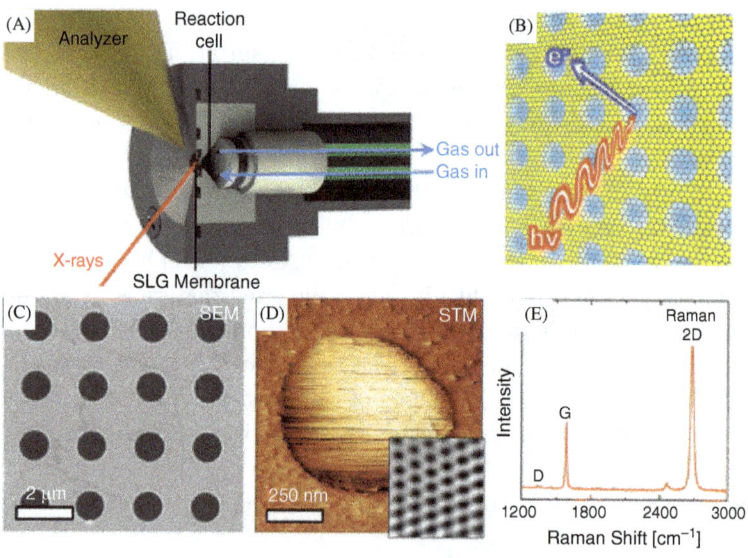

Fig. 10. (A) Cross-sectional view of the atmospheric pressure XPS setup showing the gas flow through the reaction cells and the approximate arrangement of the analyzer and X-ray beam. (B) Sketch of the graphene-based membrane illustrating the operating principle of atmospheric pressure XPS. (C) SEM image of a region of a SLG covered membrane. (D) STM image of one the holes in the membrane with SLG suspended across it ($V_S = 1.5$ V, $I_t = 300$ pA). Inset: Atomic resolution STM image of free-standing graphene measured in the hole region ($V_S = 0.18$ V, $I_t = 500$ pA, 2D-FFT filtered). (E) Representative Raman spectra of SLG transferred onto SiO_2(300 nm)/Si using the same polymer-free method used for fabricating the graphene-based membranes. Panels adapted with permission from Ref. [50], American Chemical Society.

Time resolved transient (2 sec) study of catalytic hydrogenation of CO

The catalytic hydrogenation of carbon monoxide, known as the Fischer–Tropsch process, is a technologically important, complex multipath reaction which produces long-chain hydrocarbons. In order to access the initial kinetics and the mechanism, we developed a reactor that provides information under nonsteady state conditions. We tested a CoMgO catalyst and monitored the initial product formation within 2 s of exposure to CO as well as the time dependence of high molecular weight products (in a 60 s window) and found drastic changes in the product selectivity. The probability for forming branched isomer (C_4 and C_5) peaks in the first 25 s, and within that time frame no unsaturated products were detected. The subsequent decline (at $\sim$ 35 to 40 s) of branched isomers coincides with the detection of olefins (from C_2 to C_5), indicating a change in the reaction path (Fig. 11) [51].

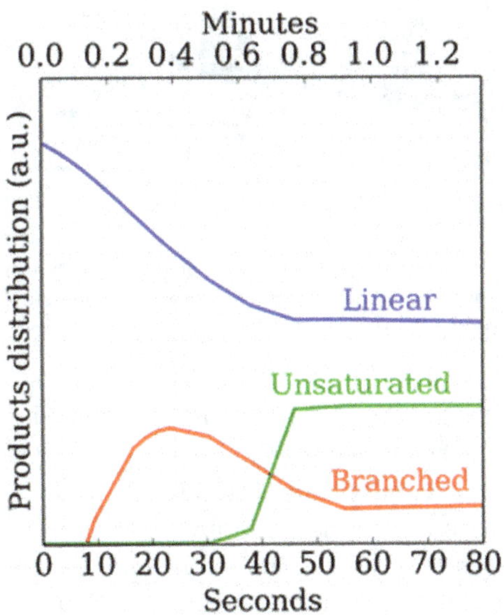

Fig. 11. Selectivity as a function of time for different products obtained using CoMgO at atmospheric pressure and 230°C for H_2:CO = 3:1, obtained by GC-MS. Panels adapted with permission from Ref. [51], American Chemical Society.

Cascade reaction in a flow microreactor

Dendrimer-encapsulated Au nanoparticles with diameter of 2 nm were loaded on mesoporous SiO_2, packed in a flow microreactor, and utilized as a heterogeneous catalyst for the cascade reaction of dihydropyran synthesis (Fig. 12A). In this reaction, propargyl vinyl ether **1** was catalytically rearranged by the Au catalyst into the primary product, allenic aldehyde **2**. Activation of the primary product **2** by the Au catalyst was followed by nucleophilic attack of butanol-d_{10}, leading to the formation of the secondary product, acetal **3**. In order to track the catalytic transformation within the flow microreactor, this multistep reaction was mapped with a spatial resolution of 15 μm, employing synchrotron-sourced IR and X-ray beams. High-resolution mapping of the catalytic reaction with IR microspectroscopy revealed the oxidation state of the catalyst along the flow reactors (Fig. 12B).

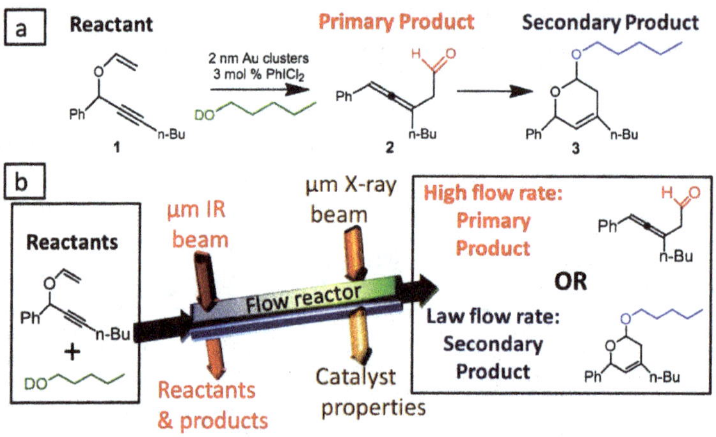

Fig. 12. (A) the cascade reaction of dihydropyran synthesis. (B) High-resolution mapping of the catalytic reaction with IR microspectroscopy revealed the oxidation state of the catalyst along the flow reactors. Panels adapted with permission from Ref. [29], American Chemical Society.

Prior to the catalytic reaction, dendrimer-encapsulated Au nanoparticles were oxidized to Au ions by a flow of an inorganic oxidizer, $PhlCl_2$, solvated in toluene. Following the formation of catalytically active, highly oxidized Au ions, the catalyst was tested toward the cascade reaction of dihydropyran formation [29].

The product selectivity was tuned by modifying the residence time of the reactants. These results are summarized in the accompanying figure along with a schematic of the reaction and spectroscopy cell. A high flow rate of 10 mL/h induced low conversion (20%) and a primary:secondary (**2,3**) product ratio of 5:1. Higher yield (75%) was obtained by decreasing the flow rate by 50-fold to 0.2 mL/h, while the product selectivity (**2,3**) was transformed to 0:100 (Fig. 2B). These results demonstrate the advantage in performing catalytic transformations in flow microreactors, enabling isolation of either the primary or the secondary products [29].

Acknowledgments

The catalysis research is supported by supported by the Director, Office of Science, Office of Basic Energy Sciences, Chemical Sciences, Geosciences and Biosciences Division of the US Department of Energy under Contract No. DE-AC02-05CH11231. The synthesis and instrumentation research is supported by the Director, Office of

Science, Office of Basic Energy Sciences, Materials Sciences and Engineering Division, of the U.S. Department of Energy under Contract No. DE-AC02-05-CH11231.

References

[1] G. A. Somorjai, Y. Li, *Introduction to Surface Chemistry and Catalysis*, 2nd Edition (Wiley, 2010).
[2] H. Song, F. Kim, S. Connor, G. A. Somorjai, P. Yang, *J. Phys. Chem. B* **109**, 188 (2005).
[3] I. Lee, F. Delbecq, R. Morales, M. A. Albiter, F. Zaera, *Nat. Mater.* **8**, 132 (2009).
[4] V. V. Pushkarev, K. An, S. Alayoglu, S. K. Beaumont, G. A. Somorjai, *J. Catal.* **292**, 64 (2012).
[5] C.-K. Tsung, J. N. Kuhn, W. Huang, C. Aliaga, L.-I. Hung *et al.*, *J. Am. Chem. Soc.* **131**, 5816 (2009).
[6] S. Alayoglu, C. Aliaga, C. Sprung, G. A. Somorjai, *Catal. Lett.* **141**, 914 (2011).
[7] G. Melaet, A. Lindeman, G. Somorjai, *Top. Catal.* **57**, 500 (2014).
[8] D. I. Enache, J. K. Edwards, P. Landon, B. Solsona-Espriu, A. F. Carley *et al.*, *Science* **311**, 362 (2006).
[9] K. Ohara, M. Kawano, Y. Inokuma, M. Fujita, *J. Am. Chem. Soc.* **132**, 30 (2010).
[10] F. T. Sejidov, Y. Mansoori, N. Goodarzi, *J. Mol. Catal. A: Chem.* **240**, 186 (2005).
[11] I. E. Markó, P. R. Giles, M. Tsukazaki, S. M. Brown, C. J. Urch, *Science* **274**, 2044 (1996).
[12] D. Gauthier, A. T. Lindhardt, E. P. K. Olsen, J. Overgaard, T. Skrydstrup, *J. Am. Chem. Soc.* **132**, 7998 (2010).
[13] K. Ishihara, S. Nakagawa, A. Sakakura, *J. Am. Chem. Soc.* **127**, 4168 (2005).
[14] G. L. Shearer, K. Kim, K. M. Lee, C. K. Wang, B. V. Plapp, *Biochemistry* **32**, 11186 (1993).
[15] J. L. Kofron, P. Kuzmic, V. Kishore, E. Colon-Bonilla, D. H. Rich, *Biochemistry* **30**, 6127 (1991).
[16] F. Bjorkling, S. E. Godtfredsen, O. Kirk, *J. Chem. Soc., Chem. Commun.* 934 (1989).
[17] R. Ye, T. J. Hurlburt, K. Sabyrov, S. Alayoglu, G. A. Somorjai, *Proc. Natl. Acad. Sci. USA* **113**, 5159 (2016).
[18] S. J. Tauster, S. C. Fung, R. T. K. Baker, J. A. Horsley, *Science* **211**, 1121 (1981).
[19] J. Y. Park, G. A. Somorjai, *J. Vac. Sci. Technol. B* **24**, 1967 (2006).
[20] Y. Huang, C. T. Rettner, D. J. Auerbach, A. M. Wodtke, *Science* **290**, 111 (2000).
[21] A. Hervier, J. R. Renzas, J. Y. Park, G. A. Somorjai, *Nano. Lett.* **9**, 3930 (2009).
[22] K. An, S. Alayoglu, N. Musselwhite, K. Na, G. A. Somorjai, *J. Am. Chem. Soc.* **136**, 6830 (2014).
[23] S. Alayoglu, S. K. Beaumont, G. Melaet, A. E. Lindeman, N. Musselwhite *et al.*, *J. Phys. Chem. C* **117**, 21803 (2013).
[24] M. E. Grass, Y. Zhang, D. R. Butcher, J. Y. Park, Y. Li *et al.*, *Angew. Chem. Int. Ed.* **47**, 8893 (2008).
[25] R. M. Crooks, M. Zhao, *Adv. Mater.* **11**, 217 (1999).
[26] Y. Li, J. H.-C. Liu, C. A. Witham, W. Huang, M. A. Marcus *et al.*, *J. Am. Chem. Soc.* **133**, 13527 (2011).
[27] J. Kleis, J. Greeley, N. A. Romero, V. A. Morozov, H. Falsig *et al.*, *Catal. Lett.* **141**, 1067 (2011).
[28] R. M. Crooks, M. Zhao, L. Sun, V. Chechik, L. K. Yeung, *Acc. Chem. Res.* **34**, 181 (2001).

[29] E. Gross, X.-Z. Shu, S. Alayoglu, H. A. Bechtel, M. C. Martin *et al.*, *J. Am. Chem. Soc.* **136**, 3624 (2014).
[30] W. Huang, J. N. Kuhn, C.-K. Tsung, Y. Zhang, S. E. Habas *et al.*, *Nano. Lett.* **8**, 2027 (2008).
[31] C. A. Witham, W. Huang, C.-K. Tsung, J. N. Kuhn, G. A. Somorjai *et al.*, *Nat. Chem.* **2**, 36 (2010).
[32] E. Gross, J. H.-C. Liu, F. D. Toste, G. A. Somorjai, *Nat. Chem.* **4**, 947 (2012).
[33] W. Huang, J. H.-C. Liu, P. Alayoglu, Y. Li, C. A. Witham *et al.*, *J. Am. Chem. Soc.* **132**, 16771 (2010).
[34] E. Gross, J. H. Liu, S. Alayoglu, M. A. Marcus, S. C. Fakra *et al.*, *J. Am. Chem. Soc.* **135**, 3881 (2013).
[35] J.-J. Velasco-Velez, C. H. Wu, T. A. Pascal, L. F. Wan, J. Guo *et al.*, *Science* **346**, 831 (2014).
[36] F. Jiao, H. Frei, *Angew. Chem. Int. Ed.* **48**, 1841 (2009).
[37] J. Oliver-Meseguer, J. R. Cabrero-Antonino, I. Domínguez, A. Leyva-Pérez, A. Corma, *Science* **338**, 1452 (2012).
[38] E. Gross, H.-C. Liu Jack, F. D. Toste, G. A. Somorjai, *Nat. Chem.* **4**, 947 (2012).
[39] R. R. Ye, B. Yuan, J. Zhao, W. T. Ralston, C.-Y. Wu *et al.*, *J. Am. Chem. Soc.* (2016).
[40] H. Wang, A. Sapi, C. M. Thompson, F. Liu, D. Zherebetskyy *et al.*, *J. Am. Chem. Soc.* **136**, 10515 (2014).
[41] K. S. Palla, T. J. Hurlburt, A. M. Buyanin, G. A. Somorjai, M. B. Francis, *J. Am. Chem. Soc., submitted.*
[42] O. Nureki, D. G. Vassylyev, M. Tateno, A. Shimada, T. Nakama *et al.*, *Science* **280**, 578 (1998).
[43] D. D. Boehr, D. McElheny, H. J. Dyson, P. E. Wright, *Science* **313**, 1638 (2006).
[44] G. Bhabha, J. Lee, D. C. Ekiert, J. Gam, I. A. Wilson *et al.*, *Science* **332**, 234 (2011).
[45] D. R. Glowacki, J. N. Harvey, A. J. Mulholland, *Nat. Chem.* **4**, 169 (2012).
[46] S. Hay, N. S. Scrutton, *Nat. Chem.* **4**, 161 (2012).
[47] P. Torres-Salas, A. del Monte-Martinez, B. Cutiño-Avila, B. Rodriguez-Colinas, M. Alcalde *et al.*, *Adv. Mater.* **23**, 5275 (2011).
[48] S. Matosevic, N. Szita, F. Baganz, *J. Chem. Technol. Biotechnol.* **86**, 325 (2011).
[49] U. Roessl, J. Nahálka, B. Nidetzky, *Biotechnol. Lett.* **32**, 341 (2009).
[50] R. S. Weatherup, B. Eren, Y. Hao, H. Bluhm, M. B. Salmeron, *J. Phys. Chem. Lett.* **7**, 1622 (2016).
[51] G. Melaet, W. T. Ralston, W.-C. Liu, G. A. Somorjai, *J. Phys. Chem. C* **118**, 26921 (2014).

TOWARDS A THEORY OF HETEROGENEOUS CATALYSIS

JENS K. NØRSKOV

SUNCAT Center of Interface Science and Catalysis
Department of Chemical Engineering, Stanford University
Stanford, CA94305, USA

Some challenges to research in the field of heterogeneous catalysis

The drive for sustainable energy solutions requires new focus on catalysis. It is very desirable to store energy in chemical bonds and that makes it essential to be able to efficiently transform energy to and from a chemical form, or from one chemical form into another. In many sustainable energy technologies, the lack of sufficiently efficient catalysts is a primary factor limiting their use. The same arguments hold for the development of a sustainable chemical industry where resources are used more efficiently and the energy input is from a sustainable source.

Heterogeneous catalysis is a very complex phenomenon and progress in catalyst design has been held back by a lack of an understanding why one catalyst is better than another. Catalyst discovery could be greatly facilitated if we had a predictive theory of heterogeneous catalysis. One could imagine a machinery that allowed us to calculate activation energies and entropies of all elementary steps in a reaction. This would in principle provide a complete theoretical description of the reaction (neglecting non-adiabatic effects and dynamic corrections to transition state theory). In general, however, such an endeavor is prohibitively demanding. For a given set of reactants and products, we need calculations of a large number of elementary processes and for a large number of possible catalysts. This is not feasible nor is it in fact desirable. It would be much more useful to have a set of concepts that can help us single out the most important parameters characterizing a catalyst.

Research contributions to a theory of heterogeneous catalysis

Our approach is to establish a systematic framework based on scaling relations, which are relationships between adsorption energies of different intermediates and transition states in a reaction. The finding that scaling relations are ubiquitous in surface chemistry forms the basis for a mapping of the many reaction and transition state (TS) energies that are needed to describe a full catalytic reaction onto a few bond energies or descriptors [1]. The principle is well-illustrated in Fig. 1.

Given the scaling relations we can express the rate and selectivity of different catalysts in terms of a limited number of descriptors allowing a rationalization

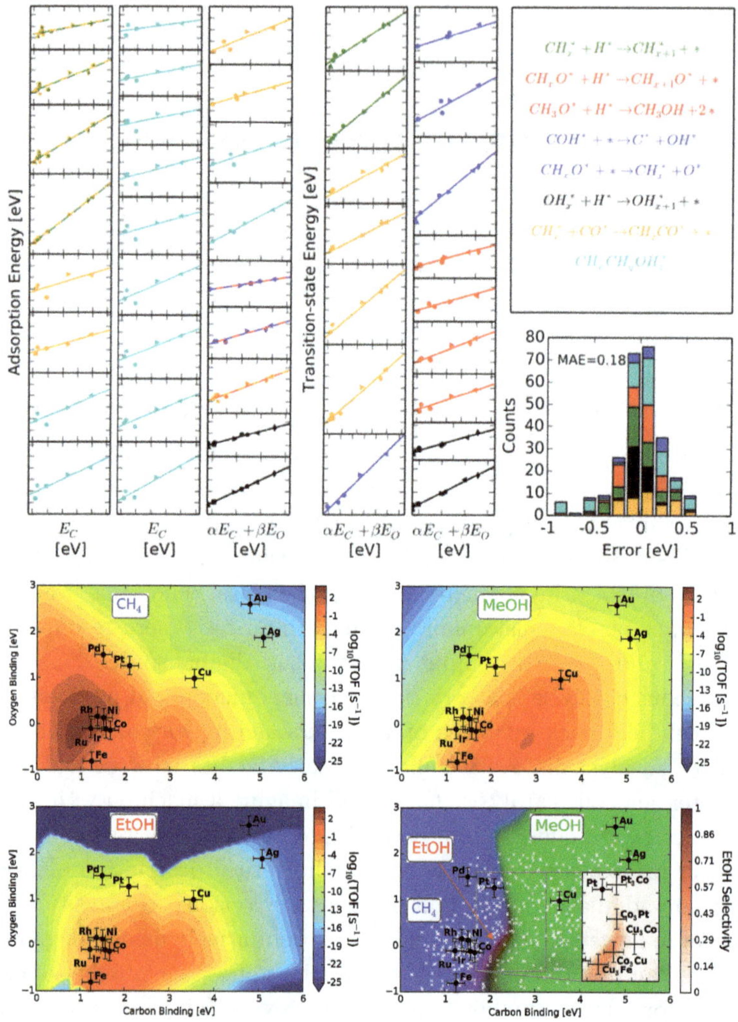

Fig. 1. Top: Scaling relations for intermediates and transition states in CO hydrogenation as a function of the adsorption energy of atomic C and O, E_C and E_O. Bottom: Turnover frequencies as a function of carbon and oxygen binding energies for (a) methane, (b) methanol and (c) ethanol production. The position of different transition metals is indicated. The resulting selectivity to ethanol is shown in (d) along with the most promising alloys. From Ref. [2].

of decades of experimental results in terms of the activity maps like in Fig. 1. The figure shows how we can understand why Cu-based catalysts are the best for methanol synthesis, why Ru, Rh and Ni are the best catalysts for methanation, and why it has so far not been possible to find good catalysts for ethanol synthesis. What is more, we have an understanding of the origin of the scaling relations and of the variation of descriptors from one catalyst surface to the next in terms of the electronic structure in terms of the d-band model [3]. This constitutes the first step towards a comprehensive theory of the intrinsic catalytic properties of transition metal surfaces.

Since the theory predicts the optimum descriptor values, we can also use it to make predictions of new catalysts. This is illustrated by the discovery of a number of new catalysts, most recently for thermal CO_2 reduction to methanol [4], as well as electrochemical hydrogen evolution [5].

Outlook to future developments of the theory of heterogeneous catalysis

The theoretical approach we have taken is a reductionist one. We first understand trends in the intrinsic catalytic activity of different surfaces and then we will need to add effects of promoters, and support. We have the best description of transition metal catalysts, and need to extend and validate the theory for other classes of catalyst materials.

An important insight that comes from the discovery of scaling relations is that they define severe limitations in catalytic activity and selectivity. The simplest possible example is shown in Fig. 2. The two most important parameters determining the ammonia synthesis rate, the N_2 dissociation transition state (TS) energy and the N adsorption energy scale for the transition metals. This scaling allows us to represent the rate as a function of only one of the two parameters (a 1D volcano), but the relationship between the two also means that we are not able to reach the real optimum for this reaction. The tens of thousands catalysts have been tested for this reaction have never shown anything better than Ru (the best so far is a "CoMo" catalyst that came out of optimizing the N adsorption energy based on the 1D volcano, as in Fig. 2 [1, 6].

The finding that scaling relations effectively limit our ability to properly optimize catalysts for a given reaction is very general. An important challenge going forward is therefore to find strategies to circumvent the scaling. Since the scaling relations stem from the fact that different intermediates and transition states bond to the active site at the surface through the same adsorbate atoms, a successful approach is likely to involve strategies to sample other parts of the reacting molecules — just like enzymes and many homogeneous catalysts do it [7].

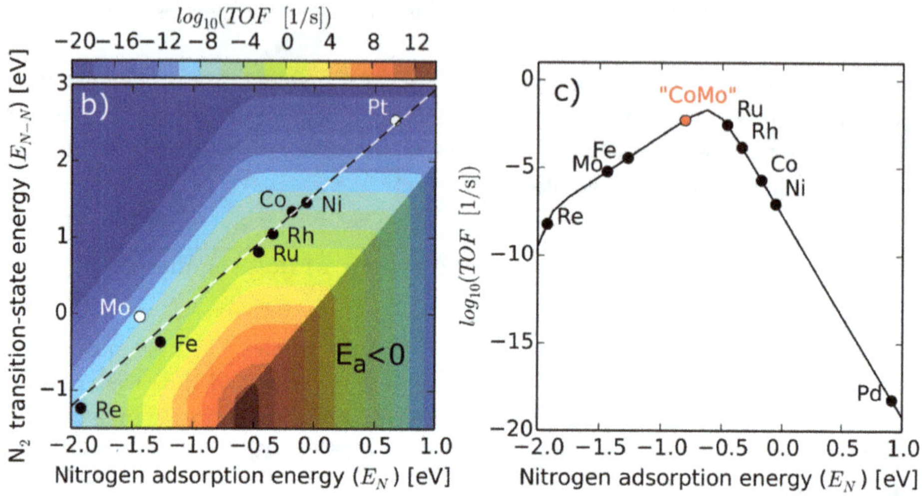

Fig. 2. Left: Color map of the ammonia synthesis rate as a function of nitrogen adsorption energy and N_2 dissociation barrier. Energetics for FCC/HCP metal step sites for different metals is included as well as the scaling line connecting the two energy parameters. A simple model is used where N_2 dissociation is always rate limiting and adsorbed N is the only intermediate. The same picture is found if these simple assumptions are relaxed. Shaded area shows the theoretical limit since the activation linear function of E_N. The red point marked CoMo denotes the expected rate on a mixed site containing both Co and Mo based on interpolation between the two nitrogen adsorption energies. From Ref. [1].

Acknowledgments

This work has been funded by the US DOE BES to the SUNCAT Center.

References

[1] A. J. Medford, A. Vojvodic, J. S. Hummelshøj, J. Voss, F. Abild-Pedersen *et al.*, *J. Catal.* **328**, 36 (2015).

[2] A. J. Medford, A. C. Lausche, F. Abild-Pedersen, B. Temel, N. C. Schjodt *et al.*, *Top. Catal.* **57**, 135 (2014).

[3] B. Hammer, J. K. Nørskov, *Advances in Catalysis*, Vol. 45, 71 (2000).

[4] F. Studt, I. Sharafutdinov, F. Abild-Pedersen, C. F. Elkjaer, J. S. Hummelshøj *et al.*, *Nat. Chem.* **6**, 320 (2014).

[5] J. Kibsgaard, C. Tsai, K. Chan, J. D. Benck, J. K. Nørskov *et al.*, *Energy Environ. Sci.* **8**, 3022 (2015).

[6] C. J. H. Jacobsen, S. Dahl, B. S. Clausen, S. Bahn, A. Logadottir *et al.*, *J. Am. Chem. Soc.* **123**, 8404 (2001).

[7] A. Vojvodic, J. K. Nørskov, *Natl. Sci. Rev.* **2**, 140 (2015).

STRUCTURE-REACTIVITY RELATIONS THROUGH CHARGE CONTROL AT THE ATOMIC LEVEL IN HETEROGENEOUS CATALYSIS

HANS-JOACHIM FREUND

Department of Chemical Physics, Fritz-Haber-Institut der Max-Planck-Gesellschaft, 14195 Berlin, Germany

My view of the present state of research on heterogeneous catalysis

The science of catalysis is the science of the 21st century. It is an exceedingly important economic and social factor in sustaining our societies and sustaining a healthy environment by controlling chemical energy conversion. From the viewpoint of fundamental research there are a number of challenges to be overcome. We need to be able, eventually, to control the vacuum-solid and solid-liquid interfaces under operation conditions at the atomic level. We are not there yet! There are still a number of other problems that we need to understand at the atomic level, before we proceed towards the in-operando problems. They are connected with the complexity of the typical catalytic material.

One approach to complexity is to use a systematic model approach [1–3]! In order to identify the active part of the material we need to be able to study a very small amount of starting material and of products at the surface of the complex material, and we need to be simultaneously able to differentiate the surface of the material from its bulk. This requires the development of specific, surface sensitive techniques and, in order to isolate the action of the various material components, a systematic variation in the complexity of the studied material [4]. This variation has to proceed from the most simple to the more complex, and not *vice versa.* Only by adding complexity will we be able to finally approach the final real system. In this sense, we set up model systems in catalysis, which may be characterized at the atomic level. There are several levels of models in catalysis, which we try to show as a flow chart in Fig. 1, both, for theory and experiment.

We look at experiment and theory separately, knowing, however, that progress in the spirit of understanding phenomena at the atomic level will only be achieved by going hand-in-hand. The left hand side of the chart refers to clusters in the gas phase in different charge states. Through a comparison or better a combination between experiment and theory it is possible to unravel details of the reactions taking place in such systems. However, the distance in complexity of the isolated cluster level with respect to supported clusters is far, and the size of clusters, possible to study, has, until now, been limited, although steadily increasing. Having said this, still very useful and unique information on details of reaction mechanisms may be unraveled. At the next level in the chart (Fig. 1), single crystal and well-ordered thin film surfaces have been used successfully in the past to start to understand fundamental phenomena at surfaces. This culminated in the Nobel Prize for Gerhard Ertl in 2007 [5, 6]. Given the enormous success of those studies, the valid question that has been asked is: What comes next? Single crystal metal studies lack important ingredients of real catalysts: One is the finite size of the metal particles, typically used in the catalyst, and, secondly, the oxide-metal interface realized by the fact that dispersed metals are typically supported on oxides, which, according to all reports so far, is of enormous importance for the reactions, their activity and selectivity, observed. Not to speak of the addition of modifiers such as poisons or promoters. So we certainly have to move to higher levels of model studies as shown in Fig. 1 before we will have a chance to reach a level, which may be directly compared and connected to situations encountered on real catalytic material. This is the point where the two arrows bottom up and top down meet. Through the interplay between experiment and theory, those additional ingredients, indeed, have been identified as essential, and it is the task of researchers in the field, both from theory and experiment, to work together to bring the field forward along the complexity axis shown in Fig. 1. We will try to exemplify the approach of experimental and computational model systems by choosing two areas:

Supported metal catalysts consist typically of small metal particles often anchored to an oxide support. How does the charge flow between the metal particle and the support influence reactivity?

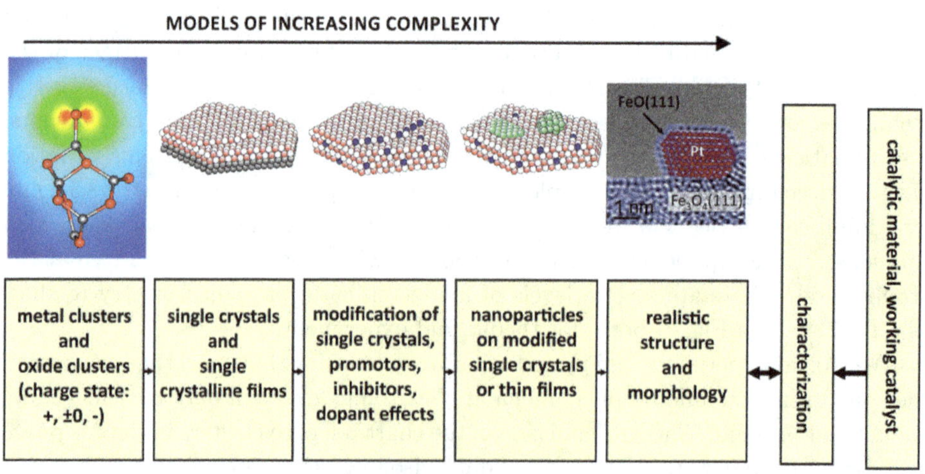

Fig. 1. Schematic and pictorial representation of model systems as a function of increasing complexity to capture the essential features of a working catalyst.

My recent research contributions to heterogeneous catalysis

Heterogenized homogeneous catalysts often consist of an amorphous, for example, silica support onto which individual metal complexes carrying a well-defined ligand sphere are bound. How do we get an atomistic view of the bonding situation between complex and amorphous support?

1. What would be the prerequisites for a model system to tackle this problem? A well defined source of electrons is needed and experimental tools to study the consequences for the particle and reactions on it. Consider the model system shown in Fig. 2: Small Au particles have been prepared by physical vapor deposition onto an ultrathin MgO(100) film, epitaxially grown on a Ag(100) metal single crystal [7]. It has been theoretically predicted [8, 9] and experimentally verified [10], that electrons are transferred from the Ag support through the MgO film onto the Au nano particle. The electron transfer causes the flat morphology of the particle [11]. Had there been no electron transfer, the particle would have the typical three-dimensional morphology. This system may be looked at as a prototypical system for a supported metal nano particle on a non-reducible oxide, and one may investigate how the transferred charge influences the electronic structure of the particle and its reactivity, and where a potential reactivity on the particle would occur. It turns out that the charge is distributed around the rim of the nano particle at the oxide metal interface [7, 11]. The reaction used to probe this is the formation of a carboxylate from CO_2 by electron transfer [12]. This particular electron transfer, however, is controlled by the amount of CO_2, since the energy needed for electron transfer depends on the number of CO_2 molecules involved in it. While an electron transfer onto a single molecule is endothermic by 0.6 eV, an electron transfer to a CO_2-dimer is exothermic by 0.9 eV, an effect known from gas phase molecular beam experiments [13–19]. After the first electron has been transferred a second electron may be transferred leading to the formation of an oxalate species, and the formation of a carbon-carbon bond concomitantly. The oxalate molecules are imaged at the rim (inset in Fig. 2), clearly supporting the proposed mechanism. This is also corroborated by IR spectra (not shown). In addition to localizing the reaction site, we may also elucidate the consequences for the electronic structure of the metal particle involved. The energies of the quantum well states of the Au particles formed from its 6s electrons can be determined using scanning tunneling spectroscopy. The electron-transfer from Au to the oxalate molecules leads to a repulsion of the electrons remaining on the Au nano particle, thus restricting their spatial extend and shifting the states in a characteristic manner [9]. The ideas deduced from such a model study may also be used to develop a strategy to design realistic powder catalyst systems, by realizing that one

needs an electron source within a catalyst support to induce charging of supported nano particles. Such an electron source may be provided by properly chosen dopants as recently demonstrated [20].

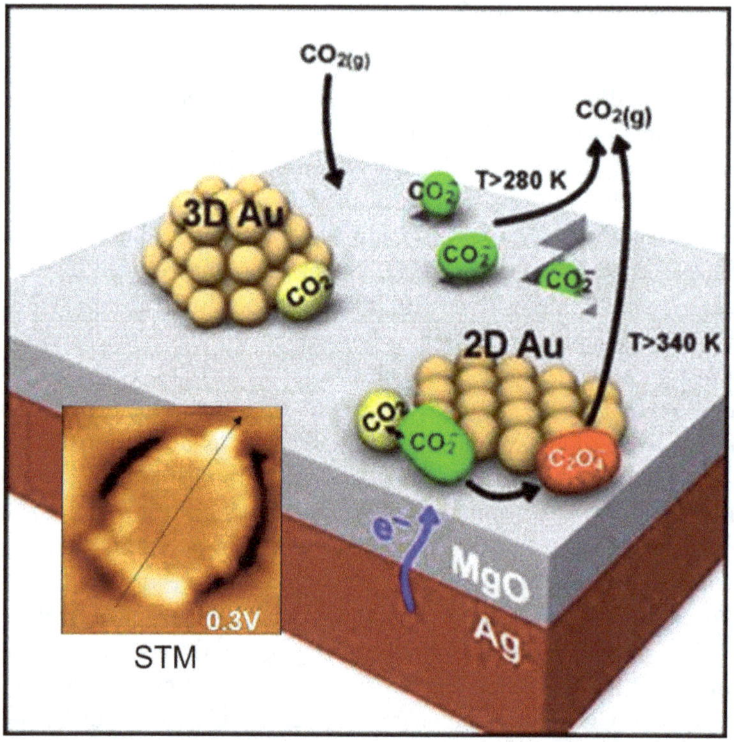

Fig. 2. Schematic showing the individual steps of oxalate formation upon chemisorption of CO_2 on the rim of 2D Au islands on thin MgO(001)/Ag(001) films [12]. Carboxylate species (green) are formed both on defect sites of the MgO film and on the rim of the 2D Au islands by electron transfer. Only on the latter, additional CO_2 (yellow) is able to solvate the carboxylate species yielding a $(CO_2)_2^-$ dimer ion, which, after an additional electron transfer, results in adsorbed oxalate C_2O_{42-} (red). This reaction does not occur on 3D Au particles [29]. The inset shows an STM image of a 2D Au particle with the oxalate molecules at the rim [12].

With respect to the second example we consider a well ordered, crystalline silica film, which may be prepared on a Ru(0001) substrate by physical vapor deposition in an oxygen atmosphere [21]. This film consists of two layers of corner-sharing SiO_4-tertahedra arranged in a hexagonal pattern bound on top of each other. The same film may also be prepared as an amorphous film exhibiting a structure predicted in 1932 by Zachariasen [22], but only experimentally verified on the basis of the present film [23]. The atomic position of the silicon and oxygen atoms have been recorded by scanning probe techniques [24, 25]. We are therefore in the position to create an amorphous silica support, that after hydroxylation, which is possible via electron bombardement, to graft complexes from solution or by chemical vapor deposition [26]. It should be possible to produce a model for a heterogenized homogeneous catalyst as, for example, the Philips catalyst based on Cr atoms bound at the oxygen atoms of the silica [27, 28].

2. If such a system could be imaged at atomic resolution, it should be possible to unravel the specific bonding situation necessary to induce the chemical reaction in question, i.e. ethylene polymerization.

Outlook to future developments of research on heterogeneous catalysis

I suggest that the described model catalyst approach may provide an avenue to unravel the complexity of catalytic material, its function and, at least, partly close the existing gap with respect to materials function. If one could develop this further to perform such investigations in-operando we would be able to provide a route to catalyst design.

References

[1] J. Sauer, H.-J. Freund, *Catal. Lett.* **145**, 109 (2015).
[2] H.-J. Freund, *Chem. Eur. J.* **16**, 9384 (2010).
[3] H.-J. Freund, N. Nilius, T. Risse, S. Schauermann, *Phys. Chem. Chem. Phys.* **16**, 8148 (2014).
[4] H.-J. Freund, N. Nilius, T. Risse, S. Schauermann, T. Schmidt, *ChemPhysChem* **12**, 79 (2011).
[5] G. Ertl, *Nobel Lecture*, 2007.
[6] G. Ertl, *Angew. Chem. Int. Ed.* **47**, 3524 (2008).
[7] C. Stiehler, F. Calaza, W.-D. Schneider, N. Nilius, H.-J. Freund, *Phys. Rev. Lett.* **115**, 036804 (2015).
[8] G. Pacchioni, L. Giordano, M. Baistrocchi, *Phys. Rev. Lett.* **94**, 226104 (2005).
[9] D. Ricci, A. Bongiorno, G. Pacchioni, U. Landman, *Phys. Rev. Lett.* **97**, 036106 (2006).
[10] X. Lin, N. Nilius, M. Sterrer, P. Koskinen, H. Häkkinen *et al.*, *Phys. Rev. B* **81**, 153406 (2010).
[11] M. Sterrer, T. Risse, M. Heyde, H.-P. Rust, H.-J. Freund, *Phys. Rev. Lett.* **98**, 206103 (2007).
[12] F. Calaza, C. Stiehler, Y. Fujimori, M. Sterrer, S. Beeg *et al.*, *Angew. Chem. Int. Ed.* **54**, 12484 (2015).
[13] K. O. Hartman, I. C. Hisatsune, *J. Chem. Phys.* **44**, 1913 (1966).
[14] R. N. Compton, P. W. Reinhardt, C. D. Cooper, *J. Chem. Phys.* **63**, 3821 (1975).
[15] J. Pacansky, U. Wahlgren, P. S. Bagus, *J. Chem. Phys.* **62**, 2740 (1975).
[16] D. Schröder, C. A. Schalley, J. N. Harvey, H. Schwarz, *Int. J. Mass Spectrom.* **185–187**, 25 (1999).
[17] A. Stamatovic, K. Stephan, T. D. Märk, *Int. J. Mass Spectrom. Ion Processes* **63**, 37 (1985).
[18] M. Knapp, O. Echt, D. Kreisle, T. D. Märk, E. Recknagel, *Chem. Phys. Lett.* **126**, 225 (1986).

[19] E. L. Quitevis, D. R. Herschbach, *J. Phys. Chem.* **93**, 1136 (1989).
[20] X. Shao, S. Prada, L. Giordano, G. Pacchioni, N. Nilius *et al.*, *Angew. Chem. Int. Ed.* **50**, 11525 (2011).
[21] D. Löffler, J. J. Uhlrich, M. Baron, B. Yang, X. Yu *et al.*, *Phys. Rev. Lett.* **105**, 146104 (2010).
[22] W. H. Zachariasen, *J. Am. Chem. Soc.* **54**, 3841 (1932).
[23] L. Lichtenstein, C. Büchner, B. Yang, S. Shaikhutdinov, M. Heyde *et al.*, *Angew. Chem. Int. Ed.* **51**, 404 (2012).
[24] L. Lichtenstein, M. Heyde, H.-J. Freund, *Phys. Rev. Lett.* **109**, 106101 (2012).
[25] L. Lichtenstein, M. Heyde, H.-J. Freund, *J. Phys. Chem. C* **116**, 20426 (2012).
[26] X. Yu, E. Emmez, Q. Pan, B. Yang, S. Pomp *et al.*, *Phys. Chem. Chem. Phys.* **18**, 3755 (2016).
[27] M. P. McDaniel, *Advances in Catalysis*, Vol. 33, D. D. Eley, H. Pines, P. B. Weisz (eds.), 47 (Academic Press, 1985).
[28] M. P. McDaniel, *Advances in Catalysis*, Vol. 53, B. C. Gates, H. Knözinger (eds.), 123 (Academic Press, 2010).
[29] C. P. O'Brien, K.-H. Dostert, M. Hollerer, C. Stiehler, F. Calaza *et al.*, *Faraday Discuss.* **188**, 309 (2016).

COMPUTATIONAL CATALYSIS: RIGOR AND RELEVANCE

JOACHIM SAUER

Institut für Chemie, Humboldt-Universität, 10099 Berlin, Germany

My view of the present state of computational research on heterogeneous catalysis

Computational quantum chemistry has contributed substantially to the atomistic understanding of heterogeneous catalysis by providing detailed information about active sites and elementary reaction steps. Microkinetic models [1, 2] are needed to examine which elementary steps of the complex reaction networks are relevant and how they are coupled together to yield a certain product distribution as a function of the type and concentration of reactants (feed) and the conditions such as pressure and temperature. Examples are the methanol-to-olefin process on acidic zeolites [3] and the oxidative coupling of methane [4].

The number of possible reactions, adsorption/desorption and diffusion steps is large, and so is the number of possible active sites (see, *e.g.*, Ref. [5] for the oxidative coupling of methane and Refs. [6–8] for the oxidative dehydrogenation of propane), but knowledge about the active sites and the elementary steps is scanty.

Often rate or equilibrium constants for individual steps have not been or cannot be measured. Missing values are either transferred from similar reactions or estimated using Arrhenius expressions with pre-exponentials (entropy terms) and activation energies (enthalpies) for rate constants and analogue expressions for equilibrium constants. Usually, there remains a rather large set of parameters that is fitted to the observed behaviour, with all the problems connected with multiple solutions and a loss of physical meaning of the parameters for individual elementary steps.

My recent research contributions to heterogeneous catalysis

Quantum chemistry can remove the ambiguity connected with parameter fitting in microkinetic models and provide rate and equilibrium constants "*ab initio*" or "*from first principles*" provided that chemical accuracy (4 kJ/mol for energies and one order of magnitude for rate constants) is reached. This is a huge computational challenge because the available methods scale exponentially with the system size and a realistic simulation may require periodic models with large simulation cells including of the order of 1000 atoms.

Recently this challenge has been met with calculations for both equilibrium constants (adsorption of methane, ethane, and propane in H-chabazite [9]) and rate constants (methylation of ethene, propene and butene in H-ZSM-5 [10]).

Following a divide-and-conquer strategy, progress has been made both for the energy and the entropy (pre-exponential) calculations. For the energy, the common calculations on periodic models based on density functional theory are augmented with the more accurate wave-function type calculations for cluster models of the reaction site (Møller–Plesset perturbation theory (MP2) and Coupled Cluster theory (CCSD(T)) [11, 12]. For the entropy, we improve on the sampling of the potential energy surface by calculating partition functions from anharmonic vibrational energies. We reduce the scaling from exponential to linear by solving 3N one-dimensional Schrödinger equations in the anharmonic potential of each normal mode separately [13].

The availability of rigorous methods for calculating kinetic parameters of crucial reaction steps proved very useful when re-considering the well-established "Lunsford" mechanism for the oxidative coupling of methane on Li-doped MgO [14],

$$2\ \mathrm{CH_4} + \mathrm{O_2} \rightarrow \mathrm{C_2H_4} + 2\ \mathrm{H_2O}$$

which proposes that the C–H bond is activated by homolytic splitting involving hydrogen atom transfer to the $O^{\bullet-}$ sites [15]:

$$\mathrm{H_3C-H} + [\mathrm{O^{\bullet-}Li^+}]_{\mathrm{MgO}} \rightarrow \mathrm{H_3C^{\bullet}} + [\mathrm{HO^-Li^+}]_{\mathrm{MgO}}.$$

For the oxygen radical sites of Li-doped MgO, our calculations yielded barriers for hydrogen abstraction between 7 ± 6 and 27 ± 6 kJ/mol, which were in obvious conflict with the much larger observed values, between 85 and 160 kJ/mol [16–19]. Microkinetic simulations yielded 139 kJ/mol [20].

From this disagreement we concluded that the $Li^+O^{\bullet-}$ site may not be the active site, and that methyl radicals released into the gas phase are not formed by hydrogen transfer to such sites [19]. Our conclusion was further supported by temperature programmed reaction experiments. They showed that the same sites are responsible for the activation of CH_4 on both Li-doped MgO and pure MgO catalysts [19]. Experiments with differently prepared MgO samples hinted at a connection between morphological defects and catalytic activity.

These findings stimulated further calculations. They have shown that the Lunsford mechanism needs to be revised and that CH_4 chemisorbs *heterolytically* on morphological defects:

$$[\mathrm{Mg^{2+}O^{2-}}]_{\mathrm{MgO}} + \mathrm{H-CH_3} \rightarrow [(\mathrm{Mg-CH_3})^+\mathrm{HO^-}]_{\mathrm{MgO}}.$$

Release of methane into the gas phase happens only when O_2 is present on the surface,

$$[(\mathrm{Mg-CH_3})^+\mathrm{HO^-}]_{\mathrm{MgO}} + \mathrm{O_2} \rightarrow (\mathrm{O_2^{\bullet-}})[\mathrm{Mg^{2+}HO^-}]_{\mathrm{MgO}} + {}^{\bullet}\mathrm{CH_3}.$$

The latter reaction yields also a superoxo surface species that is more reactive than adsorbed O_2 in the gas phase:

$$(O_2^{\bullet -})[Mg^{2+}HO^-]_{MgO} + H - CH_3 \rightarrow [HO_2^-)[HO^- Mg^{2+}]_{MgO} + {}^{\bullet}CH_3.$$

Our quantum calculations, in combination with experiments [19], suggest a new role of the oxide catalyst in the oxidative coupling reaction. They do not provide and receive back electrons as transition metal oxide catalysts do in selective oxidations (Mars-van Krevelen mechanism [21]), they rather stay inert with their own electronic system and just bring together the reactants allowing them to exchange electrons (redox equivalents) directly between themselves. Here, the role of the solid "catalyst" is to bring together the reactants by binding them onto the surface, see Fig. 1.

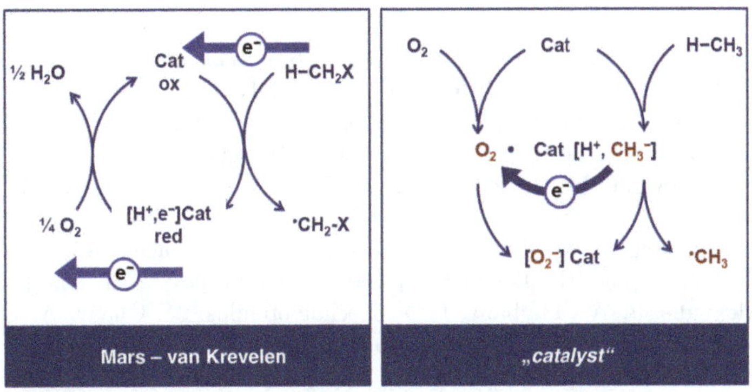

Fig. 1. Different C–H bond activation mechanisms.

Outlook to future developments of research on heterogeneous catalysis

Approximate computational methods have been proposed that make the screening of a large number of structures possible. To test and validate such approximate methods, and as a foundation for their further development, a set of benchmark systems is necessary, for which convergence has been achieved between accurate results from *ab initio* calculations and accurate experimental data for well-characterized surfaces within chemical accuracy limits (4 kJ/mol).

The real challenge in computational and joined experimental computational studies for solid catalysts is the very limited knowledge about their atomic structures. Supported metal and metal oxide catalysts are not ordered into unit cells and the size distribution of the active components is only rarely known. This is true even for the as prepared and activated catalyst, but the more for the catalyst at reaction temperature and in contact with the reactant mixture. We need to move from the active site concept to an active phase concept.

In the area of enzyme catalysis, researchers only start to think about mechanisms and computational studies on mechanistic proposals after they have got a good enough crystal structure. For studies in heterogeneous catalysis, on the one hand it is necessary to reduce complexity by looking at model catalysts [22], and on the other hand computational studies need to address catalysts under reaction conditions. Zeolites with their well-defined crystal structures are attractive objects for fundamental studies, see e.g., [3, 9, 10], although their structure is not as perfect as we would like it to be and often the distribution of active sites over different crystallographic positions is neither known nor can it be controlled.

Acknowledgements

Funding by German Research Foundation (DFG) is gratefully acknowledged.

References

[1] J. A. Dumesic, G. W. Huber, M. Boudart, *Handbook of Heterogeneous Catalysis*, G. Ertl, H. Knözinger, F. Schüth, J. Weitkamp (eds.), 1445 (Wiley-VCH, 2008).
[2] P. Stoltze, J. K. Norskov, *Handbook of Heterogeneous Catalysis*, G. Ertl, H. Knözinger, F. Schüth, J. Weitkamp (eds.), 1479 (Wiley-VCH, 2008).
[3] U. Olsbye, S. Svelle, M. Bjorgen, P. Beato, T. V. W. Janssens *et al.*, *Angew. Chem. Int. Ed.* **51**, 5810 (2012).
[4] E. V. Kondratenko, M. Baerns, *Handbook of Heterogeneous Catalysis*, G. Ertl, H. Knözinger, F. Schüth, J. Weitkamp (eds.), 3010 (Wiley-VCH, 2008).
[5] V. I. Alexiadis, J. W. Thybaut, P. N. Kechagiopoulos, M. Chaar, A. C. Van Veen *et al.*, *Appl. Catal. B* **150–151**, 496 (2014).
[6] J. Liu, F. Mohamed, J. Sauer, *J. Catal.* **317**, 75 (2014).
[7] X. Rozanska, R. Fortrie, J. Sauer, *J. Am. Chem. Soc.* **136**, 7751 (2014).
[8] X. Rozanska, R. Fortrie, J. Sauer, *J. Phys. Chem. C* **111**, 6041 (2007).
[9] G. Piccini, M. Alessio, J. Sauer, Y. Zhi, Y. Liu *et al.*, *J. Phys. Chem. C* **119**, 6128 (2015).
[10] G. Piccini, M. Alessio, J. Sauer, *Angew. Chem. Int. Ed.* **55**, 5235 (2016).
[11] C. Tuma, J. Sauer, *Chem. Phys. Lett.* **387**, 388 (2004).
[12] C. Tuma, J. Sauer, *Phys. Chem. Chem. Phys.* **8**, 3955 (2006).
[13] G. Piccini, J. Sauer, *J. Chem. Theory Comput.* **10**, 2479 (2014).
[14] J. H. Lunsford, *Angew. Chem. Int. Ed. Engl.* **34**, 970 (1995).
[15] T. Ito, J. Wang, C. H. Lin, J. H. Lunsford, *J. Am. Chem. Soc.* **107**, 5062 (1985).
[16] M. Y. Sinev, V. Y. Bychkov, V. N. Korchak, O. V. Krylov, *Catal. Today* **6**, 543 (1990).
[17] V. S. Muzykantov, A. A. Shestov, H. Ehwald, *Catal. Today* **24**, 243 (1995).
[18] S. Arndt, U. Simon, S. Heitz, A. Berthold, B. Beck *et al.*, *Top. Catal.* **54**, 1266 (2011).
[19] K. Kwapien, J. Paier, J. Sauer, M. Geske, U. Zavyalova *et al.*, *Angew. Chem. Int. Ed.* **53**, 8774 (2014).
[20] P. N. Kechagiopoulos, J. W. Thybaut, G. Marin, *Ind. Eng. Chem. Res.* **53**, 1825 (2014).
[21] P. Mars, D. W. van Krevelen, *Chem. Eng. Sci. Spec. Suppl.* **3**, 41 (1954).
[22] J. Sauer, H.-J. Freund, *Catal. Lett.* **145**, 109 (2015).

HETEROGENEOUS CATALYSIS USING SUPPORTED GOLD AND GOLD PALLADIUM NANOSTRUCTURES: UNDERSTANDING THE NATURE OF THE ACTIVE SPECIES

GRAHAM J. HUTCHINGS

Cardiff Catalysis Institute, School of Chemistry, Cardiff University, Cardiff CF10 3AT, UK

My view of the present state of research on Au catalysis

The preparation of active supported gold heterogeneous catalysts is complex. Gold being the most noble of metals readily transforms to metal gold and under reaction conditions the gold species that are present can readily sinter into large metallic particles. For this reason the observation of high catalytic activity by supported gold catalysts was not observed until the mid-1980s as at that point methods for synthesising small gold nanoparticles that were relatively stable were developed [1–3]. Subsequently, gold complexes were found to be highly active for a range of reactions and there is now an extensive field of catalysis by gold which continues to receive immense attention. Heterogeneous gold catalysts are effective for low temperature CO oxidation [2], acetylene hydrochlorination [1], and alcohol oxidation [4]; and when alloyed with palladium catalysts effective for alcohol oxidation to aldehydes [5, 6], C–H bond activation [7] and the direct synthesis of hydrogen peroxide [5, 8]. In all these reactions an outstanding feature of gold as a catalyst is the selectivity that can be achieved, *e.g.* alcohols can be selectively oxidised to aldehydes without the formation of acids [5, 6], or acetylene can be hydrochlorinated to vinyl chloride without subsequent formation of dichloroethane [9], and hydrogen peroxide can be synthesised from H_2 and O_2 with total selectivity based on H_2 [8]; something not readily achieved with other catalysts.

Heterogeneous gold catalysts are now being brought into commercial operation [9, 10] and in particular the gold supported on carbon has been commercialised in China for vinyl chloride production enabling the replacement of non-environmentally friendly mercury catalyst. This is the first time in 50 years that a catalyst composition has been completely changed for the synthesis of a commodity chemical. With the advent of commercial success there is now the remaining question of what is the nature of the active species in these catalysts, but this is a topic that has been debated since the discovery of high activity heterogeneous gold catalysts [1, 2].

There are now many preparation methods that are used *e.g.* co-precipitation, deposition precipitation, impregnation; as well as more advanced methods involving vapour deposition and indeed there are many methods for preparing elegant model systems at the mg scale. However, for supported gold catalysts that are required at the hundreds of tons scale the preparation method has to be very simple and reproducible and for this reason impregnation is currently used for carbon-supported gold [9] although co-precipitation/deposition methods could be applied to oxide-supported gold catalysts since this method is well used for the large-scale production of commercial catalysts.

My recent research contributions to Au catalysis

Supported Au as a heterogeneous catalyst

CO oxidation

A feature of supported gold catalyst is the complexity of the Au species that are present. However, this was not apparent until the advent of aberration-corrected scanning transmission electron microscopy in 2007. Before then only relatively large Au nanoparticles (2–5 nm) had been observed by the techniques available and the high activity of gold catalysts was ascribed to such particles. However, it is now known that much smaller species are present including atoms and clusters which can be monolayers or multilayer (Fig. 1). For over thirty years there has been an intense scientific debate surrounding the nature of the active species for CO oxidation on gold; however, no unequivocal identification of the active species has been reported to date and often the findings are contradictory. Bond and Thompson [11, 12] initiated the debate based solely on Au nanoparticles and Lopez *et al.* [13] demonstrated that the activity increased with decreasing nanoparticle size. Goodman *et al.* [14] explained the role played by the Au/support periphery atoms by showing that extended bi-layer Au structures on TiO_2 were extremely active. Herzing *et al.* [15] reported a study that utilised high angle annular dark field (HAADF) imaging in an aberration corrected STEM to investigate, for the first time, the full range of supported Au species present in real Au/FeO_x catalysts and they proposed that the active catalysts contained more sub-nm bilayer clusters and fewer nanoparticles $>$ 1 nm, which agreed well with Goodman [14] and also with the observations of Landman *et al.* [16] who predicted that a minimum grouping of eight Au atoms is needed to show CO oxidation activity. More recently, Schüth *et al.* [17] demonstrated that Au/FeO_x catalysts prepared by colloid immobilization methods can exhibit high CO oxidation activity while being devoid of any sub-nm clusters. Combining all these observations suggest that there is not just one distinct active site for CO oxidation over supported Au species and that particles existing over a broad size range may be effective for the reaction. Such a possibility was also recently highlighted by Haruta [18], who suggested that Au/FeO_x catalysts consist of a range of co-existing Au nanostructures each with its own characteristic activity.

Against this background we studied in detail two Au/FeO_x catalysts prepared by two different co-precipitation methods [19]. While the methods are similar there

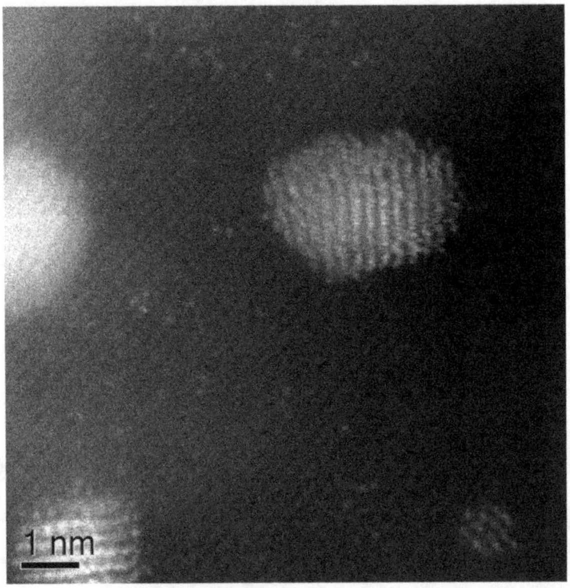

Fig. 1. Representative HAADF-STEM image of a Au/Fe_2O_3 show the co-existence of nanoparticles of various sizes, sub-nm clusters and isolated atoms.

are differences in the sequence and rate of mixing the acidic and basic precursors. In one method (denoted CP-1), the acidic solution ($Fe(NO_3)_3$ + $HAuCl_4$) was added quickly (within 2 min) into the basic solution (Na_2CO_3), whereas in the second method (denoted CP-2), the basic solution was slowly added drop-wise into the acidic solution over 30 min. These subtle preparation differences have dramatic effects on the catalytic behavior. Figure 2 shows the CO oxidation activities over a range of temperatures after drying and calcination. The two catalysts at the dried-only stage (120°C, 16 h) had similar CO conversion over the temperature range tested. However, after calcination at 300°C for 3 h the CP-2 catalyst is deactivated, whereas the CP-1 catalyst becomes more active, especially at lower temperatures. Detailed examination of the catalysts electron microscopy using a new counting algorithm showed that the two dried only catalysts had very different populations of gold species on the surface of the support, yet their activity was almost identical. Indeed, we concluded that based on the evidence, it is not possible to assign just one type of Au species as being solely active, while the others are inactive, in order to explain all the sets of data. For instance, if the sub-nm Au clusters are the only active species, then the thermal activation behavior of the CP-1 catalysts cannot be explained as there is a smaller population of these sub-nm Au clusters after calcination when this catalyst is more active. Instead, it is more logical to propose there to be an activity hierarchy for the different Au species present. This readily explains the observed behavior since the co-existence of wide range of Au nanostructures each having a different intrinsic activity needs to be considered.

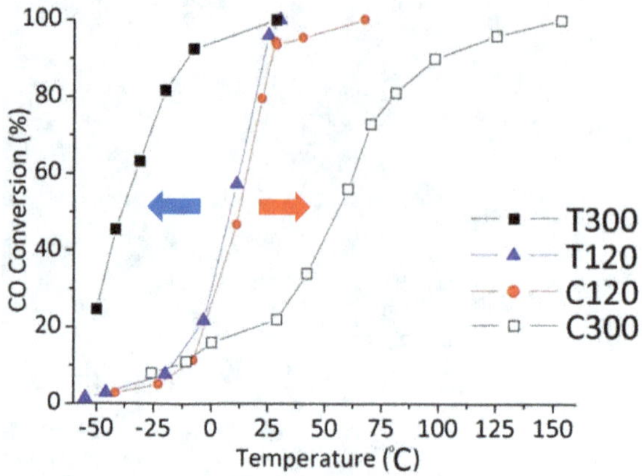

Fig. 2. CO conversion versus temperature. Catalyst mass 150 mg, Gas flow 50 ml min^{-1} 1 vol% CO in air; T120 CP-1, dried 120°C; T300 CP-1, calcined 300°C; C120 CP-2, dried 120°C; C300 CP-2, calcined 300°C. The arrows show the thermal activation behaviour (blue arrow) of the CP-1 catalyst and the thermal deactivation behaviour (red arrow) of the CP-2 catalyst.

Hence the final reported activities of the catalysts should be a weighted sum of the activity of each of the different species present, combined with their relative population densities (i.e. total activity, $A = \sum_i \rho_i \varepsilon_i$, where ρ_i and ε_i represent the population fraction and intrinsic activity for the ith active species.

Indeed, such an activity hierarchy might have a more general significance and readily exist in many well-studied supported metal systems, but remains undetected to date because the complex diversity of metal species present, spanning a range of sizes, have not been fully characterised since aberration corrected STEM imaging was not been fully exploited in their study.

Acetylene hydrochlorination

High surface area activated carbon has been found to be the only suitable support for gold for acetylene hydrochlorination. Initially, high activity catalysts were prepared by dissolving $HAuCl_4$ in *aqua regia* [3] but recently catalysts with improved activity have been prepared using Au complexes with sulfur ligands [9]. The most active catalysts require the dispersion of gold on the surface, however, the study of these catalysts is hampered by their beam sensitivity in either electron microscopy or X-ray photoelectron spectroscopy. For example in the fresh catalyst prepared using the *aqua regia* method comprises initially gold atoms but in the presence of the electron beam nanoparticles are formed. Detailed X-ray photoelectron spectroscopy [20] indicates that the activity is associated with a Au(I)–Au(III) oxidation couple, with exposure to acetylene causing the concentration of Au(I) to be enhanced and exposure to HCl leading to an increase in Au(III). Given that homogeneous Au(I) complexes are very active for reactions of alkynes it is possible that the most active species is the individual gold cation stabilized by Cl^- on the carbon surface, but as other species are present *e.g.* clusters and nanoparticles there may will be a hierarchy of activity for the species present although the catalyst is deactivated when Au is reduced to Au(0).

A very recent *in situ* study has shown this not to be the case and the active gold species in the acetylene hydrochlorination reaction is a gold cation that is fully dispersed on the carbon support [21] which is in line with the original prediction [1].

Supported AuPd as a heterogeneous catalyst

The addition of palladium to gold can influence the activity and selectivity of catalysts markedly. Bimetallic nanoparticles are relatively facile to prepare using two methods, i.e. impregnation and sol-immobilisation [22] and the materials have been well characterized. Both methods can be used to make either homogeneous alloy nanoparticles or core-shell structures, however, the interaction with the support is crucial. In these catalysts the interfacial sites between the nanoparticle and the support can be important, and the nature of the support can influence the morphology of the nanoparticles. For example, immobilizing a AuPd sol on carbon the AuPd nanoparticles do not wet the support and the nanoparticles are not faceted; whereas on MgO the AuPd nanoparticles interact very strongly with the support and can be highly faceted. The two materials behave very differently as catalysts even though they have the same precursor AuPd sol. In addition, there is a further complication with supported bimetallic nanoparticles; namely, the composition of the nanoparticles can vary systematically with particle size [22]. Hence these catalyst structures are more complex, although the addition of the second metal is a very useful tool in fine tuning activity and selectivity of gold catalysts, and this complexity increases with the addition of a third metal (*e.g.* Pt) although this provides a further way to fine tune catalyst activity [23].

As noted previously monometallic supported Au catalysts are highly effective for CO oxidation and acetylene hydrochlorination. For both the addition of palladium to gold decreases the activity significantly. In Fig. 3 the activity data are plotted against the standard electrode potential of the alloys. The high activity of gold as a catalyst for acetylene hydrochlorination was predicted on the basis of a correlation between the conversion of acetylene to vinyl chloride for a series of supported metal chloride catalysts with the standard electrode potential of the cations [1]. As Pd is added to Au there is a linear decrease in the standard electrode potential, and the alloy catalysts become less active for both reactions. This confirms that the monometallic Au catalysts are the best catalysts, and also that there is involvement of cationic Au(I) and/or Au(III) in the reaction. However, for the oxidation of benzyl alcohol and the direct synthesis of hydrogen peroxide the addition of Pd to Au, or *vice versa*, enhances the activity (Fig. 3), with catalysts comprising homogeneous alloys prepared by sol-immobilisation with Au:Pd mol ratio of *ca.* 1 having the highest activity. It is clear that these two reactions have different active species than those which correlate with the standard electrode potential. In both

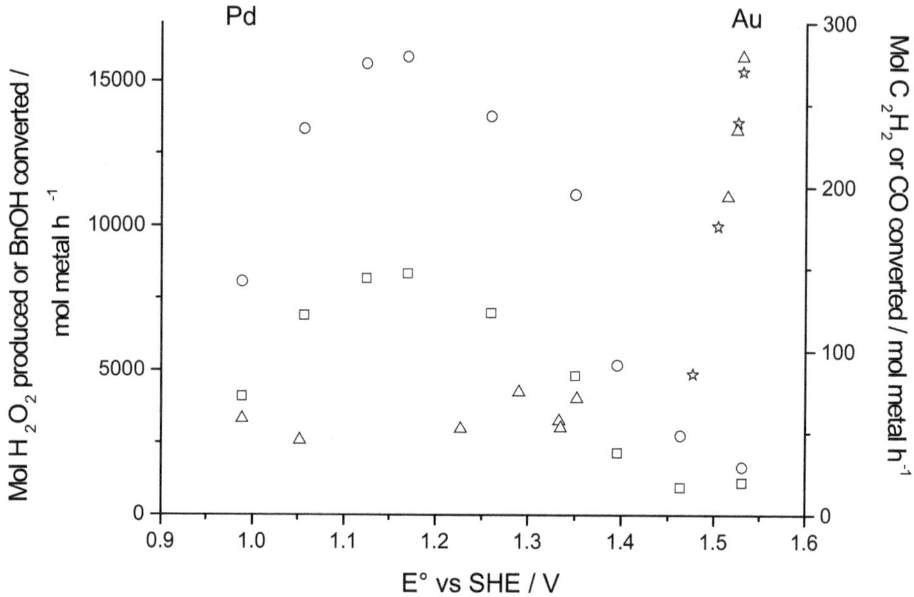

Fig. 3. □ H_2O_2 synthesis (methanol (5.6 g), water (2.9 g), 10 mg catalyst, 5% H_2/CO_2 (420 psi), 160 psi 25% O_2/CO_2 (160 psi), 2 min, stirring 1200 rpm, 2°C); ○ Benzyl alcohol oxidation (benzyl alcohol (40 ml), 0.05 g of catalyst, T =120°C, pO_2 (150 psi), stirring rate 1500 rpm, 6 h); △ CO Oxidation (5000 ppm CO in air, 20 mg catalyst, flow rate = 40 ml min^{-1}, 20°C); ★ Initial C_2H_2 hydrochloriation activity — C_2H_2 (5 $mLmin^{-1}$) and HCl (5 $mLmin^{-1}$) 200 mg of catalyst, 180°C, atmospheric pressure, GHSV = 870 h^{-1}.

cases the interfacial sites play an important role.

Using the sol-immobilisation preparation method it is very easy to prepare homogeneous alloys of AuPd or structures with a Au shell and a Pd core, or a Pd shell and a Au core [22]. This opens up possibilities of tailoring catalyst structures for specific reactions; for example the production of H_2. Currently only ~ 5% of commercial H_2 production comes from renewable sources (*e.g.* water electrolysis). Indeed photovoltaic electrolysis may become competitive in the near future and provide an effective route to H_2, but the materials used can contain toxic elements (*e.g.* CdTe) which can present a problem. Recently, there has been intense research on improving efficiency of photocatalytic process and the use of co-catalysts nanoparticles supported on the semi-conductors is proving to be a valuable approach. Au nanoparticles supported on TiO_2, can improve the photocatalytic efficiency through two methods: (i) spatial charge separation and (ii) surface plasmon resonance. However precise control of the composition and structure of the metal nanoparticle co-catalysts is essential [24, 25]. A recent example of using this approach is given by the photocatalytic generation of H_2 from glycerol or ethanol [24] using AuPd nanoparticles as co-catalysts with TiO_2. $Pd_{shell}Au_{core}$ supported on TiO_2 was found to be far more effective for the photocatalyic generation of H_2 from ethanol and glycerol than $Au_{shell}Pd_{core}$ structures, and both more effective than the supported monometallic Au and Pd co-catalysts. Hence the designing bimetallic nanoparticles with specific morphologies can provide a valuable way forward to identifying improved catalysts.

Outlook for further developments in Au catalysis

Supported gold and gold-bimetallic catalysts have immense potential in the future development of catalysis. The commercialization of gold as a catalyst for vinyl chloride manufacture has shown that stable gold catalysts can be produced in very high tonnages. However, the origin of the enhanced activity is as yet poorly understood. Going forward there needs to be more emphasis on studying both the reaction mechanisms and the nature of the catalyst morphology, composition and oxidation state under reaction conditions. Many research groups are indeed active in this area; however, the observation that there is a hierarchy of active structures in monometallic gold catalysts adds to the complexity as many *in situ* techniques operate by averaging across a cross section of all the available species present. For this reason there needs to be more emphasis placed on model systems to gain improved understanding but these need to be designed carefully and tested under realistic reaction conditions. In this way use can be made far more effectively of the full range of valuable *in situ* techniques available at this time which in combination with computational methods will be able to provide the insight needed to aid improved catalyst design. At present only a small fraction of the potential of gold as a catalyst has been realized but by combining model systems, computational modelling and *in situ* characterization its full potential could be achieved.

References

[1] G. J. Hutchings, *J. Catal.* **96**, 292 (1985).
[2] M. Haruta, T. Kobayashi, H. Sano, N. Yamada, *Chem. Lett.* **16**, 405 (1987).
[3] B. Nkosi, N. J. Coville, G. J. Hutchings, *J. Chem. Soc., Chem. Commun.* 71 (1988).
[4] A. S. K. Hashmi, G. J.Hutchings, *Angew. Chem. Int. Ed.* **45**, 7896 (2006).
[5] D. I. Enache, J. K. Edwards, P. Landon, B. Solsona-Espriu, A. F. Carley *et al.*, *Science* **311**, 362 (2006).
[6] M. Sankar, E. Nowicka, E. Carter, D. M. Murphy, D. W. Knight *et al.*, *Nature Comm.* **5**, 3332 (2014).
[7] L. Kesavan, R. Tiruvalam, M. H. Ab Rahim, M. I. bin Saiman, D. I. Enache *et al.*, *Science* **331**, 195 (2011).
[8] J. K. Edwards, B. Solsona, E. N. Ndifor, A. F. Carley, A. A. Herzing *et al.*, *Science* **323**, 1037 (2009).
[9] P. Johnston, N. Carthey, G. J. Hutchings, *J. Am. Chem. Soc.* **137**, 14548 (2015).
[10] R. Ciriminna, E. Falletta, C. Della Pina, J. H. Teles, M. Pagliaro, *Angew. Chem. Int. Ed.* **55**, 14210 (2016).
[11] G. C. Bond, D. T. Thompson, *Gold Bull.* **33**, 41 (2000).
[12] G. C. Bond, D. T. Thompson, *Catal. Rev.* **41**, 319 (1999).
[13] N. Lopez, T. V. W. Janssens, B. S. Clausen, Y. Xu, M. Mavrikakis *et al.*, *J. Catal.* **223**, 232 (2004).

[14] M. Valden, X. Lai, D. W. Goodman, *Science* **281**, 1647 (1998).
[15] A. A. Herzing, C. J. Kiely, A. F. Carley, P. Landon, G. J. Hutchings, *Science* **321**, 1331 (2008).
[16] U. Landman, B. Yoon, C. Zhang, U. Heiz, M. Arenz, *Top. Catal.* **44**, 145 (2007).
[17] Y. Liu, C.-J. Jia, J. Yamasaki, O. Terasaki, F. Schüth, *Angew. Chem. Int. Ed.* **49**, 5771 (2010).
[18] M. Haruta, *Faraday Discuss.* **152**, 11 (2011).
[19] Q. He, S. J. Freakley, J. K. Edwards, A. F. Carley, A. Y. Borisevich *et al.*, *Nature Comm.* **7**, 12905 (2016).
[20] X. Liu, M. Conte, D. Elias, L. Lu, D. J. Morgan *et al.*, *Catal. Sci. Technol.* **6**, 5144 (2016).
[21] G. Malta, S. A. Kondrat, S. J. Freakley, C. J. Davies *et al.*, *Science* **355**, 1399 (2017).
[22] G. J. Hutchings, C. J. Kiely, *Acc. Chem. Res.* **46**, 1759(2013).
[23] Q. He, P. J. Miedziak, L. Kesavan, N. Dimitratos, M. Sankar *et al.*, *Faraday Discuss.* **162**, 365 (2013).
[24] R. Su, R. Tiruvalam, A. J. Logsdail, Q. He, C. A. Downing *et al.*, *ACS Nano* **8**, 3490 (2014).
[25] R. C. Tiruvalam, J. Pritchard, N. Dimitratos, J. A. Lopez-Sanchez, J. K. Edwards *et al.*, *Faraday Discuss.* **152**, 63 (2011).

HETEROGENEOUS CATALYSIS, UNDERSTANDING THE MICRO TO BUILD THE MACRO

AVELINO CORMA

Instituto de Tecnología Química, Universitat Politècnica de València-Consejo Superior de Investigaciones Científicas, 46022 Valencia, Spain

My view of the present state of research on catalysis

The objectives of science and technology is to increase knowledge and improve the living standards of human kind, while achieving a sustainable growth. On this second aspect, social megatrends such as population growth, aging society, universalizing health care, climate change and limited resources, require, more than ever, the input of science and technology. To this respect, chemistry and more specifically catalysis can play an important role as, for instance, in the synthesis of healthcare pharmaceuticals, food security, water treatment, nitrogen fixation, generation and storage of renewable energy, saving energy by developing more efficient low carbon chemical process and enabling new routes to use CO_2 (provided that abundant and usable renewable energy is produced).

Considering the above issues, it is clear that homogeneous, enzymatic and heterogeneous catalysis with solids will play a predominant role. However, challenges such as reaching an unified approach for homogeneous, heterogeneous and biocatalysis and moving from description to prediction will be required. In other words, we should go toward science based catalyst design to attack emerging objectives related with renewables resources and energy. We will have to make the transition from an economy based only on fossil fuels into an economy in where the participation of renewable resources will constitute the backbone of the binomio energy-chemistry. During the transition period, natural gas will play an important role and key challenges for catalysis were and still are, how to directly activate alkanes in general, and methane in particular. How to activate N_2 and how to activate and react CO_2 efficiently to form hydrocarbons, assuming that renewable H_2 will be available. New technologies will have to be developed for energy storage and certainly chemical energy storage, preferably with energy reactors compatibles with existing energy infrastructure. Process intensification, i.e. "any chemical and chemical engineering development that loads to a substantially smaller, cleaner, safer and more energy efficient technology" will require structured and multifunctional catalysts with wide windows of operating conditions. More efforts on integrating homogeneous, heterogeneous and biocatalysis can be relevant. It is clear to me that more fundamental knowledge and integrative approaches between the different disciplines:

materials chemistry, physical, organic and theoretical chemistry, chemical engineering and physics will be required for achieving catalysts by design. In the present manuscript, the methodology we follow to design uni- and multi-functional solid catalysts with well-defined active sites will be presented and illustrated with some examples. We will show that it is possible to go from the concept and proof of principle in the academic lab, to the industrial application.

My recent research contributions to heterogeneous catalysis

One challenge with solid catalysts is to introduce chirality. However, since homogeneous catalysis has been very successful to carry out asymmetric reactions with large enantiomeric excess, there is the possibility to build solid catalysts with chiral catalytic centers by synthesizing hybrid organic-inorganic porous materials. Then structured mesoporous organic-inorganic solids can be made [1] by condensation of chiral organic counterparts functionalized with disilane groups, with a source of silica (tetraethyl-orthosilycate). The synthesis of the structured mesoporous solids has been achieved by using surfactant type molecules as templates [2]. However, for catalytic uses, the surfactant structure directing agent within the mesopores has to be removed. This is done by calcination at $T \geq 400°C$ or by treatment with a low pH solution. Both activations may disturb sensitive organic molecules or can affect chirality, losing the desired catalytic properties of the hybrid material. We have achieved the synthesis and successful activation of such catalysts by carrying out the preparation in absence of surfactants, working in ethanol/H_2O solution while catalyzing the condensation with NH_4F in ppm amounts. After the synthesis and filtration and washing with ethanol and water, structured mesoporous materials are formed with empty pores and the chiral organic remaining intact in the wall together with silica [3]. Good activities and enantioselectivities can be achieved for carbon-carbon bond formation by Michael type reactions. Catalysts with more than one type of active sites can be built by introducing more than one organic in the wall and active sites at the inorganic counterpart (acid, basic and/or metal sites). These types of catalysts can now be designed for intensification process and we have shown how it is possible to reduce the number of process steps during the synthesis of some pharmaceuticals. Indeed, by means of a hybrid catalyst, the number of reaction, separation and purification steps, can be reduced with respect to the conventional synthesis procedure from five to one during the synthesis of GABA derivatives. The catalyst with a chiral organic counterpart together with supported Pd metal, allows the preparation of GABA derivatives with good yield and enantioselectivities, by a one pot two reaction step [4].

In the field of hybrid materials metal organic frameworks (MOFS) have been obtained with porous crystalline structures with metals at the nodes, linked by organics with carboxylic acids or amines [5–7]. They may act as catalysts through unsaturated metals centers, functionalized organic linkers, or from both [8, 9]. One potential advantage of these materials could rely on their microporosity, flexibility

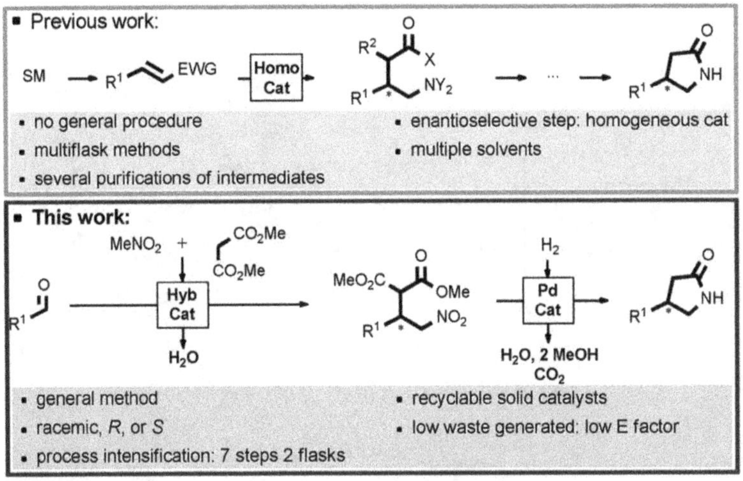

Fig. 1. Previous and current approaches toward the catalytic synthesis of GABA derivatives. SM = starting material. EWG = electron withdrawing group.

and possibility for multisite generation. The possibility to introduce chirality has also been demonstrated [10]. However, for successful industrial operation stability, turnover frequency (TOF) and turnover number (TON) have to be also considered.

Taking into account that many enzymes contain hydrophobic pockets, host-guest systems, such as for instance cyclodextines [11–13], calixarenes [14], or zeolites have been used to carry out reactions in confined spaces [15]. Self-assembled host [16–19] have shown to work as "molecular flasks" to facilitate reactions. Nevertheless, it would be of much interest to prepare more sophisticated host structures that can incorporate a variety of catalytic sites. Along this direction, micellar systems have been developed that allow to carry out reactions in aqueous media [20]. We have found that it is possible to prepare mesoscopic solid structures based on single-layer assembly of metal chains and organic alkyl spacers [21] that can be used as catalyst and reactor, and can be reused. Saturation transfer difference and two dimensional 1H nuclear Overhauser effect NOESY NMR-Spectroscopy show that non-covalent interactions can account for substrate activation. The mesoscopic character of this inorganic micelles, its hydrophobicity and chemical stability in water, makes the material able to also perform asymmetric transformations under environmentally friendly conditions.

In Fig. 2, we can see the representation of the structures made, that can work as micelles combining hydrophobicity associated to the alkylchains and catalytic active sites associated to the metals present in the structure and/or introduced by independent organic molecular species. In Fig. 2, the much higher activity of the material containing aluminium, and which we named Al-ITQ-HB, compared with MOFs or reported micellar systems for three-component condensation reactions, and for rate enhancement in enantioselective catalytic reactions is also shown [21].

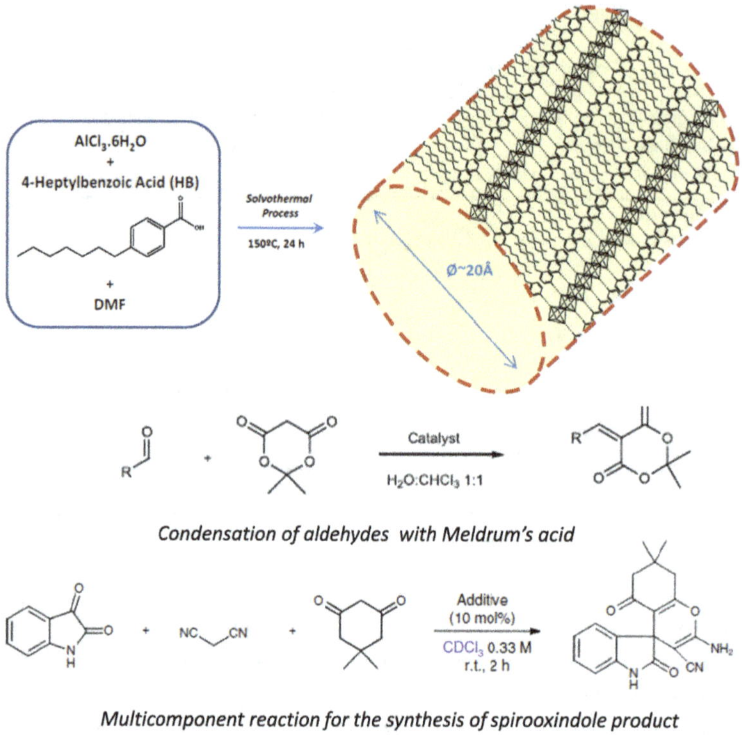

Fig. 2. Mesosocopic metalorganic material used as solid recoverable catalyst and reactor.

It appears that this type of material may open new possibilities to introduce in the system several types of active sites.

From my point of view, solid hybrid organic-inorganic materials described up to now can be of practical use in catalysis, if the TONs obtained are high (no catalyst regeneration required) and reaction temperature is moderate.

When much more robust catalytic systems, able to introduce molecular confinement effects are required, we have been working with microporous inorganic molecular sieves such as zeolites and zeotypes [22]. In this field, we have prepared materials with well-defined Brönsted and Lewis acid and basic sites, located in framework or extra framework positions. Local environment around the active sites and within the zeolite pore could be modified by changing the chemical composition. The materials can be tuned, from highly hydrophobic to hydrophilic, and the confinement effects and dispersion forces can stabilize the desired transition state of the reaction [15, 23]. The well demonstrated shape selectivity effects have made these materials widely used in petrochemistry and refining as well as for the production of chemicals and fine chemicals. Many of those zeolitic materials are made using Organic Structure Directing Agents (OSDA) to "template" the pore dimensions and topology [24, 25]. Present challenges in the field are related with the design of specific structures, avoiding or decreasing the use of expensive OSDAs, controlling active site location, synthesizing stable extralarge pore zeolites and multipore systems with controlled pore dimensions, and the introduction of different type of active sites for multistep catalytic reactions.

We have recently succeeded in the synthesis of tridimensional extralarge pore zeolites by stabilizing secondary building units such as double four and double three rings, which favor the formation of extralarge pore, even in the mesoporous range, zeolites. This was achieved by rationalizing the effect of Ge and F^- for stabilizing those secondary building units [26]. Unfortunately, the materials prepared with Si/Ge ratios lower than 5 are sensitive to H_2O and moisture. When samples with Si/Ge ratios in the range of 20 have been prepared, the zeolites are thermally and hydrothermally stable. However Ge contents have still to be further reduced for industrial applications, due to the cost of Ge. The work described above has opened new research avenues in academia and industry, to decrease and even avoid Ge during the synthesis, or to remove the Ge by postsynthesis treatments and to substitute that Ge by other M^{IV}, specially Si or Si and Al [27, 28].

Recently, the use of OSDAs has been avoided for the synthesis of some zeolites [29] and OSDAs expensive or requiring unfriendly manufacturing routes have been substituted by other more advantageous.

We would like to present here new challenges in zeolite synthesis for catalysis, such as the *ab initio* synthesis of zeolites for preestablished catalytic reactions based on molecular recognition principles [30], and the controlled location of the framework metals responsible for the active sites. The last is especially important in the case of zeolitic structures with crossing pores or pores with "cavities". We have focused our attention on two zeolites: ITQ-39 with one system of pores with 12 rings and two system of pores with 10 rings that cross, being the 10R pore system also connected to the 12R [31]; and ZSM-5 with a lineal and sinusoidal 10R pores that cross. Preferential location of framework Al in one system of pores or at the crossing of the channels can dictate the selectivity, when competing reactions with different size in the transition states can take place. This is the case of industrially relevant process such us methanol to propene or the cracking of olefins in the gasoline range during FCC operation to increase the yield of propylene. ZSM-5 is used in both processes [32, 33] (Scheme 1).

In the simplified reaction schemes presented above, propene formed is also consumed by oligomerization-cyclation-dehydrogenation (hydrogen transfer) reactions that yield aromatics, while saturating olefins. Notice that either methanol to propene or 1-hexene to propene require smaller transition states than the formation of aromatics plus alkanes by cyclization and hydrogen transfer. It appears then that there is room in the synthesis of ZSM-5 to improve the catalytic behavior. In our case by selecting the source of cations and the OSDA we have prepared ZSM-5 that improves selectivity to propene and catalyst stability for the two processes, i.e. methanol to propene and cracking of olefins. Our catalyst has achieved industrial application [34].

(a)

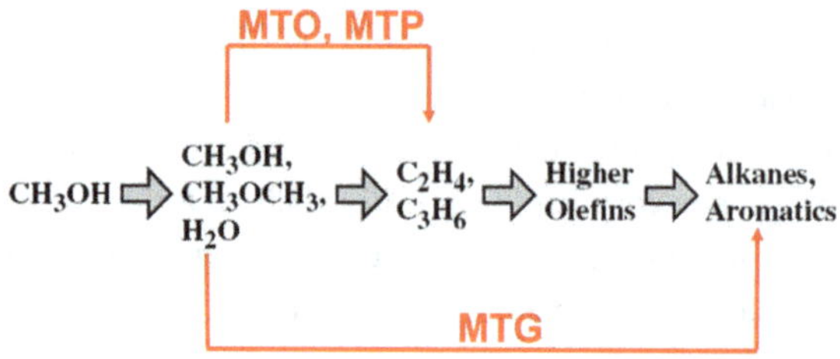

(b)

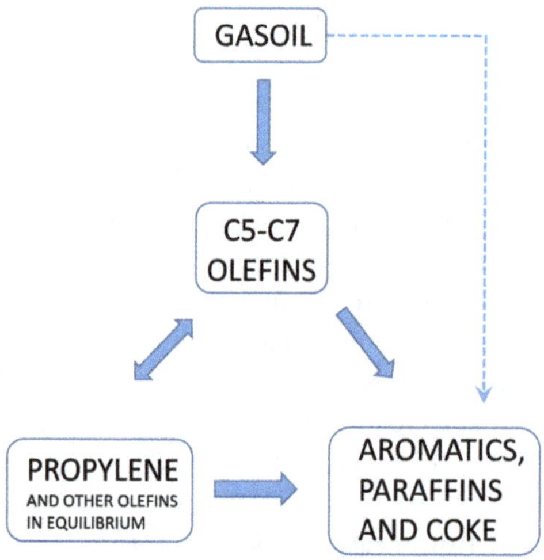

Scheme 1. Industrial relevant processes: (a) methane to propene and (b) cracking of olefins during FCC operation.

In another research line directed to the design of solid catalysts with defined catalytic sites, we have been working with single metal sites, metal nanoparticles and metal clusters. The catalyst variables to be optimized are particle size, shape and interaction with the support. Unsupported metal nanoparticles with sizes ranging from 5 to 20 nm can also be prepared and their surface and catalytic properties can be modified by introducing, carbon, nitrogen and/or a second metal [35, 36]. We will show that depending on metal site, support or the presence of other elements one can control the rate of H_2 activation, reactant and product adsorption and consequently chemoselectivities during hydrogenation. More specifically it is possible to control the chemoselective hydrogenation of substituted nitro aromatics to give substituted azoxys, substituted azos, substituted anilines and nitrones [37]. Based on the above concepts, a multimetallic catalyst that can reduce the sulfur content of natural gas below 10 ppb has been prepared and this catalyst is now being used commercially [38, 39].

Also in the field of catalysts based on metal nanoparticles, we have developed a completely new catalyst that allows chemoselective oxidations of organic molecules with air and that has been introduced in an industrial process that was before stoichiometric, and produced, at least, five times more subproducts than ours.

Metal clusters formed by a few metal atoms present a different electronic structure than metal nanoparticles ($\geq$ 25 atoms). Indeed, in the case of metal clusters formed by less than 10–12 atoms they are all exposed to reactants, are flexible and adaptable when interacting with reactants, and expose their frontier orbitals, allowing an excellent overlapping with the reactants when acting as Lewis acids [40]. We have shown that clusters with less than 10 atoms of Au, Pd, Cu and Pt are very active for a large number of reactions when in solution [41]. Furthermore, they can be stabilized on different supports, allowing, in some of them, to work at temperatures as high as 600°C without getting agglomerated [42].

Outlook to future developments of research on heterogeneous catalysis

We have still many challenges to face in heterogeneous catalysis and some of them are named below:

- Better understanding catalysis at molecular level, combining microkinetic and mechanistic studies with in-situ, operando and molecular modelling methods.
- Activation of molecules such as N_2, CO_2 and alkanes. In this sense making NH_3 at milder conditions (may also be by electrocatalysis) and, in general, directly converting the above molecules into chemicals reacting: CH_4 + CO_2 to acetic acid, benzene +CO_2 to benzoic acid, ethylene +CO_2 to acrylic acid, etc.
- Design and prepare materials:
  - Single atom and subnanometric catalysts
  - Chirality in solid catalysts
  - Multifunctional catalysts
  - 1D and 2D catalytic nanomaterials
  - "Artificial" enzimes
  - Hybrid and composite catalysts
  - Responsive-adaptive catalysts
  - Catalysis reactors with selective excitation
  - Integrating multi-catalysts-multi-reactor systems (tool box of catalytic modules).

Acknowledgments

This work has been supported by the European Union through the SYNCAT-MATCH project (Grant Agreement n° 671093), and the Spanish Government through "Severo Ochoa Program" (SEV 2012-0267).

References

[1] S. Inagaki, S. Guan, Y. Fukushima, T. Ohsuna, O. Terasaki, *J. Am. Chem. Soc.* **121**, 9611 (1999).
[2] F. Hoffmann, M. Cornelius, J. Morell, M. Frö ba, *Angew. Chem. Int. Ed.* **45**, 3216 (1999).
[3] P. García-García, A. Zagdoun, C. Copéret, A. Lesage, U. Díaz *et al.*, *Chem. Sci.* **4**, 2006 (2013).
[4] A. Leyva-Pérez, P. García-García, A. Corma, *Angew. Chem. Int. Ed.* **53**, 8687 (2014).
[5] B. F. Hoskins, R. Robson, *J. Am. Chem. Soc.* **112**, 1546 (1990).
[6] O. M. Yaghi, H. L. Li, *J. Am. Chem. Soc.* **117**, 10401 (1995).
[7] G. Férey, *Chem. Soc. Rev.* **46**, 369 (2004).
[8] A. Arnanz, M. Pintado-Sierra, A. Corma, M. Iglesias, F. Sánchez, *Adv. Synth. Catal.* **354**, 1347 (2012).
[9] A. Corma, H. García, F. X. Llabrés i Xamena, *Chem. Rev.* **110**, 4606 (2010).
[10] Y. Peng, T. Gong, K. Zhang, X. Lin, Y. Liu *et al.*, *Nature Comm.* **5**, 4406 (2014).
[11] H. Li, F. Li, B. Zhang, X. Zhou, F. Yu *et al.*, *J. Am. Chem. Soc.* **137**, 4332 (2015).
[12] A. Harada, Y. Takashima, M. Nakahata, *Acc. Chem. Res.* **47**, 2128 (2014).
[13] F. Hapiot, H. Bricout, S. Menuel, S. Tilloy, E. Monflier, *Catal. Sci. Technol.* **4**, 1899 (2014).
[14] J.-N. Rebilly, O. Reinaud, *Supramol. Chem.* **26**, 454 (2014).
[15] R. R. Cabrero-Antonino, A. Leyva-Pérez, A. Corma, *Angew. Chem. Int. Ed.* **54**, 5658 (2015).
[16] M. Yoshizawa, J. K. Klostermann, M. Fujita, *Angew. Chem. Int. Ed.* **48**, 3418 (2009).
[17] M. Yoshizawa, M. Tamura, M. Fujita, *Science* **312**, 251 (2006).
[18] T. Murase, Y. Nishijima, M. Fujita, *J. Am. Chem. Soc.* **134**, 162 (2012).
[19] C. Zhao, F. D. Toste, K. N. Raymond, R. G. Bergman, *J. Am. Chem. Soc.* **136**, 14409 (2014).
[20] L. M. Wang, N. Jiao, J. Qiu, J.-J. Yu, J.-Q. Liu *et al.*, *Tetrahedron* **66**, 339 (2010).
[21] P. García-García, J. M. Moreno, U. Díaz, M. Bruix, A. Corma, *Nature Comm.* **7**, 10835 (2016).
[22] M. Moliner, C. Martínez, A. Corma, *Angew. Chem. Int. Ed.* **54**, 3560 (2015).
[23] S. Goel, S. I. Zones, E. Iglesia, *J. Am. Chem. Soc.* **136**, 15280 (2014).
[24] B. W. Boal, M. W. Deem, D. Xie, J. H. Kang, M. E. Davis *et al.*, *Chem. Mater.* **28**, 2158 (2016).
[25] J. Sun, C. Bonneau, A. Cantín, A. Corma, M. J. Díaz-Cabañas *et al.*, *Nature* **458**, 1154 (2009).
[26] G. Sastre, J. A. Vidal-Moya, T. Blasco, J. Rius, J. Jordá *et al.*, *Angew. Chem. Int. Ed.* **41**(24), 4722 (2002).
[27] P. Eliášová, M. Opanasenko, P. S. Wheatley, M. Shamzhy, M. Mazur, P. Nachtigall, W. J. Roth, R. E. Morris, J. Čejka, *Chem. Soc. Rev.* **44**(20), 7177 (2015).
[28] P. Chlubna-Eliasova, Y. Tian, A. B. Pinar, M. Kubu, J. Cejka *et al.*, *Angew. Chem. Int. Ed.* **53**, 7048 (2014).

[29] B. Xie, J. Song, L. Ren, Y. Ji, J. Li, F.-S. Xiao, *Chem. Mat.* **20**(14), 4533 (2008). Y. Kamimura, S. Tanahashi, K. Itabashi, A. Sugawara, T. Wakihara, A. Shimojima, T. Okubo, *J. Phys. Chem. C* **115**(3), 744 (2011).

[30] E. M. Gallego, T. Portilla, C. Paris, A. León-Escamilla, M. Boronat, M. Moliner, A. Corma, *Science* **355**, 1051 (2017).

[31] T. Willhammar, J. Sun, W. Wan, P. Oleynikov, D. Zhang *et al.*, *Nat. Chem.* **4**, 188 (2012).

[32] A. Corma, C. Martinez, E. Doskocil, *J. Catal.* **300**, 183 (2013).

[33] Y. Mathieu, A. Corma, M. Echard, M. Bories, *Appl. Catal. A.* **439-440**, 57 (2012).

[34] F. Manfred, K. Stefan, B. Goetz, C. A. Corma, M. T. Joaquin, C. M. Elena, *PCT Int. Appl.* (2012), WO 2012123558. F. Manfred, K. Stefan, B. Goetz, C. A. Corma, M. T. Joaquin, C. M. Elena, *PCT Int. Appl.* (2012), WO 2012123556.

[35] L. Liu, P. Concepcion, A. Corma, *J. Catal.* **340**, 1 (2016).

[36] R. V. Jagadeesh, A. E. Surkus, H. Junge, M. M. Pohl, J. Radnik *et al.*, *Science* **342**, 1073 (2013).

[37] A. Corma, P. Serna, *Science* **313**(5785), 332 (2006).

[38] G. Wilson, N. Macleod, E. M. Vass, A. Chica, A. Corma, Y. Saavedra, *PCT Int. Appl.* (2011), WO 2011033280.

[39] G. Wilson, N. Macleod, E. M. Vass, A. Chica, A. Corma, Y. Saavedra, *PCT Int. Appl.* (2009), WO 2009112855.

[40] M. Boronat, A. Leyva-Perez, A. Corma, *Acc. Chem. Res.* **47**, 834 (2014).

[41] J. Oliver-Meseguer, J. R. Cabrero-Antonino, I. Dominguez, A. Leyva-Perez, A. Corma, *Science* **338**, 1452 (2012).

[42] L. Liu, U. Díaz, R. Arenal, G. Agostini, P. Concepción *et al.*, *Nat. Mater.* **16**, 132 (2017).

HETEROGENOUS CATALYSIS AND CHARACTERIZATION OF CATALYST SURFACES

JOHN MEURIG THOMAS

Department of Materials Science, University of Cambridge, Cambridge CB3 OFS, UK

My view of the present state of research on the above fields

(i) Present day research on heterogeneous catalysis (H^tC) is strongly influenced by societal and environmental factors, especially the need efficiently to: (a) utilize biomass and other renewable feedstocks; (b) evolve cleaner methods of utilizing the still plentiful supplies of renewable feedstocks, especially oil and gas; and (c) convert CO_2 to platform chemicals and sources of energy. I have previously summarized these issues in articles entitled: *"(H^tC) and the challenges of powering the planet and securing chemicals for civilized life"* [1] and *"Some of tomorrow's catalysts for processing renewable and non-renewable feedstocks, diminishing anthropogenic CO_2 and increasing the production of energy"* [2].

It is gratifying to note that more attention is now being paid to develop H^tCs for effecting sequential (tandem or cascade) reactions, as too much energy is wasted in effecting product separations [3, 4]. It was shown that, in a microporous framework-substituted alumino-phosphate catalyst, TAP05, the same multifunctional active site (4-coordinated Ti^{IV}) converted *c*-hexene via five distinct intermediates to adipic acid [5] using H_2O_2. But this reaction is slow. In general, it is better to use cascade (one-pot) reactions as has been done by Frechet, Corma, Lee and others. Lee *et al.* ingeniously devised a hierarchical macro/meso siliceous solid host — see results section below — to effect successive selective oxidations; and Corma has shown how to produce alkyl glycosides (from cellulose) and many other important products using cascade reactions [6].

Synthesis using structured porous polymers with bifunctional (acidic and basic sites) [7], as well as designed metal-organic frameworks (MOFs) [8], have also served as efficient, single-site H^tCs for sequential reactions involving, first, C–C bond formation (at Sc centres) followed by selective oxidation of the product (at Fe centres).

(ii) Turning to *characterization*, the work of Gladden on non-invasive, *in situ*, magnetic resonance imaging (natural abundance ^{13}C, MRI) is valuable. An early paper [9] demonstrated how conventional MRI combined with a magnetic resonance polarization enhancement technique could be used to resolve spatially both the conversion and selectivity of a heterogeneous catalytic reaction occurring within a fixed-bed reactor. Subsequent work by this group has tracked, *in situ*, the multi-phase oligomerization of ethene over a $Ni/SiO_2/Al_2O_3$ catalyst [10]. Recently [11], they made an *in situ* study of reaction kinetics using compressed sensing (CS), which provides a mathematical framework that quantifies how accurately a signal can be recovered when only a few data points can be recorded. They have used CS for other *in situ* catalytic studies in which MRI and NMR are both recorded to evaluate kinetic models, rates of reaction and selectivities.

Dramatic advances have been made in the application of aberration-corrected, high-resolution electron microscopy for *ex-situ* studies and more recently *in situ* cells have become available for investigating solid–liquid catalytic reactions [12]. Grunwaldt's joint use of both hard X-rays and electron microscopy for the *in situ* study of catalysts is a significant new development [13].

My recent research contributions to the above fields

(i) Dealing first with catalyst characterization, I have focused recently on two main areas: single-atom H^tCs and electron tomography (ET). Workers in China, US and Europe have shown that individual atoms of Pt, Pd, Ir, Au, etc., on appropriate supports, exhibit exceptional catalytic activities and selectivities. Our group's work has used aberration-corrected electron microscopy to identify situations where (Fig. 1) isolated, potentially catalytic atoms can be imaged and, aided by high-resolution electron energy-loss spectra (EELS), unmistakably identified: See Cr and Pt on MoS_2, Cu on graphene oxide and Ir on C_3N_4. The last-named is a good support, used by Perez-Ramirez *et al.* [14], because, unlike graphene (on which isolated atoms are too mobile) its shallow pockets [15] constitute favourable environments as anchors for solitary catalytically active atoms (DFT calculations and other evidence show that the energetics of forming dimer or trimer atoms in the pockets of the C_3N_4 are thermodynamically unfavourable).

My colleagues, using ET [16] have confirmed the validity of Lee's strategy in endeavouring to place nanoparticles of Pd at the inner walls of the macropores of his hierarchical siliceous host (to aerobically oxidize cinnamylalcohol to the aldehyde), and nanoparticles of Pt inside the mesopores, into which the aldehyde (unlike the alcohol) can enter to be further converted to cinnamic acid [4] (Fig. 2).

(ii) Turning to H^tCs *per se*, I have recently been involved in a collaboration on (a) designing a successful, highly selective Pt nanoparticle catalyst for the aerobic oxidation of KA oil using continuous-flow chemistry [17]. (The cyclohexanone produced is required for the production of adipic acid and

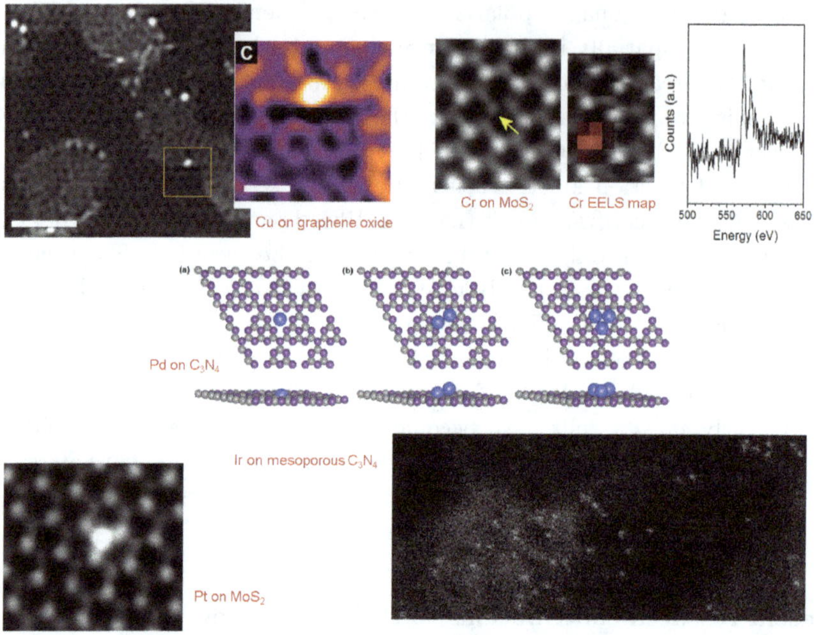

Fig. 1. Imaging and identification of potentially catalytic atoms by aberration-corrected electron microscopy and electro energy-loss spectra, respectively.

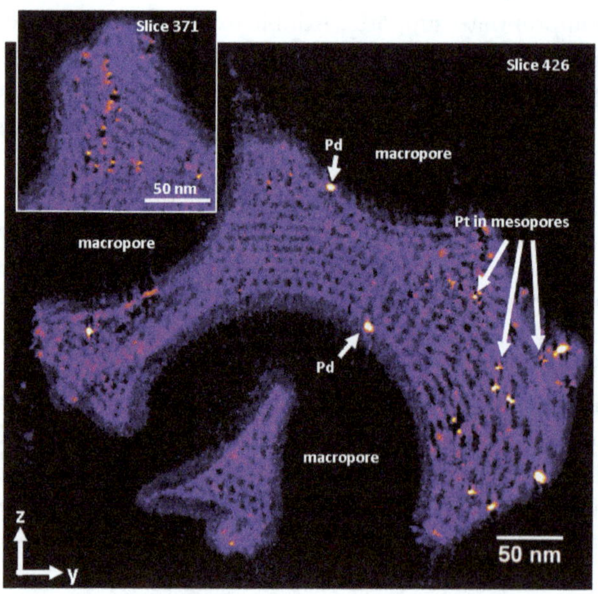

Fig. 2. Placement of nanoparticles on a hierarchical pore network for cascade reactions [4].

ε-caprolactam, which are precursors in the manufacture of nylon 6,6 and nylon 6, respectively). Also, (b) I have been involved with collaborators in Oxford, in designing a successful catalyst for the microwave-assisted rapid decomposition of a benign hydrogen storage material (long-chain hydrocarbons) to yield H_2-rich gas for use in hydrogen-powered fuel-cell vehicles [18]. (c) With the late A. H. Zewail, I examined the dynamical nature of the active centre in a single-site photocatalyst by 4D ultrafast electron microscopy [19]. In this work we could visualize the femtosecond-scale atomic movements of the Ti^{IV}-active centre in the single-site photocatalyst, known as JDFL-1 (that I had jointly designed with Chinese collaborators 25 years ago). These findings contribute fundamental insights in the development of advanced photocatalysts. (d) I have also pursued with P. A. Midgley, and his associates, new aspects of multi-dimensional electron microscopy, and, in particular, begun to explore the advantages of ET using, not images, but EELS, X-ray emission spectra and diffraction patterns.

Outlook to future developments of research in the above fields

(i) Looking to the future, there are plenty of challenges and opportunities where H^tCs can be vital. Take first, energy demand. In 2009, 16 terrawatts (TW) of energy were required by the world population. In 2050, it is estimated that 30 TW will be required. Nuclear power is not the solution since some 20,000 (gigawatt) reactors would be needed. (At present there are only *ca.* 440 operable civil reactors). All this points to the need for effective photocatalysts to harness solar radiation. Plenty of H_2 can be produced by photovoltaics and photocatalytic breakdown of H_2O. What is now needed are photocatalysts to reduce CO_2 to CH_4 (or other platform chemicals). Moreover, such photocatalysts could diminish anthropogenic CO_2 and stabilize its concentration in the atmosphere.

For the near future, here are some feasible prospects: (1) Use of homochiral MOFs (of the kind prepared by Wenbin Lin) to effect enantioselective conversions (including tandem style hydrogenations). (2) Utilize the crystalline, open-structure, hybrid organic-inorganic aluminosilicates, synthesized by Inagaki *et al.* and Bellussi *et al.*, to functionalize their C–H bonds and thereby produce novel, single-site H^tCs for a variety of transformations. (3) Improve methods of producing bulk chemicals from biomass, as was done, *e.g.* in the synthesis of terephthalic acid using oxidized variants of 5-hydroxymethylfurfural [20]. (4) Explore further the potential of single-atom catalysts, where one atom of the latter can replace tens of thousands, in a nanoparticle catalyst. (5) Perfect catalytic reactors to process wood and other plentiful sources of non-edible biomass so as to produce a wide range of products from a bio-refinery. Such work is challenging as several emulsion phases, apart from liquids, solids and gas, have to be processed.

(ii) So far as future methods of characterizing catalyst surfaces are concerned, many exciting tools have recently appeared, several of them made possible by the wide availability of synchrotron radiation, X-ray free electron lasers (XFEL) and high harmonic generation (HHG) sources. XFELs increase the coherent X-ray flux by 10^9, while table-top HHGs generate coherent radiation from UV to hard X-rays. This opens up ways of interrogating, by coherent X-ray diffraction, non-crystalline or poorly crystalline H^tCs. Femtochemistry at metal surfaces, as Wolf has shown [21], has its difficulties when dealing with nonadiabatic reaction dynamics — the whole manifold of potential energy surfaces, makes the situation fundamentally even more difficult than that faced by Zewail in his classic, progenitor experiment that investigated the transition state in the dissociation of ICN in 1988. Nevertheless, impressive progress in probing the transition states in catalytic oxidation of CO on Ru has been made by Nilsson *et al.* [22]. Although new prospects are opened up by the availability of attosecond sources [23], it is still possible to make fruitful studies of H^tCs *in situ* using modest synchrotron sources. In 1991, it was shown [24] that parallel measurements of XAFS and XRD could be made on the synthesis and subsequent performance of a Cu/ZnO methanol synthesis catalyst. Subsequent work [25, 26] has amply demonstrated the power of synchrotron radiation for exploring both short-range (interatomic) and long range order of typical, industrial-type catalysts, thereby elucidating their surface characteristics. But it is now also feasible, thanks to the work of Chergui *et al.* [27], to carry out ultra-fast time-resolved X-ray spectroscopies, as well as core-electron photoemission studies of atoms situated both at the catalyst surface and those in the reactants and products.

Note added in proof:

In a recent article entitled "The Enduring Relevance and Academic Fascination of Catalysis" by J. M. Thomas in *Nature Catalysis*, **1**, 2–5 (2018) other important future developments are described. These include (i) plasmon-enhanced photocatalysis, (ii) decarbonization of fossil fuels, (iii) elimination of oxides of nitrogen and particulates from diesel-operated vehicles.

Acknowledgments

I am grateful to the Kohn Foundation for support and the cooperation of Adam Lee, Rowan Leary, Jamie Weaver, Paul Midgley, Tom Furnival and Peter Edwards.

References

[1] J. M. Thomas, *ChemSusChem* **17**, 1801 (2014).
[2] J. M. Thomas, K. D. M. Harris, *Energy Environ. Sci.* **9**, 687 (2016).

[3] M. J. Climent, A. Corma, S. Iborna, M. J. Sabater, *ACS Catal.* **4**, 870 (2014).

[4] C. M. A. Parlett, M. A. Isaacs, S. Beaumont, L. M. Bingham, N. S. Hordow *et al.*, *Nat. Mater.* **15**, 178 (2016).

[5] S.-O. Lee, R. Raja, K. D. M. Harris, J. M. Thomas, B. F. G. Johnson *et al.*, *Angew. Chem. Int. Ed.* **42**, 1520 (2003).

[6] N. Villander, A. Corma, *ChemSusChem* **4**, 508 (2011).

[7] E. Merino, E. Verde-Sesto, E. M. Maya, M. Iglesias, F. Sánchez *et al.*, *Chem. Mater.* **25**, 981 (2013).

[8] L. Mitchell, B. Gonzalez-Santiago, J. P. S. Mowat, M. R. Gunn, P. A. Wright *et al.*, *Catal. Sci. Technol.* **3**, 606 (2013).

[9] B. S. Akpa, M. D. Mantle, A. J. Sederman, L. F. Gladden, *Chem. Commun.* **43**, 2741 (2005).

[10] S. T. Roberts, M. P. Renshaw, M. Lutecki, J. McGregor, A. Sederman *et al.*, *Chem. Commun.* **49**, 10519 (2013).

[11] Y. Wu, C. D. Agostino, D. J. Holland, L. F. Gladden, *Chem. Commun.* **50**, 14137 (2014).

[12] J. Park, H. Elmlund, P. Ereus *et al.*, *Science* **349**, 290 (2015).

[13] S. Baier, C. D. Dansgaard, M. Scholz, F. Benzi, A. Rochet *et al.*, *Microsc. Microanal.* **22**, 178 (2016).

[14] G. Vilé, D. Alham, M. Nachtegaal, Z. Chen, D. Dontsova *et al.*, *Angew. Chem. Int. Ed.* **54**, 11265 (2015).

[15] J. M. Thomas, *Nature* **525**, 325 (2015).

[16] R. Leary, P. A. Midgley, J. M. Thomas, *Acc. Chem. Res.* **45**, 1782 (2012).

[17] A. M. Gill, C. S. Hinde, R. K. Leary, M. E. Potter, A. Jouve *et al.*, *ChemSusChem* **9**, 423 (2016).

[18] X. Jie, S. Gonzalez-Cortes, T. Xian, J. M. Thomas, P. P. Edwards *et al.*, *Angew. Chem. Int. Ed.* **56**, 10170 (2017).

[19] B.-K. Yoo, Z. Su, J. M. Thomas, A. H. Zewail, *Proc. Natl. Acad. Sci. USA* **113**, 503 (2016).

[20] J. J. Pacheco, M. E. Davis, *Proc. Natl. Acad. Sci. USA* **111**, 8363 (2014).

[21] C. Frischorn, M. Wolf, *Chem. Rev.* **106**, 4207 (2006).

[22] H. Őström, H. Őberg, H. Xin, J. LaRue, M. Dell'Angelo *et al.*, *Science* **347**, 978 (2015).

[23] S. Baker, I. A. Walmsley, J. W. G. Tisch, J.-P. Marangos, *Nat. Photon.* **5**, 664 (2011).

[24] J. W. Couves, J. M. Thomas, D. Waller, R. H. Jones, A. J. Dent *et al.*, *Nature* **354**, 465 (1991).

[25] B. S. Clausen, G. Stephensen, B. Fabius, J. Villadsen, R. Feidenhansl *et al.*, *J. Catal.* **132**, 524 (1991).

[26] J. D. Grunwaldt, S. Hannemann, C. G. Schroder, A. Baiker, *J. Phys. Chem. B.* **110**, 8674 (2006).

[27] C. J. Milne, T. J. Penfold, M. Chergui, *Coord. Chem. Rev.* **227-228**, 44 (2014).

SESSION 2: HETEROGENEOUS CATALYSIS

CHAIR: GERHARD ERTL
AUDITORS: T. VISART DE BOCARMÉ[1,2], Y. DE DECKER[2,3]

[1] *Université libre de Bruxelles, Chimie Physique des Matériaux et Catalyse, Faculté des Sciences, Campus Plaine CP243, 1050 Brussels, Belgium*
[2] *Université libre de Bruxelles, Interdisciplinary Center for Nonlinear Phenomena and Complex Systems, CP231, 1050 Brussels, Belgium*
[3] *Université libre de Bruxelles, NonLinear Physical Chemistry Unit, CP231, 1050 Brussels, Belgium*

Discussion among the panel members

Gerhard Ertl: We heard now a lot of different aspects of heterogeneous catalysis. I may just ask at the beginning if there is any problem of general nature.

Graham Hutchings: There is a lot of things that draw the thread together in the heterogeneous catalysis presentations that we put forward and I think if we are going through it, there is a lot of things that build bridges between the heterogeneous catalysis community and the homogeneous catalysis community. As Avelino said: "Do we know what we are doing?" I mean the example I put forward for the commercialized gold catalysts for vinyl chloride manufacturers is in fact single gold atoms which we think are cations and are stabilized by chlorides that are on the surface and are really homogeneous supported catalysts. Gold (I) complexes are extremely well-known for acetylene-type chemistry. So perhaps something we found 30 years ago is actually a homogeneous catalyst which is made heterogeneous by the nature of the support. But I think some of the key areas that are coming out are the interplay between the theory and practice for this area and how it shows the model systems actually at the way in which I am thinking. So, these are my initial thoughts to set the frame.

Hans-Joachim Freund: That is a very important comment in order to bridge these two fields, if we want to call them two separate fields, which I think is not really appropriate. But we have now the experimental tools to actually see real heterogeneous catalysts. Things that people have previously only viewed as a model. It's extremely important for finding a common ground for discussions. So, the point I want to make is how important it is to spend time on instrument development. We need to develop instrumentations adapted to solving particulate problems. I don't want to come back to the funding situation that was raised. There is very much a need to develop instrumentations that allow us to look at a catalyst under working conditions, in other words bridging the pressure and the materials gap. I think that the examples that were shown in this round are at least pointing two directions that give us hope.

Gerhard Ertl: Let me come back to one of my questions on the last slide. I am wondering if we have a unique definition of a catalytic activity in heterogeneous catalysis. And how can we determine the activity? The turnover frequency is very often used but this requires to count the active sites and this is usually not the case. So, people use specific surface area or activity by gram of the catalyst or whatever it is. What is the activity in heterogeneous catalysis? What do you think?

Avelino Corma: It depends for what point of view you want to use that definition. From the scientific point of view, what we should do is to go and to find what are the active sites, to count them and then to calculate the turnover frequencies. Scientifically it's that. If you talk about the turnover frequencies to people from industry, they will say "Look, I want to fill up my reactor and I fill up my reactor by volume, so tell me weight, space hour velocity, or volume whatever space hour velocity" and that is the way they will look. But as a scientist, I feel that the objective has to be to try to understand the reaction, how it happens, where the reaction is happening and then to calculate the turnover frequencies.

Joachim Sauer: The turnover frequency relies on knowing the active sites and with a solid catalyst, as we have seen, there is a big problem to identify them. For my example of the oxidation coupling catalyst, it is doubtful if we really understood the catalyst active site concept and as long as this is the case, we may use the most rational definition or compare different definitions. For example, if the rate per gram and the rate per volume change differently as a function of time, this shows that the morphology has changed during the lifetime of the catalyst. So, what we need is not just one definition but we should look at several definitions.

Gerhard Ertl: I fully agree with you. I am sure that we will find a clear definition in the future but at the moment it seems to be not obvious.

Avelino Corma: Well, I think that in some way we are not disadvantaged with respect to our colleagues in homogeneous catalysis, because in homogeneous catalysis you have molecular catalysts. You can count those molecules provided that the ones that you introduced are the active sites because sometimes they are not; they evolve also then you can calculate pretty well. In our case, to calculate turnover frequencies and unless we have very well-defined systems as in the case of zeolites, it becomes difficult because we have first to know what are the exact active sites. Secondly, the active sites under operation conditions: because if we measure before, or after, they are not necessarily the ones that are working, neither the total number because you will have competitive adsorption, etc. So we have in that sense much more work to do if we want to make those definitions that our colleagues have in homogeneous catalysis.

Gabor Somorjai: We moved from a view of surfaces and catalytic surfaces as static for many decades and we know now that they are dynamic. They restructure under reaction conditions and they restructure in every way: the size, the shape, the interface. For example, if you have a solid-liquid interface, the liquid density is a thousand times greater than the gas density and so the coverages are different. So, as we look at a catalyst as a dynamic system constantly moving, maybe we have to divorce the ideas of activation or surface area with just looking at what comes out, which we can measure. We can measure quantitatively how many molecules come out, with the surface area that we defined before the reaction, the prenatal state or the surface area that, after the catalyst is dead, is again very well-defined. But what is in between is the most important form, which is exceedingly difficult to define. So, I suggest that maybe we should just only look at the products that are coming out and define what are the assumed surface area; or the working surface area is what it is. So, it is very difficult to define a dynamic system, constantly changing, with something that we can define only in a prenatal or post-mortem state. I think that this will live with us for a long time because what we find is the rise of nanomaterials. The small particles are constantly reconstructing and restructuring and absorbing things that passivate half of it, and that will be variables. So, I think that we have no real solution of defining in the old-fashioned way the surfaces that are rigid before or after use, when everything is changing. And I have no solution to this other than quantitative measurements of what is coming out.

Gerhard Ertl: Yes, that is right. In fact, the reason for that is that the strength of the bond between the adsorbed particle and the surface atom is comparable to the strength of the bond formation between the catalyst particle itself and this has been recognized by Langmuir almost 100 years ago. This is no novel insight.

Graham Hutchings: I think the one thing that we find these last few years is the fantastic possibilities of looking at catalysts under *in situ* conditions, which we didn't have before. We can find that what we thought were the active surfaces aren't what are the active surfaces. We take oxides for example. Often people model it as crystalline materials and you find the active surface is amorphous. So, it is much more complex to find out what is the net active species and if we count active species, that becomes very difficult. For amorphous materials, theory could play a very good role for what we want to do. But I am just thinking about the vanadium phosphates which are well-known industrial catalysts. They will be considered to be crystalline pyrophosphate, as they are known, with an amorphous overlayer, that is the active species. And if you make an amorphous vanadium phosphate, it is far more active than the crystalline material. The trick is how do you make it, so how do you disperse it... That's an application point of view.

Gerhard Ertl: The impression is that we more or less agree on this question. We cannot uniquely define the catalytic activity in heterogeneous catalysis and that's perhaps a good point to make a break.

General discussion

Gerhard Ertl: Is there a general question to start with?

Henk Lekkerkerker: I have a question for Professor Sauer but it's perhaps more general. We see in his presentation that there is DFT, the density functional theory. Now as far as I know, DFT still suffers from the lack of knowledge of the exact exchange correlation integrals. Many approaches have been put forward giving different results, so my question is: "how sensitive are your calculations to the choice of parameters and how transferable are your answers when you move from one system to another?"

Joachim Sauer: To the first part: we do not know the exact exchange correlation functional.

Henk Lekkerkerker: Here we agree.

Joachim Sauer: Here we agree, and I would even continue. If we would know it, it would probably be as costly to do the computation as it is now for the wave function-based methods. The alternative to density functional theory are wave function electron correlation methods. The problem is that they are scaling exponentially with the size of the problem. Applying them to systems of the order of thousand atoms is probably not possible. The advantage, however, is that we know how we can approach the right answer. So we have control about the accuracy, although we cannot reach it for systems of arbitrary size. In this situation, it is very important that there are density functionals of well-defined quality, in terms of what are the basic approximation involved. For me, it is important that we use functionals of this different type of quality. There is this one called "general gradient approximation" which is the one that is most easily applied and therefore most of the calculations with periodic boundary conditions are using this type of functional. There is a next level when you borrow something from the wave function people and you mix Fock exchange into your density functional, so called hybrid functionals, and then there are additional steps which Perdew calls the ranks of the Jacob's ladder. For those who are not daily reading the Bible, this is a ladder that leads to Heaven. I am afraid that the DFT ladder is cut at some point and that we cannot get to Heaven. So, my first statement is that we should assure ourselves how dependent the answer we are getting is on the type of functional we are going to use. Then there is a group of people who believes that you can parametrize functionals for classes of systems. But the number of parameters is increasing and every month there is a new functional that promises to do things better than the previous ones. I do not share this opinion. I would like to check what is the sensitivity of the results on the chosen functional and also look at cases for which I know the answer. But how can we go beyond? What I am doing, and this played a role today, is combining the density functional description for the whole periodic system with a wave function description of the reaction site. I have shown that this way, for important cases, we really reach chemical accuracy. This played a role in the story I was telling today because in this particular case we had to make sure that we are really reaching this level of accuracy.

Henk Lekkerkerker: Thank you very much.

Jens Nørskov: Two or three comments to this. First of all, we do know something about the accuracy by simply benchmarking against experiments and that is a very important part of those calculations and that is very important to distinguish between different classes of materials. There are classes of materials where simple density functional theory works well, transition metals for instance, and there are classes of systems, transition metal oxides, 3D oxides where we know, is doesn't work at all. So, we do know something about where we can trust it and where we need to use higher level methods. That is number one. Number two is that there are a number of methods where you can estimate the uncertainty on what to do and you can use that to assess the reliability. That is again mostly tested for transition metals, for other systems we know that there are big problems. The last thing to point out is that if you look at trend variations from one system to the next, then you are more likely to get things right than if you are looking for the absolute magnitude of anything. So, to get an absolute rate is exceedingly difficult, but to tell if something is better than another catalyst is less demanding. We know for instance that the scaling relations that I talked about are fairly functional independent. So, there are concepts you can take from the calculations that are not dependent on the actual value, if you like.

Henk Lekkerkerker: Thank you.

Joachim Sauer: Very briefly. I fully agree with what was just said.

Gerhard Ertl: Obviously, we agree on the concept of theory and on what it can do and what it cannot do.

Frank Neese: I agree with what both colleagues said but I would also like to point out that there is a large and very active area of research that brings the scaling on the wave function methods down to the linear scaling, maintaining the accuracy, they have made huge progress in recent years and you can get genuine wave functions and calculate forces in systems of hundreds of atoms now.

Henk Lekkerkerker: I can only say that I am happy to hear this, certainly from a world expert.

Gerhard Ertl: Perhaps I can address again the question of relations between heterogeneous and homogeneous catalysis.

Ben Feringa: I heard several times during the presentation "design" and "design of active sites." What do you define as an active site? Because I am a synthetic chemist and if you want to synthesize something, I would like to hear what is an active site. How do you define this? How should we target it?

Avelino Corma: If you want, for the first part that I said in a zeolite, for as it is clear, for instance, if you put titanium in the framework, you put tin, those are going to be the active sites that we have. If you put aluminum for instance, what is going to be the active site is the counter cation that you are going to put now. That can be a proton. Those cases are very well-defined. In other cases, when you go into transition metal oxides for instance, then the things are getting a little bit more difficult. And for metals, I will let Jens to come.

Ben Feringa: May I react directly to this? How far does an active site extend in a heterogeneous system, because you talk about single metal centres, or maybe bimetallic, but the whole space around it, the surfaces, all these parameters must be important, no?

Avelino Corma: You remember in my talk, I said that we want to optimize two things: not only the active site but short, medium range and long range. And that, you do it by changing the neighbors that you have there and of course, different types of pores that you have and those are the ones that are going to control in your transition state the interactions for the stabilization.

Jens Nørskov: Let me add up a couple of comments to this. Not for zeolites, but for systems like metal particles, there will be many different active sites. They will each have their own activity or activation energy, and may even have a change in mechanisms. The nice thing is of course that the rate depends exponentially on activation energies and therefore you are typically in a situation where a single or a few sites determine everything. Then you are in the following situation, this adds a little to the discussion on how we define rate. There are basically two ways to improve a catalyst in a situation like that. You can increase the number of those most active sites. So for instance, let me take an example: ammonia synthesis or NO activation or CO activation. We know that steps are the way you break the CO or N_2 bond. So, those are the sites that determine the activity and you can then change the activity in two ways: you can increase the number of active sites or you can change the nature of the active site. The first will give you a linear dependence, the other an exponential dependence. And therefore, both are important, obviously. And if you can control one and the other, you are in business. If you are stuck with a single kind of material, then obviously, you want to optimize or increase the number of those defects that do things for you but it's actually better to find another catalyst because you get the exponential dependence. Again, when you then compare different catalysts, then the exponential dependence usually dominates and that is why you can get trends as I illustrated without worrying in detail about the number of active sites. But this is an important factor.

Gabor Somorjai: This is a very important question. But it is far too complicated and it makes it impossible to define an active site because of the complexity of the neighborhood. Nevertheless, I try to suggest something: how about isotope exchange? The H_2-D_2 exchange or oxygen isotope exchange or nitrogen isotope exchange to define at least some of the active sites by a reaction that you can identify like C–H bond activation. Of course, if you look at an acid-base catalysis (in my type), which is an oxide metal interface, very often it is a much more complicated site because if you exchange the oxide from the silica to titanium oxide or to cobalt oxide and you keep the metal the same, it has a very different behavior pattern, because the charge transfer, which is crucial, — as you ionize the molecule —, is very different and complicated. But it is up to the community to define the active site by the chemistry it does, well enough that it can move that concept into a textbook stage, or something like that, because it is so important.

Jens Nørskov: There are actually examples where it has been done. Let me just say it again in metal catalysis which is what we understand the best. Gerhard Ertl showed one example of the NO dissociation on surfaces by STM, but perhaps an even more striking example is an experiment made by Ib Chorkendorff where he basically selectively poisoned the steps that we knew from the theory are the active sites. And the rate of the N_2 dissociation as a part of the ammonia synthesis dropped by eight orders of magnitude. So, there are actually examples where you can selectively poison those that you hypothesise are the active sites and you can see it directly, so I think there are examples.

Xinhe Bao: I would like to continue the discussion about the active sites. Well it is difficult to identify the nature of the active sites even in the static state. There is also a very important thing: during the reaction, the active site is changed. We have done a lot of experiments which say that. If you look at the turnover frequency, and if that turnover frequency is one thousand, it means that in one millisecond, the structure will be changed. Then, it is really very difficult to say what is the active site. There is sometimes also confusion when you talk about sites. So, I think it is a passport to the active environment. If you talk about the site, it has really a definite size for the activation. A lot of people are working now on single atoms as good catalysts. The single atom is certainly not freestanding and is maybe coordinated also to different oxides or such things. Then it is also difficult to say that one single atom is the active site. This is also the case for clusters with their surrounding environment. I think it is somewhere misleading to say "the sites" because if people talk about the sites, we should have suitable sites to do things. That is one thing. There is another thing: a lot of people use TEM, high resolution TEM that shows very beautiful pictures. Perhaps you know that the pictures; come through the differences of populations of the samples, and their structure are really different from the real catalytic reaction. Then what is the relevance between the picture we see and the real catalytic reaction? Sometimes I think it is also misleading to synthesize nanoparticles and say that this is the active site. Really, during the reaction, such a nanoparticle is not existing. I just want to have your comments on this.

Graham Hutchings: Thanks for your comments. I didn't show anything on this this afternoon, but I am sure people have got examples of it. One of the new techniques on the block is environmental TEM and STEM with aberration corrected capabilities, and so you can look at the formation of the catalyst under the reaction conditions, as has been done for CO hydrogenation and various reactions. So, I think you are right. The thing is that TEM or STEM are looking at very small parts of the sample, we all know that, and if you are looking at with XPS, then you are averaging over a whole sample, so you have got to use a whole range of techniques and use as many *in situ* techniques as you can to get to the answers so there is not one technique that tells you the answer. In 10 minutes, you don't get much chances to show what is there, so you show one micrograph, there is a lot of work behind it. I must admit we haven't done environmental STEM because I don't think anybody would like to put HCl and acetylene in a microscope. If there are volunteers, I would be very happy. But we can do CO-hydrogen and that is very instructive. You can see the defective small nanoparticles that are formed under reaction conditions. *In situ* techniques might become of age that you can really use them. *In situ* X-ray absorption techniques have been used for years but are now in combination with other techniques: you can really get important and valuable information.

Gerhard Ertl: The concept of active sites was introduced in 1925 by Hugh Taylor and it was used all the time as a concept and we have now learned that this concept is questionable. Nevertheless, we will use it further, like the term "activity." This is also something for which we cannot have a clear definition but we will use it. I think the same goes for the active sites, because Gabor has pointed out that the catalyst surface is continuously changing so the active sites will also change in their nature.

Jens Nørskov: When you say that is it continuously changing, I think one needs to include time scales in the discussion. There is an induction when, during reduction, you change the catalyst surface. If you are in a steady state or close to a steady state, then there won't be changes except very long term changes, sintering and other effects. When one talks about dynamics, one needs to invoke the time scale, and of course the time scale of the chemical reaction is quite different from those of the other time scales.

Kurt Wüthrich: Coming from the side of protein science, we have large molecules where we have usually one active site per molecule. So, that is well-defined. Now I am bit worried about the semantics used here. Should you not rather speak about the active surface and the activity of the surface being largely determined by the density of presumed active sites?

Gerhard Ertl: We were talking here about the properties of the surfaces, not of the reacting molecules. The sites at the surface that do the job, they are called active sites according to the definition by Taylor.

Hans-Joachim Freund: I wanted to make a comment concerning the possibilities to actually design a system that has a given active site, as Ben (Feringa) asked. You can use scanning tunnelling microscopy, environmental scanning tunnelling microscopy to look at a flat surface. I have shown you a silica surface, you can hydroxylate it, and you can see individual metal atoms anchored to this. If you could see an ethylene chain growing via polymerization on this site, then it would define the active site. However, as the first molecule is bound, and that is the reaction, the site is changing somehow, and then we don't know the details. But we do know which is the local active site. So, I think it is possible somehow to design systems appropriate for certain experimental techniques to answer that question in specific cases. I don't think we can do it in general, that would take decades of work, but in principle, it is possible for special cases.

Bert Weckhuysen: I would like to come back to the link between homogeneous and heterogeneous and biocatalysis and heterogeneous catalysis. Jens Nørskov and Avelino Corma referred at some point to "confinement." What is your opinion on the confinement? Is the confinement aspect something that bridges things like the solvent effects, trying to make a reaction intermediate, a transition state? How can we now link homogeneous, heterogeneous and biocatalysis? What tool do we have by which heterogeneous catalysis could help to forward homogeneous catalysis and the other way around?

Avelino Corma: I think that confinement in this case is what is going to help views on your medium range type interactions that I was talking there. It has to stabilize or to decrease the free energy of your transition state. And that happens certainly with heterogeneous catalysis or with enzymatic catalysis. In homogeneous catalysis, I am not expert at all, but I think that people are looking also at the possibilities to maximize the type of interactions not only by designing active sites but also they are finding other functionalities into ligands. Is that so? Because then it will come into the same principle.

Bert Weckhuysen: The solvent effect has already been mentioned. How many solvent molecules have to be close to the active site in a homogeneous catalyst? If you think about a zeolite pore, how many solvent molecules can you still have close to the Brønsted acid site to get a molecule you want to get reacting? Are there similarities between these things?

Gabor Somorjai: That is a very hard question that you are asking. I give you a practical example: we tried CO oxidation, a simple reaction. Keep the metal, platinum, always the same, and we do that on cobalt oxide, iron oxide, nickel oxide and we measure the turnover rate. By far the platinum-cobalt interface is about seven thousand times more active for the same reaction under the same conditions than the platinum-silica interface. The silica doesn't give anything and I assume that is because of a big band gap and a charge transfer is very difficult. But iron oxide and nickel oxide have oxide band gaps which are in the right range to allow the electron to go over the kinetic energy barrier that have Schottky devices, and it's two orders of magnitude less active. Now I have no explanation for this. In the acid-base things, both the oxide and the metal are part of the active site but in a way that they feel very different.

JoAnne Stubbe: I wanted to come back to what Kurt said, that all enzymes have a sort of single active site, they don't really. What you have is a conformational landscape, and if you do single molecule experiments and look at the turnovers, they can differ by a factor of a thousand. I think you will hear more about the importance of dynamics in proteins in the session on Friday morning. I think there are striking parallels between these systems in many ways.

Avelino Corma: If I can make a comment on what Bert was saying before. This morning, when I was hearing our colleagues from homogeneous catalysis, and they were talking about the effect of the solvent, about impurities, they are not different to what we have in heterogeneous catalysis. We deal with that by adsorption, we can do it, but also by kinetics. And it is possible to distinguish when you have physical adsorption and when you have chemical adsorption. And because if you do it kinetically you will see that the heats of adsorption are going to be different in one case or the other and that should be quite similar. I would like to know why you don't deal in that way in the case of homogeneous catalysis. Simply adsorption, and you can have physical adsorption, low heat of adsorption, or chemical adsorption, so large that sometimes it can be a poison, to tell like that.

Christophe Copéret: I wanted to add on that. If you look at bio- hetero- homo-, what we can do is we can measure a rate. I think today no one will say we cannot measure a rate in any discipline. The problem is to relate a rate to an active site and I believe in all fields it is nearly impossible. In heterogeneous catalysis, the question is what is an active site. We will spend hours debating on what is an active site, because it is a model. And so basically we should be very careful between the accuracy of measurements, which we can do very well and to relate a measurement to a turnover frequency which needs a model. And the model will be difficult in heterogeneous catalysis. In biocatalysis, as Professor Stubbe said very clearly, we know that the enzymes are on and off. What are we measuring? Are we measuring the rate of a single enzyme? Are we measuring the rate of many enzymes? In homogeneous catalysis also the catalyst can be on and off, that is the resting state. What do we observe? Actually it is very difficult. What you need to do is an accurate measurement of elementary steps. So you said adsorption in heterogeneous catalysis. You can imagine calorimetry or coordination — de-coordination in homogeneous catalysis. I am sure that in biocatalysis people are also studying these elementary steps. We should be very careful. As an author of a paper we should be very careful to always state clearly what we measure, and then to derive a model to try to relate a measurement to some active sites and this is a model that can be refined in 50 years. The active site of today will not be the active site of tomorrow, because we don't know what it is. So, we should be very careful with that.

Judith Klinman: I just want to make a comment on enzymes. There is such a huge literature defining the active sites in enzymes. A lot of it is chemical modification and so, this idea of poisoning for example in sites is seen. But in proteins, we have a crystal structure, we can make covalent modifications and we can quantify the number of active sites. That is a pretty easy thing to do. The confinement is very straightforward because we know the turnover numbers of enzymes, we know they are of rates similar to the turnover numbers, so we pretty much know the lifetime of the complexes between the enzyme and the substrate. That is all very straightforward, that is traditional biochemistry but there is also the issue that JoAnne brought up which is that, even after you form an enzyme substrate complex, if you have a conformational landscape, only some fraction of those sites are the active ones. You are sampling for the geometries, the active site geometries that are most active, and that is another level. It is subtler. I still think it is much easier to define a real turnover number of an enzyme than what we have been hearing so far. But there, there are a lot of complications.

Christophe Copéret: What I was saying is not turnover number but rates. Turnover frequency, if you want to relate what Professor Sauer does when he is trying to do computation of the rate or to evaluate a rate, then the rate is actually for your model, which is often a single atom in a single conformation and that is, in most cases, a turnover number which is — I totally agree with you — easy to measure or to evaluate. But a turnover frequency is something which is much more complicated to obtain.

Rutger van Santen: The bottom-line of this discussion is of course: "do we know the structure?" That is one. And very often we don't know really the structure of the surface that is doing the catalysis. Second point is that even if we would know the structure, then the number of atoms or the actual complex that is being used of that structure can be very different depending on the reaction. Take the example of the activation of the C–H bond, where we often believe that it's a single atom that will be reactive and that of course, it will depend on the number on neighbors. But if we take for instance hydrogenolysis, carbon–carbon cleavage, there we know that several atoms will be involved in that particular reaction. So the same structure will actually use different numbers of atoms. Of course you can normalize the whole thing if you know the structure. But as long as we don't know that precisely, there is an issue.

Jens Nørskov: I am not sure that this discussion is very productive for exactly the reasons that Rutger is mentioning, and I think in many ways the variations from one catalyst to the next is more interesting and more important than the absolute magnitude or even how we normalize the rates. If you ask an industry, what they are interested in is what is better than the other. You can measure the rate per gram or per volume which is really what they care about, and then you would like to know how you could make that better. I think there are good reasons that we may never totally be able to nail it down, but it may not also be the most important thing that we nail down.

Melanie Sanford: I wanted to come back to that sort of parallels or similarities and differences between homogeneous and heterogeneous catalysis. I think one of the big differences, just looking at the presentations this morning but those also of this afternoon, is the reactions. On one level, I looked at talks and I have seen a lot of reactions that, as an organic chemist, I would really want to do: C–H activation, C–C bond cleavage and forming reactions. But the substrates are much simpler than the many other substrates that we have seen this morning. I guess my question and I have two possible answers and maybe there another one too, is: "Why are there such differences in substrates?" or "Why can these catalysts, that do these beautiful chemistries, be used for the kind of substrates we saw this morning?" Some of the things that seem possible are the temperature — that the temperatures are too high — or poisoning from the nitrogen atoms and things like that in the substrates, but what do you view as the biggest challenges for taking these catalysts and using them to form kinds of bonds in more complex molecules?

Gerhard Ertl: Who wants to answer that?

Avelino Corma: In more complex molecules, we can do it also with solid catalysts. Not as complex as some that we saw before this morning, but yes, we can do it. We can do it from the point of view of acid-base solid catalysis, but also from the point of view of the metal catalysis. Yes, that is possible to be done. And we can do it also at low or relatively low temperature. However, we don't have the freedom that you have in more defined sites, with the ligands for instance, to be able to do it under milder conditions sometimes. But it is possible, some of them.

Gabor Somorjai: It is a very interesting question and I debated that for a long time. It seems to me that, ever since Wilkinson, homogeneous catalysis was defined: single transition metal ions surrounded by ligands. This is the definition of the whole field. If you go to heterogeneous catalysis, we sometimes have a single atom but very rarely. You look at clusters. I tried to measure catalytic activity as a function of the size of a cluster. Above 0.8 nm, which is thirty atoms, I can go higher and higher and higher, and I find changes and I think I understand them. But below 30 atoms, I could not stabilize the cluster so far. It would be very nice to try to answer you but with only two atoms, four atoms, six atoms, and how the chemistry would change. And that, to my understanding, I cannot do at this point.

Jens Nørskov: A part of the answer is that if we try to do CO bond breaking or N_2 bond breaking, thermally at low temperature, we can do it either. But we have the luxury being able to increase the temperature quite a bit, and if you want to break C–C bonds, if you do it at a low enough temperature, everything just turns into carbon and it's gone. I think the temperature is a very important factor. If we could find systems that don't follow these scaling relations that I talked about, then we could actually lower the temperature quite a bit. That is one of the big challenges. Or we can try to do it in redox or electrochemistry, and then of course, you totally change the chemical potentials and things can change as well. That's at least part of the answer, I think.

Gerhard Ertl: I agree with you. The temperature you need for bond breaking is an essential point that will make a difference. We didn't talk about another possibility to break bonds, namely by electrochemistry, by electron transfer. This is some aspect we didn't cover so much, so I hope that tomorrow we will talk more about that.

Steven Boxer: In homogeneous catalysis, if new students come into my lab, they might make a system that we have made before and then again after a little bit of experience they should get, within a couple of percent, the same results. And if it is transferred to medicinal chemists at Merck who have some experience, have a PhD, they should get within five percent the results that we get. So what happens — that is a completely naïve question — in heterogeneous catalysis if somebody in Gabor's lab comes up with a system? How easy is that for somebody in Graham's lab to do the same thing? Or is it just because it is so complicated or is it simple? I don't have any idea.

Graham Hutchings: I think in a lot of systems it is quite easy to transfer. You have got to make sure the details are in the preparation. We do the same things by the way...

Steven Boxer: And with what precision?

Graham Hutchings: With what precision? I would expect PhD students coming in to get the same results, pretty much the same as yours, and if Avelino wishes to take something of my lab results, which I know you have... then you will get the same result. Basically when there is something which is extremely exciting we have to make sure that we reproduce it before we tell the world, otherwise...

Steven Boxer: And could somebody of my lab do it? Do we need some more fancy equipment?

Graham Hutchings: No, the solid immobilized catalysts are very easy to make. We could all make one of these, and we should all be able to see some catalytic activity. It will take a little bit of time but you should be able to do it, yes. The systems that I described are not difficult to make and they can be made. And the one for the gold commercialized catalyst is made on the multiton scale by very simple methodology. So, I think it is easy to transfer and the point that I was going to address in the earlier discussion is that we all chose today to speak on relatively simple systems. We could have chosen cascades with quite complex molecules to talk about. We haven't done that.

Gabor Somorjai: I totally agree with you. The synthesis technique is crucial and it has undergone major changes from the beginning of the seventies to now. The colloid science techniques now dominate heterogeneous catalysis preparation with roughly between two nm and ten nm, about five percent change of the particle distribution. This is very easy. What is a variable is the capping you use in organic solvents, so some of the metals, because of much higher surface energy, are coated with PVP, organic. I can name five or ten of these, and these have an effect. But this is a synthetic problem. But over the last 15 years the definition between Graham's lab and mine allowed us to go to sizes within five percent that would change the product distribution of the turnover rates. This is the way it is. But just to remind this group, gold was not supposed to be a catalyst at all for a very long time, but it turns out that the size makes all the difference. Because the size means the change of the electronic structure. So the gold maybe stabilizes in 2+ or 4+ states if it is too small. There are variables that came to light that are crucial, but it is a time-dependant change of understanding.

Donald Hilvert: It seems to me that one of the distinctions of biological catalysts is their ability to use remote interactions between the catalyst and substrate to steer reactivity and selectivity. That would seem to be less the case for homogeneous catalysis and even less the case for heterogeneous catalysts. Is that impression correct or is that not so? It would explain one of the reasons why, as you generate a more complex molecular environment, that you can't work on more complex substrates.

Jens Nørskov: Yes, that is absolutely correct in my view. Here is the problem we have in heterogeneous catalysis. Let's say we want to make ammonia again, just to stay with that simple example. Then, as I showed, the energy of the transition state for the N_2 dissociation scales very well with the nitrogen binding energy. So you can easily find a catalyst with a lower transition state energy, but then it binds nitrogen so strongly that there are no active sites anymore because it is all covered by nitrogen. You have to get beyond that problem. One way of doing that would be if you had not just the surface to stabilize the transition state but had something else to help to stabilize it and in fact the potassium that Gerhard Ertl was talking about does exactly that. That actually stabilizes the transition state because the transition state is a little charged relative to the final state, and that is the reason why you add it. But if we could more systematically do that, that was exactly the point I tried to make about confinement and effects like that, we would be much better and I think that enzymes are very good at that, and some homogeneous systems as well. And that is a place where we should meet and try to get better.

Kyoko Nozaki: In relation to the complexity issues I just heard, I heard that there are differences with homogeneous and heterogeneous cases. The homogeneous players use ligands to modify the activities. You do know if examples exist of heterogeneous catalysis which utilizes like anchored ligands for asymmetric catalysis. So my question is: "Will this ligand usage become more and more popular in heterogeneous catalysis or not?"

Jens Nørskov: Let me just make one more comment there. In transition metal heterogeneous catalysis, the analogue of ligands is when you make intermetallic compounds or alloying. The metals that do the catalysis, they get other neighbors and that changes the electronic structure and that changes the reactivity. We do that all the time. The problem is that it is a much cruder method than by changing ligands on a single transition metal. When it comes to put ligands on the surface, then of course if they bind, you have a less accessible active metal atom and therefore it is much more difficult. But perhaps there are examples, I don't know.

Ben Feringa: Let me come back to the previous discussion, the point that was raised by Melanie (Sanford) and also by Don (Hilvert) about these multifunctional molecules and confined space. When I do a total synthesis, for instance a drug as synthetic organic chemists do, it is typically a multistep synthesis. Half of the steps are protection-deprotection steps. Enzymes (that's why I admire enzymes) typically don't need protecting groups. So could we take heterogeneous catalysis as an advantage? The surface, the multiple sites that are on the surface, etc, to use this for doing protection group-free catalysis? Or, is per definition the condition when you have multiple functional groups in a big molecule? Are they sticking to a surface so much that it is impossible to do this in catalysis? I try to get a feeling for that because it would solve a major problem in synthetic chemistry.

Graham Hutchings: Ben, that is a very good point and I think that this is a thing that a number of us are trying to do because protection-deprotection is a very expensive process. In certain reactions we seem to be able to do it. It's the case of the complexity of the molecules we can do it with. One example is: we can easily take alcohols to do aldehydes, so primary alcohols to aldehydes without making the acids. We have known it for a very long time. And then you can put that in another molecule and you can just functionalize that. We have a very close collaboration with physical organic chemists and synthetic organic chemists. That is one of the driving forces. So, the answer is: in the future, I very much hope so.

Ben Feringa: That is what you see now. People make the very complex supramolecular catalyst where it takes advantage of multiple interactions, noncovalent interactions, etc. But, these catalysts are often (I admire these catalysts) quite complex. Now, when you take advantage of the surface or these multiple sites on the surface, maybe there are great opportunities there. I try to get a good feeling for that.

Graham Hutchings: I agree Ben. The other point on confinement and also ligands to answer to together: there are classic examples from the literature for many years ago where you put cinchona alkaloids onto a platinum surface and you increase the rate of hydrogenation of an α-keto-ester by a factor of 300. And that is telling you something, and that is something that we have never really exploited, I would say. Because it is very specific for a very specific chemistry but it can be made far more general.

Christophe Copéret: The problem is a cultural problem. Melanie was addressing the question of very complex molecules and often, as catalysts. Because heterogeneous catalysis is much more on the physical chemical engineering side, we spend a lot of time trying to make better catalysts, trying to understand the catalyst. And obviously, it takes a lot of effort to bridge the gap between trying to make what Melanie would probably call "useful molecules," you know, discovering molecules. And you say "I want to address this problem in a very complex system." This is what I consider myself doing more at the heterogeneous side. We spend so much time trying to characterize, understand, discover other reactions, make them better. I think the problem is just culture. If homogeneous people and heterogeneous people work together and they have an organic mind, I am 100% sure that, as said Graham, we would discover new stuff. We know that if we add ligands on nanoparticles which are supported, you often do not decelerate the rate; you accelerate the rates and you get selectivity. It is known for a long time in asymmetric catalysis but is also known in semi-hydrogenation of alkynes to alkenes. So, there are many many examples but the problem is to work together. And that is probably the purpose of this type of meeting to realize that. If people work together, they might actually be able to achieve new transitional chemistry. That's what I think.

Avelino Corma: What about doing more work together and making hybrid catalysts? Hybrid catalysts you were talking today about cascade-type reactors and reactions. So, in hybrid catalysts, maybe we can get the best of the two worlds. In fact, we can get the molecular catalysts, we can get the inorganic framework, and with that we can change adsorption properties. We can work in one medium that, before, neither one or the other can work, we can get sites in one or the other or in both of them and in fact, modestly, we try to do that and it works. Because for making γ-aminobutyric acid (GABA), or derivates of GABA, you need between five and even seven steps. One is enantiomeric, and then you can do it in one pot. Two steps in one pot making one hybrid catalyst. We should work more on that because for instance, for enantioselectivity, for us it is very difficult to get. But we can get other types of reactions there. Now, putting the two together, we can go for it.

Kurt Wüthrich: I am afraid that the discussion is still going on but we have to come to a close rather soon. Would someone want to make a clear statement as to the possibility today to rationally design heterogeneous or homogeneous catalysis systems? I am unclear after the discussions of this morning and the discussions of this afternoon whether or not we are anywhere near such state.

Jens Nørskov: At least I can say we have examples where we have done it but there are few, on the heterogeneous side. There is a long way to go before we do it routinely.

Avelino Corma: I agree on that and on different aspects. There are examples, clear examples.

Gerhard Ertl: I think we should come to a close now. I think everybody agrees that the two fields are different but they come closer to each other. We can both learn from each other, and this is a fruitful result of this day. So I thank you for all the discussions and I thank you very much for attending and for staying so long.

Session 3

Catalysis by Microporous Materials

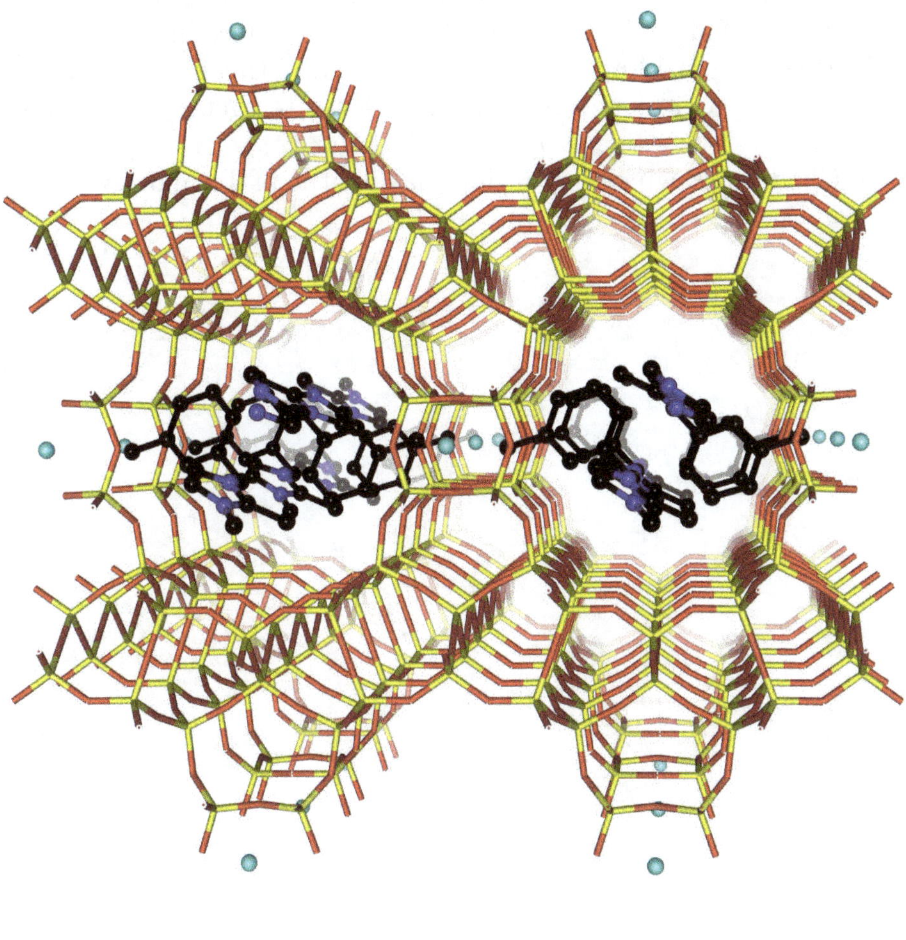

Schematic representation of the as-synthesized, molecular sieve CIT-13. This solid is the first to contain intersecting pores constructed from 14 and 10 oxygen atoms, and this highly desired arrangement provides new opportunities for heterogeneous catalysis.
Image by: Jong Hun Kang,
Chemical Engineering, California Institue of Technology.

CATALYSIS BY CRYSTALLINE, MICROPOROUS MATERIALS

MARK E. DAVIS

Chemical Engineering, California Institute of Technology, Pasadena, CA 91125, USA

Reaction pathways enabled by crystalline, microporous materials

Crystalline, microporous materials are a class of solids that have found widespread application as heterogeneous catalysis. These materials include zeolites and zeolite-like solids. Because of the uniformity in the structures of these materials, the expression of macroscale properties can be controlled by the materials chemistry at the atomic/molecular scale. Thus, these solids provide some of the best opportunities for understanding structure-property relationships in heterogeneous catalysts. New materials with new properties are being discovered, and as such, their use in fundamental studies as well as commercial applications continues to expand.

Historically, zeolites and zeolite-like solids have been exploited for use as shape-selective, heterogeneous catalysts. That is because they have uniform pore sizes and shapes that are on the same length scale as small molecules (0.3–1.5 nm). Early applications of zeolite catalysis employed shape-selectivity to discriminate between reactant molecules as well as product molecules. However, today, there are many features that can be controlled in the synthesis of the materials that enable more than shape-selectivity. For example, interior surfaces of the pore spaces can be tuned over a very broad spectrum from highly hydrophilic to highly hydrophobic. Element substitutions for the framework Si and/or Al atoms by elements like Ti, Sn, etc., have opened new chemistries and new reaction pathways (see below). Additionally, active sites can be confined in the microporous spaces, and localization there provides opportunities such as: (i) protection from components that could cause their loss in catalytic activity (*e.g.*, Ti and Sn centers in hydrophobic pore spaces that do not allow for adsorption of liquid water to enable Lewis acid catalysis in aqueous solvents), (ii) stabilization of atomic configurations within the pores spaces (*e.g.*, 6–7 atom Pt clusters for the aromatization of n-hexane to benzene) that would not be so in non-confined spaces, and (iii) their combination with other types of catalysts that would normally not be compatible (*e.g.*, strong acid site within a zeolite combined with a soluble organometallic or enzyme catalyst). In addition to providing real catalysts (can be scaled-up, can be regenerated, can be produced at acceptable costs, etc.) for many reactions in the petrochemicals and pollution abatement industries, these materials are enabling reaction chemistries that have not been realized previously (*e.g.*, non-metal center-based carbonylation,

Diels–Alder-dehydration reactions with ethylene and substituted furans). Thus, diversity of reactions and applications continues to expand.

Examples of recent contributions from M. E. Davis and collaborators

Over the past few years, the Davis group has been involved in creating crystalline, microporous solids that catalyze a number of different reaction chemistries with sugar substrates. The emphasis of that work has been on finding new catalytic routes for converting biomass-derived sugars into monomers for polymers [1]. The initial goal was to develop a solid catalyst to isomerize glucose into fructose. By understanding the mechanism and essential features of how metalloenzymes accomplish this reaction, the Davis lab and collaborators developed reaction pathways for glucose transformation over molecular sieves with isolated Lewis acid centers (via Sn and Ti substitutions for Si) that are localized in hydrophobic pore spaces. It was first demonstrated that these materials catalyze glucose isomerization to fructose in aqueous media [2]. Leading on from this discovery, it was shown that the reaction pathway in the microporous material is analogous to that of the metalloenzyme [3]. In further investigations, it was learned how to design and control the types of reactions that can be performed (selective intramolecular hydride shift (glucose to fructose) or intramolecular carbon shift (glucose to mannose)) [4]. In addition, new reaction pathways were discovered (*e.g.*, glucose to sorbose [5] and Diels Alder-dehydration reactions of substituted furans and ethylene [6]). More recent efforts involved the combination of retro-aldol and isomerization catalysts to convert hexoses into alkyl lactates [7].

Examples of potential future developments with crystalline, microporous materials

Chiral Zeolites: In a 1992 review, Lobo and I discussed the possibility of creating chiral zeolites [8]. To date, no one has synthesized a sample of a powder that is an enantiomerically pure, chiral zeolite. This goal remains elusive. Chiral, crystalline frameworks and layered materials do exist. With zeolites or zeolite-like materials, we first reported a slight enantioselectivity from a zeolite beta that was synthesized using a chiral structure directing agent [8]. Zeolite beta is an intergrowth of two polymorphs (A and B), and one (A) has enantiomorphs. Thus, to prepare an enantiomerically pure powder of zeolite beta, one would have to synthesize pure polymorph A, and then one enantiomorph of polymorph A. Lobo and I pointed out that there are frameworks that should have single crystals that are enantiomerically pure (analogous to the dense silica, quartz). When synthesizing these in powder form, there should be a 50-50 mixture of the two enantiomorphs. These structure types are likely to be more amenable to creating enantiomerically pure powders as one would not have to first solve the problem of obtaining a pure polymorph. In order to obtain a pure sample of a chiral zeolite, chiral organic molecules will be needed in the synthesis to direct the formation of a bulk sample of a pure enantiomorph [8]. Several chiral molecules have been used in the synthesis of zeolites [9]. The question arises as to why the chiral molecules used have not led to chiral framework structures. Part of the answer involves the mobility of the organic in the formed inorganic structure. Thus, the use of a rigid organic that fits into the zeolite pore structure without the ability to have any motion (even rotational motion of any portion of the molecule) will be necessary. Recently, materials have been synthesized that have enantiomorphs [10, 11]. First, the germanosilicate material was prepared (structure code STW) that has helical pores [10]. Since germanosilicates do not have the thermal stability of silicates, it was desirable to prepare the STW structure in silica alone. Rojas and Camblor synthesized the pure-silica material they called HPM-1 that has the STW structure [11]. This pure-silica material was prepared using an achiral organic molecule and the individual crystals do contain optical activity [11]. However, the bulk powder must be a racemic mixture of the two enantiomorphs. It would be quite interesting to perform the synthesis of HPM-1 with chiral organic molecules in the attempt to obtain a single enantiomorph, as separation of the individual chiral crystals into enantiomerically pure powdered samples is unlikely to be feasible.

Framework Positioning of Atoms by Design: It is well-known that the position of atoms such as aluminum can have dramatic effects on the properties of the zeolite. Therefore, it is of great scientific and technological interest to create synthetic methods that would allow for the placement of atoms into specific crystallographic sites within framework materials. At present, there are synthetic efforts aimed at altering the framework positioning between large and small cages, and the resulting materials have shown interesting catalytic behaviors. We have attempted to create site-specific heteroatom incorporation by the use of Zn as a substitute for Si. Since the Zn–O–Si angle is *ca.* 130 degrees, Zn has a tendency to order into those sites with tighter angles. While we have been able to produce ordered zincosilicates, we have thus far been unable to prepare framework materials that show good microporosity and stability and have Zn ordered into specific crystallographic positions.

The first example of an enantioenriched, polycrystalline molecular sieve has now been reported [12].

References

[1] M. E. Davis, *Top. Catal.* **58**, 405 (2015).

[2] M. Moliner, Y. Roman-Leshkov, M. E. Davis, *Proc. Natl. Acad. Sci. USA* **107**, 6164 (2012).

[3] R. Bermejo-Deval, R. S. Assary, E. Nikolla, M. Moliner, Y. Roman-Leshkov *et al.*, *Proc. Natl. Acad. Sci. USA* **109**, 9727 (2014).

[4] R. Bermejo-Deval, M. Orazov, R. Gounder, S. J. Hwang, M. E. Davis, *ACS Catal.* **4**, 2288 (2014).

[5] R. Gounder, M. E. Davis, *ACS Catal.* **3**, 1469 (2013).
[6] J. J. Pacheco, M. E. Davis, *Proc. Natl. Acad. Sci. USA* **111**, 8363 (2014).
[7] M. Orazov, M. E. Davis, *Proc. Natl. Acad. Sci. USA* **112**, 11777 (2015).
[8] M. E. Davis, R. F. Lobo, *Chem. Mater.* **4**, 756 (1992).
[9] M. E. Davis, *Top. Catal.* **25**, 3 (2003).
[10] L. Q. Tang, L. Shi, C. Bonneau, J. L. Sun, J. J. Yue *et al.*, *Nat. Mater.* **7**, 381 (2008).
[11] A. Rojas, M. A. Camblor, *Angew. Chem. Int. Ed.* **51**, 3854 (2012).
[12] S. K. Brand, J. E. Schmidt, M. W. Deem, F. Daeyaert, Y. Ma *et al.*, *Proc. Natl. Acad. Sci. USA* **114**, 5101 (2017).

CAN WE PREDICT THE REACTIVITY OF THE ZEOLITE CATALYST?

RUTGER A. VAN SANTEN[1,2], CHONG LIU[2], EVGENY A. PIDKO[1,2] and EMIEL J. M. HENSEN[2]

[1] *Institute of Complex Molecular Systems, Eindhoven University of Technology, 5612 AJ Eindhoven, The Netherlands*

[2] *Laboratory of Inorganic Materials Chemistry, Faculty of Chemistry and Chemical Engineering, Eindhoven University of Technology, 5600 MB Eindhoven, The Netherlands*

My view of the present state of research on zeolite catalysis

Here we will discuss shortly current understanding of reaction mechanism in zeolite catalysis and relationship between catalyst reactivity, catalyst composition and nanopore architecture. Computational studies have contributed largely [1]. Essential to the computational advances in zeolite catalysis has been the critical evaluation of quantum-chemical methods and development of new approaches [2, 3] and the availability of periodical DFT codes as VASP [4] that made calculation of transition states of elementary reaction steps in complex zeolitic structures possible.

General aspects of zeolite catalysis

Major insights on zeolite use and reactivity are:

— Zeolites are solid acid catalysts mainly of use in oil refinery related processes [5]. Proton activation by solid acids is mechanistically different from hydrocarbon activation in liquid super acids. Because of the low dielectric constant of the zeolite and of its rigidity, formation of intermediate carbenium or carbonium ions by proton transfer from zeolite to organic molecule is endothermic. Energy cost of proton transfer varies largely from 40 kJ/mol to 200 kJ/mol. The carbonium ion and carbenium ion chemistry that evolves upon protonation has many similarities to that in the super acid, but occurs as part of transition states or activated reaction intermediates [1].

— By incorporating reducible or non-reducible cations in the zeolite framework or locating cationic clusters in the zeolite micropore also selective oxidation or carbohydrate conversion catalysts have become available [6]. Such materials have unique reactivity, because they are often single site catalytic systems. For instance nanoporous silicalite activated by locating Ti^{4+} cations in its framework selectively activates H_2O_2, whereas a cationic $Ti_xO_y^{n+}$ cluster containing several Ti ions will decompose H_2O_2.

— A third important characteristic of zeolitic systems is the constraint on access of reactant molecules or formation of reaction intermediates due to steric limitations of the nanopores. The relative position of cations located in the zeolite nanopore leads to molecular recognition type reaction selectivity [7].

— Compared to enzyme catalysts zeolites are less selective and active. This is due to the rigidity of the zeolite frame work and the single site nature of the catalytic reaction centers. On the other hand, zeolites are more robust and less critical with respect the substrates.

—In zeolites, because of the floating motion of molecules through their nanopores when not sterically hindered the rate of diffusion is fast compared to inter particle Knudsen diffusion. The latter tends to dominate diffusion limitation of zeolite catalyzed reaction [8, 9].

Reactivity performance indicators of protonic zeolites

The activation energy of protonation will depend on the strength of the zeolite hydroxyl bond as well as on the relative stability of the transition states of intermediate carbenium or carbonium ions. Both will vary as a function of nanopore structure and zeolite framework composition. A scaling rule exists for the relative stability of the carbenium ions that increases from the primary to secondary and tertiary carbenium ions respectively.

In contrast the energy change of π adsorbed unsaturated hydrocarbon to a protonated σ alcoxy intermediate does not vary substantially between different zeolites, Therefore the Brønsted–Evans Polanyi linear activation energy-reaction energy rule, that is found to apply in transition metal catalysis is not of general use to solid acid zeolite catalysis.

Catalyst reactivity also relates to the adsorption energy of reactants and reaction intermediates, since this influences the surface state of the catalyst. For instance the apparent activation energies of the proton catalyzed direct monomolecular cracking reaction of light linear alkanes decreases linearly with hydrocarbon chain length [10, 11]. The adsorption energies of linear alkanes are incremental in the number of carbon atoms [12]. At high temperature nanopore loading is low, so that alkane concentration increases proportionally to their length. The match of intermediate molecule shape and size with nanopore structure determines the adsorption free energies largely [13].

Another example is the volcano plot dependence on zeolite nanopore structure found for the selectivity of lower temperature hydrocracking reaction of long chain alkanes [12]. The selectivity of the reaction then is determined by the adsorption equilibria of the reaction intermediates within the nanopore.

Catalytic reaction cycles

In zeolite catalysis it is essential to consider the three stages of reaction initiation, stable operation and catalyst deactivation together. Reaction initiation may activate organic molecules to form organo catalyst intermediates that propagate the reaction. Reaction can also be activated by changes of the inorganic chemistry of the catalysts surface.

We will use the catalytic cracking reaction catalyzed by solid acidic zeolites to illustrate formation and importance of the organocatalyst cation intermediates [14]. The activation energies we will quote in this subsection derive mainly from quantum-chemical calculations of the past decade.

Reaction is initiated by the formation of an unstable carbonium ion, that results from reaction of zeolite proton and alkane. Its formation requires an activation energy between 150–200 kJ/mol. The carbonium ion decomposes into a smaller carbenium ion and small alkane molecule. This carbenium ion can also generate a shorter alkane product when a hydride transfer reaction occurs with reactant molecule. The latter is by this reaction step converted directly into a carbenium ion. The activation energy of the hydride transfer process is estimated to be 30 kJ/mol less than that of carbonium ion formation. Propagation of the reaction network through carbenium ions circumvents slow carbonium ion formation of the reaction initiation phase. The carbenium ions can be considered organocatalysts that are part of a complex reaction network that form also olefins through the β C–C bond cleavage reaction. The activation energies for β C–C bond cleavage can be as low as 80 kJ/mol.

Hydride transfer competes with olefin oligomerization. Olefin oligomerization reactions lead amongst others to formation of heavier molecules or cations that will deactivate the catalyst. This is one of the key problems to be overcome in solid acid catalysis.

In the hydroisomerization process a solution has been found to this competition between hydride transfer and detrimental olefin oligomerization. The selectivity and stability of the reaction is dramatically increased when reaction is done in the presence of excess H_2 and transition metal particles as Pt or Pd are deposited on the protonic zeolite. These particles catalyze hydrogenation of the alkene intermediates, so that oligomerization of the olefins becomes suppressed. Relatively slow olefin removal by hydride transfer is now replaced by rapid alkane-alkene equilibration. The isomerization reaction is propagated by the olefins formed by dehydrogenation of alkane reagent. The olefins, that have an extremely low steady state concentration react through carbenium ion intermediates.

Interestingly again an initiation process is responsible for high selectivity of the hydroisomerization reaction. The highly active transition metal particles become rapidly carbided [15] that suppresses the hydrogenolysis reaction that otherwise would lead to methane production from the alkanes. This is an example where initial catalyst activation is due to modification of the inorganic chemistry of the catalysts surface.

When in another example instead of transition metal particles small Ga oxide clusters are used to promote CH activation of alkanes, that are more selective than proton activation, the competitive deactivating reaction is decomposition of the oxycations, that can be suppressed by the addition of water as a moderator the reactant feed [7].

In the Panov reaction the difference in deactivation catalysis is key to understand the difference in reactivity of different Fe based cationic complexes. Only Fe^{2+} gives selectively phenol whereas oxycation in clusters lead to deactivation of the catalyst by phenol oligomerization [16].

My recent research contributions to the prediction of catalyst performance, the example of the alkylation of isobutane and propylene [17]

Zeolite catalyst performance is due to the interplay between reaction intermediate nanopore occupation and the rates of rate controlling elementary reaction steps. We will show here how small differences in adsorption free energies affect catalyst performance largely.

The reaction to be simulated is the alkylation of isobutane and propylene catalyzed by a protonic zeolite of the faujasite structure. Simulated reaction rate data of the alkylation reaction of isobutane and propylene, based on a complete set of quantum-chemical data of the elementary reaction rates of the complex catalytic cycles (see Fig. 1) of this reaction are compared in Fig. 2. In the simulations results are compared for selected different values of the adsorption energies of propylene and isobutane.

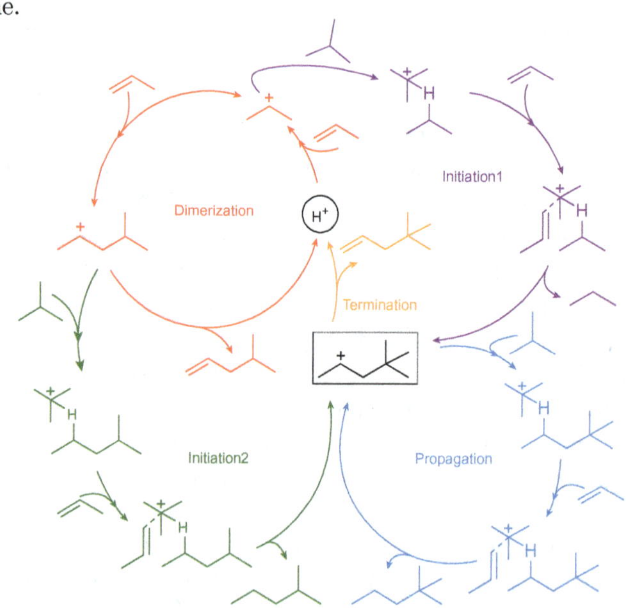

Fig. 1. Catalytic cycles of the elementary reaction steps considered in the alkylation reaction. The propagation cycle is at the bottom right. Cartoons relate to calculated reaction intermediates [17].

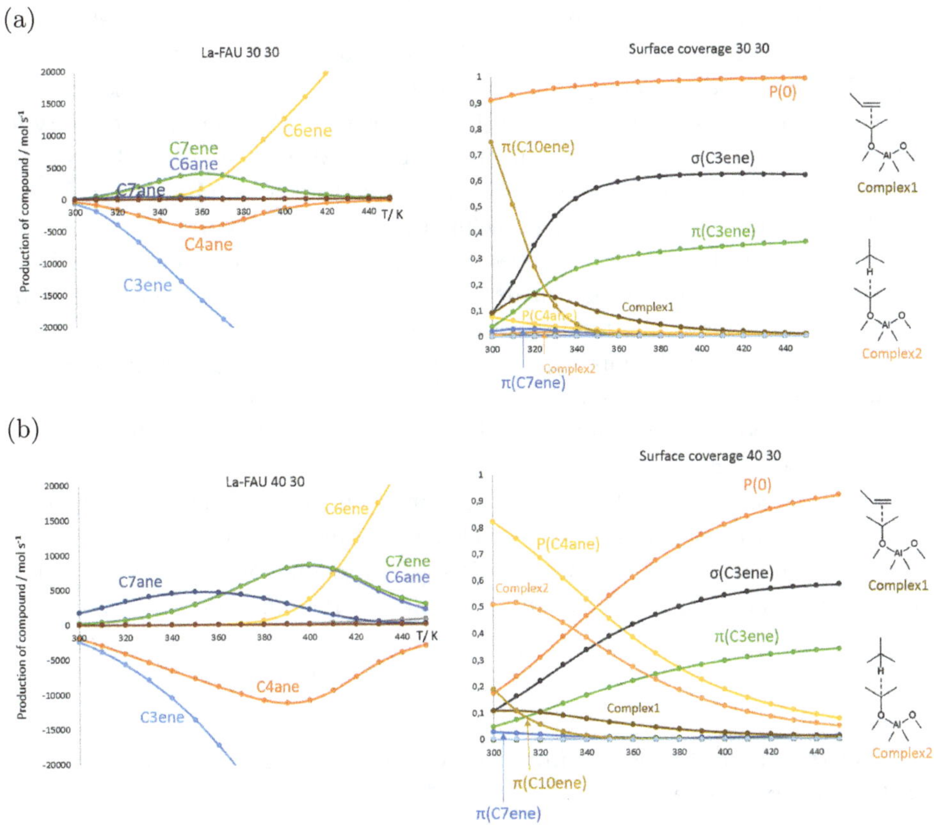

Fig. 2. Microkinetics simulated conversion and production rates of the alkylation reaction (Fig. 1) as a function of temperature. Comparison of results for two values of adsorption energies of isobutane and propylene in the proton free part of the zeolite nanapores. (a) Eads (isobutane) = 30 kJ/mol, Eads (propylene) = 30 kJ/mol), (b) Eads (isobutane) = 40 kJ/mol, Eads (propylene) = 30 kJ/mol [17].

When reactant adsorption energies vary there is a dramatic difference in simulated activity and selectivity of the reaction. The key competing reaction steps are hydride transfer of isobutane with a $C7^+$ carbenium ion to give the desired C7 alkylate versus $C7^+$ (in the propagation cycle) or $C4^+$ deprotonation (in the initiation cycles) to give olefins that will lead to catalyst deactivation. When the nanopore is empty (Fig. 2a) only the latter reactions occur and C7 olefin and hexane (or propane) are the main products.

When by increase of the adsorption energy the nanopore becomes preferentially occupied with isobutane (Fig. 2b) the hydride transfer reaction rate with isobutane increases and at low temperature C7 is formed selectively. The large differences in activity and product selectivity is found for changes in adsorption energies of only 10 kJ/mol.

In solid acid alkylation catalysis deactivation is not only relatively fast, but it also changes selective production of alkylate to production of olefins. Deactivation is due to consecutive and parallel oligomerization reactions of the olefins that form stable carbocations that consume the protons through a paring reaction type mechanism [18]. This will change selectivity because it enhances the relative rate of $C7^+$ deprotonation, due the increased charge of the zeolite frame work. When alkylate production selectivity is high olefin oligomerization reactions are relatively slow, because hydride transfer competes with carbon–carbon bond formation of the intermediate carbenium ions as can be deduced from the higher temperatures in Fig. 2 where oligomerization products are observed.

Whereas in the simulations initiation and propagation cycles are included deactivation is not explicitly treated, because of its substantially longer time scale. Often also site blocking will happen, so that the rate of deactivation will also relate to nanopore topology.

Simulations explicitly including deactivation is a multiscale problem, that so far has not been solved based on molecular elementary reaction rate data. Molecular mechanistic and computational information on the details of zeolite deactivation processes are scarce, but are becoming available [18, 19] and are highly needed.

Outlook to future developments of research on the prediction of catalyst performance

The results of the previous sections illustrate that microkinetics simulations should include next to calculated activation free energies of elementary reaction steps also accurate adsorption free energies as a function of pore occupation.

Currently calculation of adsorption equilibria and elementary reaction rates is treated separately. Coarse graining methods and force fields are used to calculate adsorption isotherms, whereas activation barrier calculation requires an atomistic treatment. There is a need to refine existing methods so as to be able to treat both with the same accuracy. Improved methods to calculate entropies beyond the commonly used harmonic approximation are also in need. It is for instance of particular importance to transition states involving carbonium ions that show complex dynamics. For a discussion of recent advances we refer to Ref. [20].

The multiscale aspect of the reaction due to the different time scales of initiation, propagation and deactivation pose an important further challenge to the development of modeling tools.

Whereas accuracy of calculated activation barriers and interaction energies computed with current state of the art quantum-chemical approaches usually suffices to discriminate between different mechanistic proposals, there is also a need to improve for complex systems quantum-chemical accuracy to at least 1 kJ/mol to reach kinetic predictability.

Acknowledgments

C. Liu acknowledges support from China Scholarship Council (CSC). Support from the NWO-MCEC program is also gratefully acknowledged.

References

[1] C. R. A. Catlow, V. van Speybroeck, R. A. van Santen, *Modelling and Simulation in the Science of Micro- and Meso-porous Materials* (Elevier, 2018); C. R. A. Catlow, R. A. van Santen, B. Smit, *Computer Modelling of Microporous Materials* (Elsevier, 2004).

[2] G. Piccini, M. Alessio, J. Sauer, *Angew. Chem. Int. Ed.* **55**, 5235 (2016); P. Nachtigall, J. Sauer, *Studies in Surface Science and Catalysis*, Vol. 168, J. Čejka, H. van Bekkum, A. Corma, F. Schüth (eds.), 701 (Elsevier, 2007).

[3] A. T. Bell and M. Head-Gordon (2011), Quantum Mechanical Modeling of Catalytic Processes, *Annu. Rev. Chem. Biomol. Eng.* **2**, 453–477. Available at: http://www.annualreviews.org/doi/10.1146/annurevchembioeng-061010-114108.12.

[4] J. Hafner, L. Benco, T. Bučko, *Top. Catal.* **37**, 41 (2006).

[5] A. Corma, *Chem. Rev.* **95**, 559 (1995).

[6] G. Sankar, J. M. Thomas, C. R. A. Catlow, C. M. Barker, D. Gleeson *et al.*, *J. Phys. Chem. B* **105**, 9028 (2001).

[7] E. A. Pidko, E. J. M. Hensen, R. A. van Santen, *Proc. R. Soc. A.* **468**, 2070 (2012).

[8] J. Kärger, *Diffusion and confinement.* In *Sitzungsberichte der Sächsischen Akademie der Wissenschaften zu Leipzig, Mathematisch-Naturwissenschaftliche Klasse*, Vol. 128 (2003).

[9] J. Kärger, S. Vasenkov, *Microporous Mesoporous Mater.* **85**, 195 (2005).

[10] W. O. Haag, *Studies in Surface Science and Catalysis*, Vol. 84, W. Hölderich, H. G. Karge, J. Weitkamp, H. Pfeifer (eds.), 1375 (Elsevier, 1994).

[11] T. F. Narbeshuber, H. Vinek, J. A. Lercher, *J. Catal.* **157**, 388 (1995).

[12] B. Smit, T. L. M. Maesen, *Chem. Rev.* **108**, 4125 (2008).

[13] R. Gounder, E. Iglesia, *Acc. Chem. Res.* **45**, 229 (2012).

[14] S. Kotrel, H. Knözinger, B. C. Gates, *Microporous Mesoporous Mater.* **35-36**, 11 (2000).

[15] R. A. van Santen, B. G. Anderson, R. H. Cunningham, A. V. G. Mangnus, L. J. Van Ijzendoorn *et al.*, *Angew. Chem. Int. Ed. Engl.* **35**, 2785 (1997).

[16] G. Li, E. A. Pidko, R. A. van Santen, Z. Feng, C. Li *et al.*, *J. Catal.* **284**, 194 (2011).

[17] C. Liu, R. A. van Santen, A. Poursaeidesfahani, T. J. H. Vlugt, E. A. Pidko and E. J. M. Hensen, Hydride Transfer versus Deprotonation Kinetics in the Isobutane-Propene Alkylation Reaction: A Computational Study, *ACS Catal.*, 2017, DOI: 10.1021/acscatal.7b02877.

[18] A. Feller, A. Guzman, I. Zuazo, J. A. Lercher, *J. Catal.* **224**, 80 (2004).

[19] D. M. McCann, D. Lesthaeghe, P. W. Kletnieks, D. R. Guenther, M. J. Hayman *et al.*, *Angew. Chem.* **120**, 5257 (2008).

[20] V. Van Speybroeck, K. Hemelsoet, L. Joos, M. Waroquier, R. G. Bell *et al.*, *Chem. Soc. Rev.* **44**, 7044 (2015).

CONSEQUENCES OF CONFINEMENT FOR CATALYSIS WITHIN VOIDS OF MOLECULAR DIMENSIONS

ENRIQUE IGLESIA

Laboratory for the Science and Applications of Catalysis, Department of Chemical Engineering, University of California, Berkeley, CA 94720, USA

Present state of research on confinement effects and transition state selectivity in catalysis by microporous solids

Shape selectivity concepts have sought to capture the ability of voids of sub-nanometer size to control molecular traffic via diffusional constraints and to stabilize transition states, via specific contacts between organic guest moieties and the inorganic confining voids with the intent to control reactivity and selectivity in chemical transformations on microporous solids. H-forms of zeotypes, most prominently of aluminosilicate composition, act as Brønsted acids with conjugate anions of stability that is essentially independent of the framework type. van der Waals stabilization of bound species, intermediates, and transition states combine in practice to bring remarkable catalytic diversity to zeotype frameworks that differ in structure, but are similar otherwise in composition and acid strength. Each framework type contains distinct crystallographic sites, in which the presence of Al and charge-balancing cations lead, in turn, to different confining environments. Differences in size and shape among transition states, bound precursors, and the confining voids, taken together with the ~200 known zeotypes continue to provide the design flexibility that underpins the spectacular success of zeotypes in the practice of catalysis.

Recent experimental and theoretical developments that address the mechanism of chemical transformations and the structure and binding properties of inorganic solids have led to a shift from heuristic geometric constructs towards firmer conceptual grounds; in doing so, they have created unexpected practical successes. This account describes some examples of how the preferential stabilization of a transition state brings forth specific enhancements in reactivity or selectivity, mediated by entropic or enthalpic effects that lower the activation free energies of elementary steps. In doing so, the examples chosen illustrate how the placement and detection of binding sites at specific locations, the mechanistic expression of reactivity in terms of elementary steps, and the interpretative and design prowess of modern theoretical treatments that describe dispersion forces and entropic effects combine to provide predictive guidance and, quite often, unprecedented quantitative assessments. As a result, reactivity descriptions based on geometry or acid strength are being replaced by their more relevant energetic descriptors — van der Waals confinement energies, proton affinities of organic molecules, and deprotonation energies — to account for reactivity, here for different reactants on diverse solid acids, but in general for catalysis within confined spaces. The focus here on confinement effects, to the exclusion of other features also relevant for reactivity, merely reflects the focus of the session and the limitations of space.

Recent research contributions to catalysis by microporous materials

Enthalpic stabilization of transition states within siliceous voids of molecular dimensions without specific binding sites [1, 2]

The concept of "physical catalysis" denotes the ability of a confining pocket of molecular dimensions to enhance the rate of homogeneous reactions without effecting specific interactions of transition states or intermediates with any type of binding site. Mesoporous silicas and purely siliceous zeolites (SIL, CHA, BEA) catalyze reactions between NO and O_2 at near ambient temperatures with kinetic trends identical to those in homogeneous routes [1]. NO oxidation rates are 10^2–10^4 times higher within the intracrystalline void space than in the extracrystalline gas phase; such rates are similar on SIL materials with and without silanol defects. Such remarkable rate enhancements reflect the stabilization of termolecular transition states (TS) by confinement, mediated by van der Waals forces and without the involvement of specific binding sites. Voids enthalpically stabilize TS structures similar to those prevalent in homogeneous routes, leading to negative apparent activation energies and to unprecedented reactivity at the near ambient temperatures that favor NO_2 formation thermodynamics. The concomitant entropy losses upon confinement become inconsequential because of the preeminent role of enthalpy for Gibbs free energies at these low temperatures. These data and their mechanistic interpretation provide clear evidence for the mediation of molecular transformations by mere confinement of TS structures, without the specific chemical binding of any reactants or transition states at active sites. These interpretations and the non-chemical nature of the stabilizing interactions are consistent with density functional theory treatments that accurately describe measured rate enhancements when using density functionals accounting for dispersion forces, but which predict no detectable rate enhancements with functionals that exclude these interactions [2]. This example of catalysis without specific binding finds practical outlets in the management of NO_x emissions via lean deNO_x traps, during fast selective catalytic reduction strategies mediated by NO_2, and in the synthesis of nitrates and nitric acid. The underpinning phenomena, however, are likely to prevail and to impact other low-temperature associative or radical-mediated chemical transformations that proceed, albeit slowly, as homogeneous reactions at low temperatures, such as acetylene trimerization, triene cyclizations, and Diels–Alder reactions in general.

Remarkable specificity of active sites within eight-member ring (8-MR) structures for the carbonylation of methyls to acetyls [3, 4]

The most useful (and only known) selective route for the formation of a single C–C bond between C_1 molecules is the carbonylation of methanol to acetic acid, a reaction currently practiced in large scale using organometallic complexes of precious metals and CH_3I intermediates. Persistent efforts to react CO with CH_3OH on acid-forms of zeotypes led to very low rates and fast deactivation, symptomatic, as shown later [3], of the competing hydrolysis of bound methyls with H_2O molecules formed via fast CH_3OH homologation and dehydration reactions, of the weak nucleophilic nature of CO, and of voids that did not preferentially stabilize the carbonylation TS in the chosen zeotypes. These hurdles were surmounted by using DME instead of CH_3OH, thus ensuring anhydrous conditions during catalysis, and by the discovery that 8-MR structures provide voids that are uniquely capable of enforcing effective van Waals contacts with the carbonylation transition state [3, 4]. In doing so, the paradigm of a given framework type as a unique descriptor of its catalytic properties was revisited, making the placement of heteroatoms, Al in the case of aluminosilicate acids, at crystallographic sites with diverse local environments for a given silicate framework a design criterion for reactivity and selectivity. Such findings led the field to seek synthesis protocols that purposely place heteroatoms at specific locations, as well as chemical or spectroscopic means to assess the success of these protocols.

Carbonylation occurs on acid forms of zeotypes via a kinetically-relevant step consisting of backside attack by CO on DME-derived bound methyls at Al sites, as shown by kinetic and isotopic data, infrared spectra, and transient placement and scavenging of bound CH_3 groups within different crystalline aluminosilicates [3, 4]. The relevant transition state structures are preferentially stabilized (relative to their bound CH_3 precursors) at 8-MR motifs that prevail in MOR and FER frameworks, leading to the formation of methyl acetate with $> 99\%$ selectivity at 450–500 K. Al sites within larger voids (10-MR, 12-MR motifs) lead to much lower rates and to unselective homologation reactions.

Such inferences about mechanisms and structural motifs required the emergence of methods to assess the number, type, and location of active sites within environments that differed markedly in accessibility and confining properties. The use of organic titrants of different size and the deconvolution of location-dependent O–H infrared bands led to the accurate assignment of reactivity at each specific location, while exchanged cations and controlled dealumination created novel materials with diverse distributions of Al and protons among the different void environments. The correlation between 8-MR Al sites and carbonylation turnover rates led, in turn, to a conceptual framework for seeking even more reactive void structures. Such specificity of methyl reactivity through confinement within small channels was not anticipated; it was considered unprecedented at the time for microporous inorganic solids, which typically influence reactivity by mere size exclusion of bulkier transition states or by controlling molecular access to active sites. These confinement effects depend on an appropriate "fit" that causes van der Waals forces to preferentially stabilize the TS relative to its relevant precursors. The preeminent role of dispersive forces is evident in DFT-derived carbonylation TS enthalpies that are not influenced by location when using functionals that neglect such forces, but become favored at 8-MR locations when using either higher levels of theory or functionals that describe non-specific induced dipole interactions. As in NO oxidation, the enthalpic benefits of confinements in carbonylation prevail over the concomitant entropic penalties at low temperatures, for which the right fit becomes a tighter fit for TS structures than for their bound precursors.

Transition state entropies and the benefits of looser fits and of partial confinement at higher reaction temperatures [5–8]

As the previous two examples show, confinement imposes enthalpy-entropy compromises that determine the free energies of TS and precursor structures and thus turnover rates. The key-lock visual constructs of enzymatic and host-guest chemistries rely on "shape matching" formalisms that enhance enthalpic stabilization through effective contacts between organic moieties and confining voids. Inorganic voids, with their remarkable structural integrity, significantly extend the useful temperature range, thus becoming practical in catalysis of endothermic reactions that are thermodynamically infeasible at lower temperatures. At higher temperatures, looser fits and the concomitant higher entropies of transition states can offset the enthalpic benefits of confinement and lead to lower activation free energies. These preeminent effects of entropy at high temperatures became evident in alkane dehydrogenation and cracking on microporous solid acids [5–8], for which the marked effects of alkane size on reactivity were shown to reflect the higher entropy of late transition states, at which alkane or H_2 leaving groups acquire significant mobility at ion-pair transition states. Yet, their turnover rates are higher in 8-MR pockets than in larger 12-MR channels in MOR, in spite of seemingly tighter environments. In fact, the relevant transition states are confined only in part within the shallow 8-MR pockets; as a result, they retain higher entropies than fully-confined structures, but also the ability to interact with the locally confined conjugate anion formed upon proton transfer to form the transition state. Partial confinement and the concomitant entropic benefits cause the preferential stabilization of the looser ion-pair TS structures that prevail for transition states with higher enthalpies. Such TS structures tend to occur later along the reaction coordinate, causing partial confinement to favor the more enthalpically demanding reactions among competing paths; for instance, dehydrogenation of linear alkanes (versus cracking) and cracking of isoalkanes (versus dehydrogenation) are favored at protons in 8-MR pockets over those in larger 12-MR channels in MOR frameworks. Partial confinement accounts for ubiquitous phenomena often described as "pore mouth catalysis"; it also provides predictive guidance for altering selectivities by placing protons at specific locations within crystalline aluminosilicate frameworks otherwise indistinct in acid strength [9]. The examples discussed so far underscore the role of temperature as the arbiter of enthalpy-entropy compromises in determining activation free energies and thus rates and selectivities in acid catalysis within confining voids.

Mechanism-based descriptors of confinement effects — from geometry-matching to activation free energies and from static to dynamic constructs of host–guest interactions [10–13]

Mechanistic interpretations of measured turnover rates enable the assignment of precise chemical origins to kinetic and thermodynamic constants in rate equations, as well as the identification of specific transition states and intermediates that determine activation free energies. In acid catalysis, the free energy differences responsible for reactivity and selectivity reflect the (relative) stabilization of transition states and their relevant precursors via electrostatic interactions that depend on acid strength, a property that is similar for protons in all aluminosilicate frameworks [9], and through van der Waals interactions imposed by confinement within voids of molecular dimensions. The two examples in this section illustrate the evolution of descriptors of confinement from those based on rigid shapes and sizes to ones involving estimates of van der Waals energies and dynamic structures in order to describe geometry-matching in a manner that reflects the rigorous kinetic and energetic contexts of confinement.

The remarkable ability of small inorganic voids to stabilize molecular guests has been assessed until recently using formalisms based on geometric similitude and on heuristic topological descriptions of the void space [12], often as visual descriptors based on the largest spheres that can be either inscribed or translated without obstructions throughout a porous solid [13]. These geometric descriptors are being replaced with experimental and theoretical physisorption energies for molecular proxies of the relevant transition states [10] and with theoretical estimates of interaction energies and specifically of their dispersive components. The latter are enabled by dispersion-corrected density functionals that detect how inorganic frameworks distort in response to molecular guests in order to enforce more effective van der Waals contacts with guest species [11].

The first-order rate constants for acid-catalyzed CH_3OH dehydration to DME reflect the free energies of bimolecular transition states relative to their smaller H-bonded CH_3OH precursors. They depend exponentially on measured adsorption energies of n-hexane probe molecules and on the ensemble-averaged dispersive components of DME adsorption enthalpies derived from DFT or Lenard-Jones potentials on FAU, SFH, BEA, MOR, MTW, MFI, and MTT zeolites [10], indicating that van der Waals forces predominantly account for the diverse reactivity of these different frameworks. In contrast, zero-order rate constants reflect the stability of the same transition state, but relative to similarly-sized CH_3OH dimers; as a result, measured (and DFT-derived) rate constants depend only weakly on the nature of the void space, because dispersive forces stabilize guests of similar size to the same extent. These results, taken together, demonstrate the preeminent effects of confinement on zeolite reactivity and the manner by which the local voids within different intracrystalline environments give rise to the remarkable catalytic diversity of aluminosilicates. Enthalpic stabilization prevails over entropy losses upon confinement at these low temperatures (400–450 K), in a manner reminiscent of the effects of solvents and of "pockets" in catalysis by molecular complexes and enzymes.

Kinetic and infrared data and theoretical treatments that account for dispersive interactions also show that alkene dimerization rate constants on aluminosilicates (TON, MFI, BEA, FAU, mesoporous MCM-41) reflect the free energy of the bimolecular C–C bond formation TS referenced to alkene-derived monomers [11]. These activation barriers decrease as voids approach the size of transition states given by theory, thus causing TS structures to be preferentially stabilized over smaller bound precursors via contacts with the confining framework. Such heuristic geometric analogies are being replaced by the dispersive components of DFT-derived energies for TS and intermediates; these descriptors account for size and shape and for the most effective "fit" for a given reaction among different frameworks. The dispersion forces recently built into density functionals allow structure optimizations of TS and bound intermediates to balance the energy costs of framework distortions with any concomitant enhancements of van der Waals contacts with guest molecules. These distortions lead to global energy minima and optimal reaction paths and cause framework O-atom displacements to vary systematically with the size and shape of TS and precursor structures. It seems plausible that such lattice distortions upon adsorption of guest molecules are larger and most consequential near the external surfaces of zeolite crystals; they may therefore shed light on phenomena, such as "skin effects" and "pore mouth catalysis," typically invoked with limited mechanistic evidence in order to account for some unexpected effects of external crystal surfaces on diffusion and catalysis.

Perspectives, advances and opportunities in catalysis by microporous solids

The examples and concepts described herein have relied on mechanistic details and practical advances made evident by advances in (i) the synthesis of materials with well-defined structures, high crystallinity, and isolated O–H species acting as Brønsted acids of uniform strength; (ii) methods that determine the number (and in some cases the location) of accessible protons during catalysis; (iii) measurements of rate and selectivity data in the absence of (or with known) diffusional constraints; and (iv) mechanistic interpretations underpinned by theoretical methods that account for van der Waals interactions and which allow increasingly accurate estimates of the entropy and enthalpy components of activation free energies. In all aspects of these enabling advances, the merits and the impact remain largely unexploited and the hurdles become more difficult as the easier challenges are met. Yet, the conceptual and practical benefits of confinement and solvation by van der Waals forces are likely to move forward undeterred, as the methods of interrogation emerge to match the tasks at hand.

A single zeolite framework itself contains a range of local void environments, each with distinct reactivity and selectivity; as a result, varying the distribution of protons among these locations within a given framework, selectively adding or removing sites within specific locations, or modifying a given location by partial occlusion of the void space can significantly extend the range of catalytic opportunities for zeolites. Yet, synthetic protocols that purposely direct the location of framework heteroatoms and their balancing cations remain sparse and far from general in applicability. Inferences about the location, and thus the success of synthesis strategies, are possible today only for very distinct environments, allowing knowledge to be brought only into the most egregious differences in environment. The local and selective modification of the void environment near active sites without the introduction of unintended diffusional hurdles remains anecdotal and accidental for those few examples that have seemingly succeeded in practice.

Knowing what to make and for which reactions to make it remains the enduring challenge that precedes the process of learning how to make a structure for a specific purpose; being able to know what one has made remains the ultimate arbiter of success or failure in making that structure. The ubiquitous deployment of a mere handful of zeolites for many practical reactions, among the several hundred known structures attest to the challenges of matching function with structure, and serves to remind us, that solutions may lie undiscovered amongst materials already at hand without always requiring the discovery of new frameworks. These sharper methods for searching are based on answering questions about what limits reactivity, and thus selectivity among competing paths; in more colloquial terms, one asks what one measures when we measure a rate. Here, mechanistic interpretations, even incomplete or speculative ones, provide the only reliable method in the search for improvements. These require the dissection of chemical reaction rates into elementary steps, a task that starts with the collection of reliable rate data, the exclusion of (or accounting for) diffusional effects, the spectroscopic and isotopic interrogation of plausible steps and intermediates, and the elimination of implausible mechanistic proposals using theory at the state-of-the-art. The impact of theoretical methods in systems where reactivity is driven by confinement effects has been rather recent, as density functionals and higher levels of theory have become increasingly adept at describing dispersion forces and at handling systems of increasing chemical and structural complexity, a trend that will certainly continue.

Acknowledgments

The author gratefully acknowledges the efforts and contributions of his research group and specifically those of Nancy Artioli, Aditya Bhan, Eric Doskocil, Rajamani Gounder, Stanley Hermann, Andrew Jones, Josef Macht, Matteo Maeshi, Michele

Sarazen and Stacey Zones. The financial support of BP, p.l.c., Chevron Energy and Technology Co., and the U.S. Department of Energy is also acknowledged with thanks.

References

[1] N. Artioli, R. F. Lobo, E. Iglesia, *J. Phys. Chem. C* **117**, 20666 (2013).
[2] M. Maestri, E. Iglesia, *J. Phys. Chem. C*, submitted.
[3] A. Bhan, A. Allian, G. Sunley, D. Law, E. Iglesia, *J. Am. Chem. Soc.* **129**, 4919 (2007).
[4] A. Bhan, E. Iglesia, *Acc. Chem. Res.* **41**, 559 (2008).
[5] A. Bhan, R. Gounder, J. Macht, E. Iglesia, *J. Catal.* **253**, 221 (2008).
[6] R. Gounder, E. Iglesia, *J. Am. Chem. Soc.* **131**, 1958 (2009).
[7] R. Gounder, E. Iglesia, *Acc. Chem. Res.* **45**, 229 (2012).
[8] R. Gounder, E. Iglesia, *Chem. Commun.* **49**, 3491 (2013).
[9] A. Jones, S. Zones, E. Iglesia, *J. Phys. Chem. C* **118**, 17787 (2014).
[10] A. Jones, E. Iglesia, *ACS Catal.* **5**, 5741 (2015).
[11] M. Sarazen, E. Doskocil, E. Iglesia, *J. Catal.* **344**, 553 (2016).
[12] A. Jones, C. Oustrouchov, M. Haranczyk, E. Iglesia, *Microporous Mesoporous Mater.* **181**, 208 (2013); M. L. Sarazen, E. Iglesia, *J. Catal.* **354**, (2017); M. Sarazen, E. Iglesia, *Proc. Natl. Acad. Sci.* **114**, (2017) E3900; S. T. Herman, S. Iglesia, *J. Catal.* **346**, (2017).
[13] M. D. Foster, I. Rivin, M. M. J. Treacy, O. D. Friedichs, *Microporous Mesoporous Mater.* **90**, 32 (2006).

ADVANCED ZEOLITES WITH HETEROATOM SITE DISTRIBUTION IN THE FRAMEWORK CONTROLLED

TAKASHI TATSUMI

Tokyo Institute of Technology, Midori-ku, Yokohama 226-8503, Japan

My view of the present state of research on control of heteroatom site distribution in zeolites

Zeolites are commonly crystalline aluminosilicates and take on the properties of solid acids. The catalytic performance of zeolites should depend on various factors such as the zeolite structure, namely the size and shape of the channels and cavities, as well as acid strength and acid amounts. Besides these factors, the location and distribution of Al atoms in tetrahedral (T) sites of the framework have been recognized as a vital factor for activity, selectivity, and life, because they would profoundly affect the accessibility of molecules to acid sites and the spatial constraints on the reaction in the pores [1–10]. Furthermore, the location of Al atoms in the framework could affect the acid strength. Although zeolite researchers have seriously tackled the estimation of the arrangement and distribution of framework Al atoms and their control in the pores, these issues still remain challenging.

My recent research contributions to control of heteroatom site distribution in zeolites

Strategy for controlling the Al siting in zeolites

As is generally known, the choice of structure-directing agents (SDAs) has a powerful effect on the self-assembly process of the zeolite structure. However, there are only limited studies on SDA structure influencing the Al site distribution in zeolites. It is well-recognized that, to balance the charge, Al species are located near cations comprising inorganic cations, *e.g.*, Na^+ or K^+, or organic ones such as quaternary ammonium ions as organic SDAs (OSDAs). Hence, the number of the cations affects the Al content, and the location of Al atoms may be dependent on their size and type. If the SDAs occupy specific positions within the void volume of the framework structure, they could drive Al atoms to be placed in certain T positions.

A synthesis strategy for controlling the Al siting in the FER type zeolite by using different organic SDAs in the presence or absence of Na cations has been developed [2–5]. It has been revealed that the size of cyclic amine in the synthesis gel affects the distribution of acid sites over the FER zeolites. We have also found that the Al distribution over the RTH-type framework was clearly dependent on the type of the cations [11]. Methods for evaluating the distribution of the acid site in the pores have been extensively investigated [12]. They include selective adsorption of probe molecules on acid sites [3], crystallographic techniques based on the Rietveld refinement [5], and EXAFS analysis and ^{27}Al MAS NMR spectroscopy supported by DFT-based molecular dynamics simulations [9]. The distribution of Al atoms in the zeolite framework between single Al atoms and Al pairs was also monitored on the basis of Co(II) ion exchange capacity in combination with ^{29}Si NMR and UV-Vis spectroscopies [6–8, 10].

Fig. 1. Acid site distribution depending on the structure-directing agent.

Recently, we have tackled the control of the distribution of the Al atoms in the MFI-type zeolites (ZSM-5), which were synthesized by using tetrapropylammonium (TPA) cations or some amines as OSDAs with or without Na cations [13]. In addition to high-resolution ^{27}Al MAS NMR and ^{27}Al MQMAS NMR techniques, "constraint index (CI)", which is defined as the cracking rate of n-hexane to that of its isomer 3-methylpentane, was used to estimate the location of Al sites in the pores [4, 14, 15]. The Al atoms on the H-ZSM-5 sample synthesized by using TPA cations in the absence of Na cations proved to be predominantly located in large spaces, namely channel intersections (Fig. 1, left) [13]. The activities of the H-ZSM-5 zeolites in the reactions that proceed via bulky transition states were greatly affected by the distribution of acid sites, while the activity in the reactions that proceeds without considerable constraints in the pores of MFI-type zeolites was not. Furthermore, it was found that the distribution of acids sites affects the catalytic lifetime for the cracking of methylcyclohexane; the H-ZSM-5 zeolite with Al atoms located at the intersection showed a shorter catalytic lifetime, suggesting that the coke formation in the intracrystalline space principally occurs at the intersection in the case of 10-membered ring (MR) zeolites. In contrast, the Al atoms on the H-ZSM-5 sample synthesized by using TPA and Na cations appears to be unselectively distributed in the straight and sinusoid channels and at the intersection (Fig. 1, center).

It is noteworthy that the Al distribution also influences the susceptibility of framework Al atoms to dealumination. Kubota *et al.* found that OSDA-free ZSM-5 was resistant to dealumination during severe acid treatment while extensive dealumination occurred to TPA-assisted ZSM-5 [16]. It has long been known that steam dealumination of FAU zeolites preferentially removes Al atoms adjacent to Q^2 sites in ^{29}Si MAS NMR spectra [17].

Very recently, we have discovered that the use of bulky alcohol, *e.g.*, pentaerythritol, in the combination with Na cations led to the preferential distribution of Al atoms in narrow straight and/or sinusoidal channels of the MFI framework with the framework T-sites facing intersection unoccupied with Al (Fig. 1, right). This distribution led to the long-life catalysts in the hexane cracking and also methanol to olefins (MTO) reaction, probably because the coke formation at the intersection of two channels was efficiently prevented [18]. These findings will contribute to the development of a new class of acidic zeolite catalysts with the location of Al atoms in the pores controlled to improve the catalytic performance not only in the cracking reaction but also for other various acid-catalyzed reactions.

For the MTO reaction a large number of zeolites were screened and we have found CIT-1 (CON) zeolite is promising, exhibiting long catalytic life and high selectivity for C3–C4 olefins [19]. Interestingly, the CON-type aluminosilicate zeolite synthesized by the direct-synthesis method showed a much longer catalytic lifetime than the one obtained by the post-synthesis method. The ^{27}Al MQMAS NMR spectra suggest that the states of tetrahedrally-coordinated Al species in the framework were dependent on whether they were synthesized directly or by the post-synthesis method. Thus, it has been demonstrated that the location of Al atom strongly affected the catalytic performance. We have also found that CON-type aluminosilicate zeolites put up very strong resistance to steaming under severe conditions compared to ZSM-5, making them promising for being used for industrial catalysts.

Control of Ti sites possible?

Compared to the distribution of Al sites in the zeolite framework, the distribution of Ti sites in titanosilicate zeolites should be hard to control; Ti^{4+} produces no negative framework charge in the Si-based framework and therefore, strong interaction between the SDA cations and the Ti species is not expected. We have established a method for synthesizing Ti-MWW zeolite in the presence of a large amount of boron species to the mother gel [20]. Ti-MWW can be synthesized by a different manner; from the B-MWW zeolite crystallized according to a conventional method in the absence of Ti, the B atoms were almost completely removed by acid washing to give deboronated MWW-type materials. We were able to introduce Ti atoms into the defect sites resulting from deboronation. Thus obtained Ti-MWW-PS (synthesized by the post-synthesis method) showed much higher activity in the alkene epoxidation and ketone ammoximation than Ti-MWW-DS (directly synthesized) [21]. The difference in the activity is probably due to the difference in the location of

Ti atoms in the framework and the site-dependent specific activity of Ti in the oxidation; Supposedly, in Ti-MWW-PS, Ti was located in the sites that had been originally occupied by B, while in Ti-MWW-DS, Ti was placed in the sites that had not been occupied by B. In this way, the distribution of Ti sites in the zeolite could be changed by the synthetic methods.

Unfortunately, it is difficult to distinguish a difference in the Ti siting in the framework because Ti atoms are not NMR-sensitive and Ti substitution is severely limited in extent. We have attempted to detect a difference in the location of Ti atoms between Ti-MWW-DS and Ti-MWW-PS by comparing the epoxidation rates of olefins different in bulkiness around the C–C double bonds [22].

Outlook for future developments of research on control of heteroatom site distribution in zeolites

The distribution of heteroatom sites changes according to the synthetic methods. However, it is not easy to clearly determine which T sites in the zeolite framework the heteroatoms occupy. We assume complete control of zeolite synthesis if we succeed in developing the methods for incorporating desired amounts of desired heteroatoms into desired specific T sites. It is our goal to accurately estimate the location of heteroatoms and properly evaluate the performance of zeolite catalysts to establish the structure-activity relationship, leading to effective development of the zeolite catalysts that give excellent performance.

References

[1] R. Gounder, E. Iglesia, *Acc. Chem. Res.* **45**, 229 (2012).
[2] A. B. Pinar, L. Gomez-Hortiguela, J. Perez-Pariente, *Chem. Mater.* **19**, 5617 (2007).
[3] A. B. Pinar, C. Marques-Alvarez, M. Grande-Casas, J. Perez-Pariente, *J. Catal.* **263**, 258 (2009).
[4] Y. Roman-Leskov, M. Moliner, M. E. Davis, *J. Phys. Chem. C* **115**, 1096 (2011).
[5] A. B. Pinar, R. Verel, J. Perez-Pariente, J. A. van Bokhoven, *Chem. Mater.* **25**, 3654 (2013).
[6] J. Dedecek, V. Balgova, V. Pashkova, P. Klein, B. Wichterlova, *Chem. Mater.* **24**, 3231 (2012).
[7] V. Pashkova, J. Dedecek, *Chem. Eur. J.* **22**, 3937 (2016).
[8] A. Janda, A. T. Bell, *J. Am. Chem. Soc.* **135**, 19193 (2013).
[9] A. Vjunov, J. L. Fulton, T. Huthfelker, S. Pin, D. Mei *et al.*, *J. Am. Chem. Soc.* **136**, 8296 (2014).
[10] J. R. Di Iorio, R. Gounder, *Chem. Mater.* **28**, 2236 (2016).
[11] M. Liu, T. Yokoi, M. Yoshioka, H. Imai, J. N. Kondo *et al.*, *Phys. Chem. Chem. Phys.* **16**, 4155 (2014).
[12] J. Dedecek, Z. Sobalik, B. Wichiterlova, *Catal. Rev.* **54**, 135 (2012).
[13] T. Yokoi, H. Mochizuki, S. Namba, J. N. Kondo, T. Tatsumi, *J. Phys. Chem. C* **106**, 6115 (2002).
[14] V. J. Frilette, W. O. Haag, R. M. Lago, *J. Catal.* **67**, 218 (1992).
[15] J. R. Carpenter, S. Yeh, S. I. Zones, M. E. Davis, *J. Catal.* **269**, 64 (2010).

[16] S. Inagaki, S. Shinoda, Y. Kaneko, K. Takechi, R. Komatsu *et al.*, *ACS Catal.* **3**, 74 (2013).
[17] K. U. Gore, A. Abraham, S. G. Hegde, R. Kumar, *J. Phys. Chem. B* **115**, 1096 (2011).
[18] T. Biligetu, Y. Wang, T. Nishitoba, R. Otomo, S. Park, H. Mochizuki, J. N. Kondo, T. Tatsumi, T. Yokoi, *J. Catal.*, 2017, **353**, 1 (2017).
[19] M. Yoshioka, T. Yokoi, T. Tatsumi, *ACS Catal.* **5**, 4268 (2015).
[20] P. Wu, T. Tatsumi, T. Komatsu, T. Yashima, *J. Phys. Chem. B* **105**, 2897 (2001).
[21] P. Wu, T. Tatsumi, *Chem. Commun.* 1026 (2002).
[22] T. Yokoi, T. Fujii, R. Otomo, J. N. Kondo, T. Tatsumi, The 3rd Euro-Asia Zeolite Conference, Book of abstracts, p. 63 (OP20, January 22–25, 2017, Bali, Indonesia, 2017).

EVOLUTION TRENDS IN ZEOLITE-BASED CATALYSTS

GIUSEPPE BELLUSSI and ROBERTO MILLINI

Research & Technological Innovation Dept., Eni S.p.A., 20097 San Donato Milanese, Italy

Zeolites in catalysis

Zeolites form a fascinating family of crystalline microporous materials of primary scientific and technological importance. The modern era of the science and technology of zeolites began in years 1950s, when, thanks to the pioneering work of R. M. Barrer (Imperial College, London) and R. Milton (Union Carbide Corp., USA), the first syntheses and applications of zeolites were reported. In analogy with the natural phases, zeolites were firstly considered as aluminosilicates only. Their physical and chemical properties directly derive from the characteristics of the 3D frameworks. The organization of the corner-sharing $[TO_4]$ (T = Si, Al) tetrahedra generates a microporous system, whose characteristics vary from a structure to another, being composed by regular channels and/or cages opened to the exterior of the crystal and with free dimensions in the range 3–12 Å. This allows the use of zeolites as molecular sieves, for separating molecules of different dimensions (*e.g.* *n*-alkanes from branched ones) from a mixture. On the other hand, the chemical characteristics of the frameworks are determined by the presence of $[AlO_4]$ tetrahedra, which impart negative charges to the framework compensated by extra-framework cations (usually alkali or earth-alkali metal ions) located in the micropores. These cations are loosely bound to the framework and can be easily exchanged rendering zeolites ion-exchangers for application, such as the water softening. When the extra-framework cation is the proton, a zeolite assumes acidic properties finding applications as heterogeneous catalyst for several acid-catalyzed reactions. The use of zeolites as solid acid catalysts has had a strong development in the last few decades, in particular to replace the dangerous mineral acids or the unstable and low-performing non-zeolite solid acids in several refining and petrochemical processes. As an example, we can examine the evolution of the catalytic cracking, one of the most important oil-refining processes. The first cracking catalyst was $AlCl_3$ introduced by A. McAfee around 1920 [1]; few years later, E. Houdry discovered the catalytic properties of activated clays [2] and then of amorphous silica-aluminas [3]. One of the main problems in catalytic cracking was the deposition of coke on the catalyst, which required frequent regenerations. This was the spring for the development of new reactors and processes, the last being the continuous regeneration cracking in fluid bed reactors (FCC, Fluid Catalytic

Cracking [4]). Only 20 years later appeared patents claiming the use of zeolites in FCC catalyst [5]. Since then, zeolites have become an essential component of catalysts for FCC, either as active phases (coke selective zeolite RE-HY and high-coke selective USY and RE-USY) or as pro-olefins and octane boosting FCC additives (ZSM-5). It is worth noting that ~95% of the ~270 kton of zeolites yearly produced for the formulation of catalysts are consumed in the FCC plants, the remaining being used in other (mainly petrochemical) processes. Interesting statistical surveys of industrial processes using solid acid-base catalysts were recently presented [6, 7]. They highlighted that 74 of the 180 solid catalysts employed in 127 different industrial processes are based on zeolites. Most of these catalysts are employed in petrochemical processes that, differently from oil-refining ones, consist of well-defined reactions on pure substrates (*e.g.* alkylation, transalkylation and disproportionation of aromatics). In these cases, the selectivity towards the desired product(s) is fundamental and zeolites provide significant advantages respect to the homogeneous and to non-zeolite catalysts. In fact, the active sites are located within the pores with dimensions comparable to those of the different species involved in the reaction. Therefore, the zeolite structure may impose a steric control to the reaction, favoring the formation of the target product(s), limiting in the same time the undesired ones. These revolutionary observations were made from the years 1960s when the zeolites started to be defined as *shape selective* catalysts. In particular, the classical concepts of *reactant*, *product* and *transition state shape selectivity* were proposed for explaining the behavior of zeolites in different reactions [8–11]. This has had several implications in the research on the microporous solids both with regard to their applications as catalysts and, above all, for the synthesis of new structures able to satisfy the demands of new materials.

Trends in the synthesis of zeolite catalysts

What emerges from the statistical surveys cited above is that all the zeolite catalysts currently employed are based on only a few framework types (actually 13 of the 231 known today, the most widely used being MFI, MOR, FAU and Beta). Moreover, all belong to the groups of *medium-* and *large-pore* zeolites, with 10- and 12-ring openings, respectively [6, 7]. When considering that the corresponding free dimensions of the pore openings are in the range 5–8Å, it is clear that the applicability of these zeolites is limited to reactions involving small molecules, i.e. those able to diffuse in the *medium-* and *large pores*. The problem that the scientific community has faced was therefore to find solutions that can extend the use of zeolites in reactions involving larger molecules. In this regard, the preparation of the M41S family of ordered mesoporous materials, reported in 1992 by researchers at Mobil, opened interesting opportunities for new application in catalysis [12, 13]. However, these new materials showed important drawbacks since it was soon demonstrated that their acidic strength and thermal/hydrothermal stability are much lower than that of the most used zeolites. Therefore, the attention moved to the preparation of new structures with pore openings formed by 14 or more tetrahedra, the so-called *extra-large pore* zeolites. Taking advantage of the increasing knowledge of the phenomena involved in the zeolite nucleation and growth as well as of the influence of the different synthesis parameters, interesting results were achieved [14, 15]. Several *extra-large pore* zeolites were obtained in the presence of elements different from Si and/or Al, such as P [16] and B [17]. Decisive, in this regard, is the use of Ge, since it stabilizes the D4R (double-four-ring) secondary building unit, in turn predicted to favor the formation of low-density zeolite frameworks. Among the *extra-large pore* zeolites, ITQ-37 [18] and ITQ-43 [19] are worthy to mention because, unique among the crystalline porous materials known so far, they possess pore openings close to 20 Å, i.e. at the border of the lower dimensions of mesopores. In spite of the relative large number of *extra-large pore* zeolites available today, they did not find any industrial application yet. In some cases, however, the catalytic tests demonstrated the high potential of these zeolites. For instance, ITQ-33 with 18-ring openings, proved to have superior performances in the catalytic cracking of Arabian Light vacuum gasoil respect to the conventional active phases (*e.g.* USY, with or without ZSM-5) [20]. In spite of that, their high costs (determined by the use of expensive reagents, including GeO_2 and complex non-commercial Structure Directing Agents (SDA), and low thermal/hydrothermal stability prevent their use.

As a consequence of these results, new pathways and new strategies are under development to increase the accessibility of the active sites located within the zeolite crystals. One of them is based on the use of the so-called *bidimensional* (or 2D) *zeolites* [21]. This term defines layered phases composed by thin zeolite sheets, which can be intermediates (precursors) in the crystallization pathway of given zeolites or prepared by specifically designed approaches. A well-known example of 2D zeolite is the MWW-type precursor whose characteristics were reported in the mid years 1990s [22]. Upon calcination of this precursor, composed by randomly stacked loosely bound MWW-type layers, the ordered 3D zeolite structure forms. On the other hand, 2D zeolites can be prepared also for phases that do not involve layered precursors. An example concerns the preparation of thin (2–3 nm thick) MFI layers using specifically designed bifunctional SDAs [23] or by post-synthesis treatment of 3D zeolites with the UTL-type framework type [24]. Regardless of the synthesis method, the zeolite layers can be considered as periodical building blocks used for the preparation of new materials through delamination, intercalation, pillaring, etc. [21]. In all cases, the materials thus obtained have very different characteristics from conventional 3D structure, especially with regard to the easy accessibility to active sites. In fact, due to the small thickness of the zeolite layers, virtually all the active sites are exposed to the external surface of the layers themselves. Consequently, the control of the reaction does not take place through one or more of the classic concepts of shape selectivity described above, but it is necessary to invoke that known as *nest effect*: the reaction occurs at the pore mouth, under a steric control but without any diffusion limitation [25].

The elimination (or, at least, reduction) of the diffusion limitations in zeolite crystals is one of the hottest topics in zeolite science and one of the solutions concerns the preparation of the so-called *mesoporous zeolites*, i.e. zeolites possessing a *hierarchical micro- mesoporous system* [26]. In this way, the mesopores present in the zeolite particle facilitate the mass transport to the crystalline nano-domains where the reaction takes place. The reduced length of the micropores assures from one side the easy elution of the products and from the other the full exploitation of the zeolite crystals with undoubted advantages on the entire catalytic process. Several different approaches for preparing mesoporous zeolites have been proposed, each of them being characterized by a different degree of complexity and success [26]. *Mesoporous zeolites* are already employed in the formulation of industrial catalysts. We can cite, for instance, the zeolite beta nanocrystals, which are the active phase of the PBE-1 catalyst for the alkylation of benzene with propylene (cumene process) [27] and the mesoporous USY as a component of the FCC catalyst [28]. Recently, we have evidenced that it is possible to combine the catalytic cracking performances of these materials with the high hydrogenation, HDS and HDM properties of the dispersed MoS_2 catalyst in a slurry type hydrocracking reactor [29]. Besides the excellent catalytic properties, MoS_2 has the ability to protect the cracking catalyst against the fast decay due to coke deposition and metal poisoning [30].

Conclusions and perspectives

Synthetic zeolites are among the most important class of heterogeneous catalysts used since more than fifty years in refinery and petrochemistry. Their main properties are due to the presence of an ordered pores architecture and acid sites in their structure. With the aim of expanding the use of zeolites to the transformation of larger molecules, new materials and sophisticate preparation techniques have been developed in the last two decades.

However, notwithstanding the potential expressed by the new materials, the cost issues and the thermal and hydrothermal stability are still hampering the practical applications. For the new materials the improvement of the thermal stability and the development of cheaper and industrially affordable production technologies are the challenges that has to be faced in the next years for expanding the use of this extraordinary class of heterogeneous catalysts.

References

[1] A. McD. McAfee, *US Patent* 1,405,054 (1922).
[2] E. Houdry, *US Patent* 1,957,648 (1934).
[3] E. Houdry, *US Patent* 2,078,945 (1937).
[4] E. V. Murphree, C. L. Brown, H. G. M. Fischer, E. J. Gohr and W. J. Sweeney, *Ind. Eng. Chem.* **35**(7), 768 (1943).
[5] C. J. Plank, E. J. Rosinski, *US Patent* 3,140,249 (1964).
[6] K. Tanabe, W. F. Hölderich, *Appl. Catal. A* **181**, 399 (1999).
[7] W. Vermeiren, J.-P. Gilson, *Top. Catal.* **52**, 1131 (2009).

[8] P. B. Weisz, V. J. Frilette, *J. Phys. Chem.* **64**, 382 (1960).
[9] S. M. Csicsery, *J. Catal.* **19**, 394 (1970).
[10] S. M. Csicsery, *J. Catal.* **23**, 124 (1971).
[11] S. M. Csicsery, *Pure Appl. Chem.* **58**, 841 (1986).
[12] C. T. Kresge, M. E. Leonowicz, W. J. Roth, J. S. Beck, *Nature* **359**, 710 (1992).
[13] J. S. Beck, J. C. Vartuli, W. J. Roth, M. E. Leonowicz, K. D. Schmidt *et al.*, *J. Am. Chem. Soc.* **114**, 10834 (1992).
[14] J. Jiang, J. Yu, A. Corma, *Angew. Chem. Int. Ed.* **49**, 3120 (2010).
[15] G. Bellussi, A. Carati, C. Rizzo, R. Millini, *Catal. Sci. Technol.* **3**, 833 (2013).
[16] M. E. Davis, C. Saldarriaga, C. Montes, J. Garces, C. Crowder, *Nature* **331**, 598 (1988).
[17] A. Burton, S. Elomari, C. Y. Chen, R. C. Medrud, I. Y. Chan *et al.*, *Chem. Eur. J.* **9**, 5737 (2003).
[18] J. Sun, C. Bonneau, A. Cantin, A. Corma, M. J. Diaz-Cabanas *et al.*, *Nature* **458**, 1154 (2009).
[19] J. Jiang, J. L. Jorda, J. Yu, L. A. Baumes, E. Mugnaioli *et al.*, *Science* **333**, 1131 (2011).
[20] A. Corma, M. J. Diaz-Cabañas, J. L. Jorda, C. Martinez, M. Moliner, *Nature* **443**, 842 (2006).
[21] W. J. Roth, J. Cejka, *Catal. Sci. Technol.* **1**, 43 (2011).
[22] R. Millini, G. Perego, W. O. Parker Jr., G. Bellussi, L. Carluccio, *Microporous Mater.* **4**, 221 (1995).
[23] M. Choi, K. Na, J. Kim, Y. Sakamoto, O. Terasaki *et al.*, *Nature* **461**, 246 (2009).
[24] W. J. Roth, O. V. Shvets, M. Shamzhy, P. Chlubná, M. Kubů *et al.*, *J. Am. Chem. Soc.* **133**, 6130 (2011).
[25] E. G. Derouane, Z. Gabelica, *J. Catal.* **65**, 486 (1980).
[26] J. Garcia-Martinez, K. Li (eds.), *Mesoporous Zeolites. Preparation, Characterization and Applications* (Wiley-VCH, 2015).
[27] G. Bellussi, G. Pazzuconi, C. Perego, G. Girotti, G. Terzoni, *J. Catal.* **157**, 227 (1995).
[28] B. Speronello, J. Garcia-Martinez, A. Hansen, R. Hu, *Refinery Operations* **4**, 1 (2011).
[29] D. Molinari, G. Bellussi, A. Landoni, P. Pollesel, *WO Patent* 2013/034642 (2013).
[30] G. Bellussi, G. Rispoli, D. Molinari, A. Landoni, P. Pollesel *et al.*, *Catal. Sci. Technol.* **3**, 176 (2013).

MOLECULAR SIEVE ZEOTYPES — TARGETING STRUCTURE TYPES THROUGH SYNTHESIS DESIGN

JOHN L. CASCI[1], RAQUEL GARCIA[1], ALESSANDRO TURRINA[1,2], PAUL A. WRIGHT[2] and PAUL A. COX[3]

[1] *Johnson Matthey Technology Centre, Billingham TS23 1LB, UK*
[2] *EaStCHEM School of Chemistry, University of St Andrews, St Andrews KY16 9ST, UK*
[3] *School of Pharmaceutical and Biomedical Sciences, University of Portsmouth, Portsmouth PO1 2DT, UK*

My view of the present state of research on molecular sieve zeotypes

In recent years there has been a tremendous increase in the number of known zeolite topological types [1]. While some of this dramatic increase in numbers has been driven by advances in structure solution it is largely as a result of the success in discovering new materials. The most successful approach to new materials generation has been the use of templates or Organic Structure Directing Agents (OSDAs) typically amines or quaternary ammonium compounds [2, 3]. Novel materials are usually the result of a screening study in which the OSDA is employed in a scan of the experimental free-space: gel composition, reaction conditions (temperature, time) and reagent sources. However, there have also been a number of successful syntheses of specific zeolite topological types by designing OSDAs to target the synthesis of the desired framework type [4–8].

The widespread use of zeolite molecular sieves in a range of catalytic, sorptive (separation) and ion-exchange applications is well-known. The commercialization of CHA topological type zeolites in Methanol-to-Olefins [9] and Selective Catalytic Reduction (SCR) of NOx (Automotive applications) [10] in aluminosilicate and silicoaluminophosphate compositions has given added impetus to the search for new structure types, especially those that have some structural relationship to Chabazite.

This work builds on previous studies on co-templating where more than one OSDA is employed with each template (or OSDA) occupying a specific location (cage or channel) in a structure [11]. In this work the aim is to target particular structures by designing OSDAs to fit in specific cages [12] — targeting structure types through synthesis design.

My recent research contributions to molecular sieve zeotypes

The target structures chosen was a sub-set of the ABC-6 family: those made up from D6R units and containing *gme* cages as one of the two types of cavity/void. This selection was made because the *gme* cages looked a suitable target to template independently. Even within this sub-set there exists a number of known frameworks: GME, AFX, SFW and AFT topological types.

Once target structures had been selected a modelling study was conducted initially looking at potential OSDAs for the *gme* cages and thereafter looking for suitable templates for the other (defining) cages or channels that distinguished the various topological types of interest (AFX, SFW, GME).

For the *gme* cages a series of small amine (and protonated alkylammonium cations) were examined with trimethylamine (TriMA) and trimethylammonium (HTriMA+) showing the best binding energies (determined by Molecular Dynamics and Energy-Minimised Calculations) in the *gme* cages. For the larger cages/channels a series of OSDAs based on DABCO (bisdiazabicyclooactane) were studied. Specifically, where two DABCO units were linked by polymethylene chains of differing lengths (3 to 8 $-CH_2-$ groups). For the GME structure a series of $[DABCO\text{-}(CH_2)_6]_n$ oligomers were chosen.

Syntheses were carried out in Teflon-lined 30 mL autoclaves at 160 or 190°C for between 4 and 7 days. Solid products were recovered by filtration, washed repeatedly with water and then dried overnight in air at 80°C for 12 hours. Overall gel compositions which gave the desired materials as pure phases are given below.

gel composition[a] Al : P : Si : R : TrMA : TBAOH : H_2O	R[b]	seed AFX (wt %)	temp. (°C)	time (days)	product[d] (topology type)
1.0 : 0.77 : 0.23 : 0.10 : 0.21 : 0.23 : 40	diDABCO-C4	4	190	4	SAPO-56 AFX
1.0 : 0.71 : 0.29 : 0.10 : 0.21 : 0.23 : 40	diDABCO-C5	4	190	4	SAPO-56 AFX
1.0 : 0.70 : 0.30 : 0.10 : 0.13 : 0.28 : 40	diDABCO-C6	2	190	7	STA-18 SFW
1.0 : 0.70 : 0.30 : 0.10 : 0.13 : 0.28 : 40	diDABCO-C7	2	190	7	STA-18 SFW
1.0 : 0.70 : 0.30 : 0.10 : 0.11 : 0.30 : 40	diDABCO-C8	2	190	7	STA-18 SFW
1.0 : 0.70 : 0.30 : 0.15 : 0.42 : 0.12 : 40	$[DABCO\text{-}C5]_3$[c]	4	160	7	STA-19 GME
1.0 : 0.70 : 0.30 : 0.15 : 0.21 : 0.24 : 40	$[DABCO\text{-}C6]_3$[c]	4	160	7	STA-19 GME
1.0 : 0.70 : 0.30 : 0.15 : 0.21 : 0.24 : 40	$[DABCO\text{-}C6]_5$[c]	4	160	7	STA-19 GME
1.0 : 0.70 : 0.30 : 0.15 : 0.21 : 0.24 : 40	$[DABCO\text{-}C6]_7$[c]	4	160	7	STA-19 GME

[a] Al, P, Si correspond to $Al(OH)_3$, H_3PO_4, SiO_2.

[b] All DABCO SDAs were added as dibromides.

[c] Quantities of $[DABCO\text{-}C5]_3$, $[DABCO\text{-}C6]_x$ is on the basis of the repeat units ($C_{11}N_2H_{22}Br_2$), ($C_{12}N_2H_{24}Br_2$).

[d] SAPO-56 (AFX), STA-18 (SFW) and STA-19 (GME).

Detailed structural and compositional analysis was carried out on samples to demonstrate that the OSDAs were present, and intact, within the frameworks. Specifically, ^{13}C sold-state nmr was used to determine OSDA integrity while a combination of XRF and TGA analysis provided (overall) volatiles (water and OSDA) content.

Outlook to future developments of research on molecular sieve zeotypes

A retrosynthetic method has been developed to design the synthesis of target silicoaluminophosphates (SAPOs) whose frameworks belong to the ABC-6 family and which have at least one *gme* cage per unit cell. This strategy allowed the preparation of SAPO versions of AFX (SAPO-56), SFW (STA-18) and GME (STA-19) types of framework.

The "double template" method opens up the possibility of synthesising other new materials possessing two different cage types, whether of a known topology type but with a different composition or with topology types as yet only predicted hypothetically. On a more fundamental level, the cooperative way in which the two SDAs act during nucleation and crystal growth should be studied in depth to enable the full potential of the method to be realised.

Acknowledgements

The authors gratefully acknowledge the contribution of the EPSRC UK National Solid-State NMR service at Durham. Mrs. S. Williamson, University of St. Andrews, for TGA, AAS and N_2 adsorption isotherms, Mr. S. Boyer, London Metropolitan University, for elemental analysis and Johnson Matthey PLC for permission to publish.

References

[1] http://www.iza-structure.org/databases/
[2] B. M. Lok, T. R. Cannan, C. A. Messina, *Zeolites* **3**, 282 (1983).
[3] R. F. Lobo, S. I. Zones, M. E. Davis, *J. Incl. Phen. Macrocycl. Chem.* **21**, 47 (1995).
[4] D. W. Lewis, D. J. Willock, C. R. A. Catlow, J. M. Thomas, G. J. Hutchings, *Nature* **382**, 604 (1996).
[5] D. W. Lewis, G. Sankar, J. K. Wyles, J. M. Thomas, C. R. A. Catlow *et al.*, *Angew. Chem. Int. Ed. Engl.* **36**, 2675 (1997).
[6] R. Pophale, F. Daeyart, M. W. Deem, *J. Mater. Chem. A* **1**, 6750 (2013).
[7] M. Moliner, F. Rey, A. Corma, *Angew. Chem. Int. Ed.* **52**, 13880 (2013).
[8] J. E. Schmidt, M. W. Deem, M. E. Davis, *Angew. Chem.* **126**, 8512 (2014).
[9] J. Haw, W. Song, D. M. Marcus, J. B. Nicholas, *Acc. Chem. Res.* **36**, 317 (2003).
[10] L. Ma, Y. Cheng, G. Cavataio, R. W. McCabe, L. Fu *et al.*, *Chem. Eng. J.* **225**, 323 (2013).
[11] M. Castro, R. Garcia, S. J. Warrender, P. A. Wright, P. A. Cox *et al.*, *Chem. Commun.* 3470 (2007).
[12] A. Turrina, R. Garcia, P. A. Cox, P. A. Wright, J. L. Casci, *Chem. Mater.* **28**, 4998 (2016).

CATALYSIS BY MICROPOROUS METAL ORGANIC FRAMEWORKS

CHRISTOPHER W. JONES, LALIT DARUNTE, KIWON EUM, SIMON PANG, GUANGHUI ZHU, STEPHANIE A. DIDAS and SANKAR NAIR

School of Chemical & Biomolecular Engineering, Georgia Institute of Technology, Atlanta, GA 30332, USA

MOFs — Discovery & Development as Catalysts

Metal-organic frameworks (MOFs) are crystalline porous materials first discovered and popularized in the mid-1990s [1, 2]. Because many MOFs are microporous materials, natural comparisons with zeolites quickly evolved, and over the ensuing two decades, MOFs were developed as catalysts as well as separation and gas storage materials.

Whereas zeolites have profoundly changed science and society due to their impact as catalysts, the impact of MOFs as practical catalysts that offer cost-effective solutions to synthetic problems is comparatively poorly developed. At this stage of development, after two decades of research, one can say that MOF catalysis has grown out of the infancy stage and is now in adolescence. In this presentation, the factors that will determine if MOFs can be applied as practical catalysts as the field grows into adulthood are discussed.

MOFs

MOFs are constructed from metal or metal oxide cluster ions interconnected by organic linkers with permanent porosity after solvent removal. As crystalline materials, they have controllable pore structures with high topological diversity imparted via unlimited metal and linker combinations. While MOFs and zeolites share some of the same crystalline topologies, their differing building blocks lead to different properties that affect their utility as catalysts [3, 4].

MOFs versus Zeolites as Catalysts

Zeolites have developed as practical, large scale catalysts used in numerous applications over a period of almost 50 years. In this time, over 200 topologies have been prepared, and extensions into other compositional families such as aluminophosphates and silicoaluminophosphates have resulted in an array of versatile catalysts [5]. In many cases, zeolites are unquestionably the best catalysts ever developed for a given reaction, when considering unique reactivity patterns and cost/practicality. In contrast, while the diversification of MOF topologies has grown much faster than that of zeolites (> 5000 topologies known today) [6], and similarly the array of building blocks is much more diverse in MOFs than in zeolites, examples of MOFs as catalysts where they offer significant advantages over other catalysts are very rare. Nonetheless, MOFs can offer some unique advantages not matched by zeolites.

Potential Advantages and Disadvantages of MOFs as Catalysts

As noted above, MOFs have enhanced tunability of structure and composition compared to zeolites [3]. This versatility makes creation of a wide array of types of active sites possible, including relatively facile creation of chiral structures, compared to zeolites. Additionally, MOFs have been easily extended into the mesopore regime [7], while maintaining excellent crystallinity, allowing for conversion of large molecules.

However, MOFs currently have key disadvantages compared to zeolites as well. The most important disadvantage is stability to water and/or humidity, with thermal stability being a secondary factor [8–10]. Water is a ubiquitous component on earth and is contained in many chemicals and process streams. Furthermore, it is produced as a byproduct or consumed as a reactant in many important reactions. Many MOFs are thermally stable over a wide range of temperatures, allowing their use in gas phase reactions, though zeolites are generally more thermally stable, overall.

Structural Diversity and Tunability of MOFs

As an example of the great tunability of MOFs, a class of MOFs called zeolitic imidazolate frameworks (ZIFs), is considered. ZIFs, which are built from zinc cations and imidazolate linkers, share many of the same topologies as zeolites [11]. ZIF-8, which is composed of methylimidazole linkers, and ZIF-90, built with carboxaldehyde-2-imidazole linkers, both share the same sodalite (SOD) topology. As a demonstration of the wide degree of tunability of these ZIFs, we have successfully created a family of mixed-linker ZIFs containing a mixture of the two ligands, covering the whole compositional space from pure ZIF-8 to pure ZIF-90, while maintaining the cubic SOD topology [12]. This allows fine control of the porosity of the material, which can be tuned in 0.1–0.2 Å increments across wide effective aperture sizes of ~4.0–5.0 Å [13]. In simple tests of diffusional selectivity using *n*-butane and *i*-butane, selectivities can be tuned over three orders of magnitude by fine control of the pore size. By comparison, zeolites cannot generally be tuned over a full compositional range (*e.g.* Si/Al ratio) without changing the crystal structure of the material. With this being only one example of the great versatility of MOF structures, one can begin to envision how MOFs may offer significant advantages over zeolites.

Potential for breakthroughs in MOF catalysis

The huge diversity in structure and composition that MOFs offer for catalysis with crystalline microporous materials offers great potential for breakthroughs in catalysis. Given the strengths and weaknesses of MOFs, building on the advantages (topological diversity including chiral structures, potential for larger pore sizes, diversity of building blocks, ability to fine-tune pore size, etc.) and considering the disadvantages (limited kinetic stability to humidity/water, modest thermal stability, cost of building blocks) one may surmise that MOFs are most likely poised to yield breakthroughs in liquid phase catalytic reactions of large molecules of relatively high value, where the chemistry can leverage the advantages and limit the impact of the disadvantages.

Nonetheless, today, after two decades of research and development, with many thousands of papers published, such breakthroughs have not yet been realized. While there are numerous examples of effective MOF catalysts [14–17], there are very few examples of MOF catalysts that offer significant advantages over other types of (often simpler) catalysts, such as traditional homogeneous or heterogeneous catalysts.

To this end, we will close by considering the question: in 20+ years, will a hypothetical Solvay Conference on Chemistry featuring a session on "Catalysis with Microporous Materials" be dominated by examples of MOF catalysis that have changed the world? The answer largely depends on whether MOFs with greatly enhanced kinetic stability to water can be developed, and the realization of systems demonstrating unique catalytic performance or function not attainable in any other way. Advances in either or both of these areas could lead to commercial successes using MOFs as catalysts that, to date, are largely absent.

Acknowledgments

Funding from Phillips 66 and the King Abdullah University of Science and Technology for research on MOF materials is gratefully acknowledged.

References

[1] O. M. Yaghi, H. Li, *J. Am. Chem. Soc.* **117**, 1041 (1995).

[2] H. Li, M. Eddaoudi, M. O'Keeffe, O. M. Yaghi, *Nature* **402**, 276 (1999).

[3] S. Bhattacharjee, Y.-R. Lee, P. Puthiaraj, S.-M. Cho, W.-S. Ahn, *Catal. Surv. Asia* **19**, 203 (2015).

[4] M. Hara, K. Nakajima, K. Kamata, *Sci. Technol. Adv. Mater.* **16**, 034903 (2015).

[5] Ch. Baerlocher, L. B. McCusker, Database of Zeolite Structures, http://www.iza-structure.org/databases/.

[6] Y. G. Chung, J. Camp, M. Haranczyk, B. J. Sikora, W. Bury *et al.*, *Chem. Mater.* **26**, 6185 (2014).

[7] S. Chavan, J. G. Vitillo, D. Gianolio, O. Zavorotynska, B. Civalleri *et al.*, *Phys. Chem. Chem. Phys.* **14**, 1614 (2012).

[8] N. C. Burtch, H. Jasuja, K. S. Walton, *Chem. Rev.* **114**, 10575 (2014).

[9] T. Devic, C. Serre, *Chem. Soc. Rev.* **43**, 6097 (2014).
[10] M. Bosch, M. Zhang, H.-C. Zhou, *Adv. Chem.* **2014**, 182327 (2014).
[11] K. S. Park, Z. Ni, A. P. Côté, J. Y. Choi, R. Huang *et al.*, *Proc. Natl. Acad. Sci. USA* **103**, 10186 (2006).
[12] J. A. Thompson, C. R. Blad, N. A. Brunelli, M. E. Lydon, R. P. Lively *et al.*, *Chem. Mater.***24**, 1930 (2012).
[13] K. Eum, K. C. Jayachandrababu, F. Rashidi, K. Zhang, J. Leisen *et al.*, *J. Am. Chem. Soc.* **137**, 4191 (2015).
[14] F. Vermoortele, B. Bueken, G. Le Bars, B. Van de Voorde, M. Vandichel *et al.*, *J. Am. Chem. Soc.* **135**, 11465 (2013).
[15] R. C. Klet, S. Tussupbayev, J. Borycz, J. R. Gallagher, M. M. Stalzer *et al.*, *J. Am. Chem. Soc.* **137**, 15680 (2015).
[16] K. Manna, T. Zhang, F. X. Greene, W. Lin, *J. Am. Chem. Soc.* **137**, 2665 (2015).
[17] A. M. Fracaroli, P. Siman, D. A. Nagib, M. Suzuki, H. Furukawa *et al.*, *J. Am. Chem. Soc.* **138**, 8352 (2016).

SESSION 3: CATALYSIS BY MICROPOROUS MATERIALS

CHAIR: MARK E. DAVIS
AUDITORS: E. M. GAIGNEAUX[1], C. GOMMES[2]

[1] *Université catholique de Louvain, Institute of Condensed Matter and Nanosciences, Division Molecules, Solids and Reactivity, Place Louis Pasteur L04.01.19, 1348 Louvain-la-Neuve, Belgium*
[2] *Université de Liège, Department of Chemical Engineering, Allée du Six Août 3, 4000 Liège, Belgium*

Discussion among the panel members

Mark Davis: We will try to keep everything on time, so we will only go about 15 or 20 minutes before the coffee break. I would like to suggest that the speakers maybe try to address some comments that would tie into yesterday. If you have some comments, say, for example on identifying active sites, turnover frequencies, relationships to homogeneous enzyme catalysis, I think it would be appropriate to do it now in this limited time that we have. Any takers?

Enrique Iglesia: So, as I sort of re-lived the discussion of yesterday on the subject of active sites, from what you saw today, it's quite likely that you have seen an active site in a zeolite; that it is a proton; that it can be titrated selectively; that it can be identified as to where it may be by the size of the titrant; that the rates (provided that they are measured away from any sort of concentration gradients) are true measures of turnover rates; that you can change those turnover rates without refusing to call an active site any more, by just putting it in a different place. It is not that in a 12-membered ring Mordenite a proton is not active. It is much less active than it is in eight-membered ring structures, but it is too an active site. They're countable, you can measure their reactivity. And one of the things that were truly missing in this area, in my view, is to try to locate them on purpose in certain places and to be able to detect where they are.

Rutger van Santen: I'd like to challenge this view. It's true that we understand very well the location of the protons. But of course, in the course of the reaction, after an initiation period, very often, intermediates have been found that actually are the catalytic species. So there is an indirect relation, I believe, between the protons and the actual reactivity of the zeolite, that relates to the location, and also the position and the particular channel that is reactive, catalytic reactive, organo-catalytic species will be located. One particular example is the MTO system, where we now know of course that the catalytic active species is this methylated cyclopentylcarbenium cation. So I think there is an issue, also here.

Giuseppe Belussi: Just picking up another discussion on the real site and the turnover number. In cracking, in the catalysis of cracking, in refinery, we are used talking about an equilibrium catalyst. Why? Because the equilibrium catalyst (that is the catalyst that has to stand at the same condition for a long time) has not the same catalytic activity than the starting catalyst. The catalyst is modified, in the first period on stream, by the formation of amounts of coke, which are distributed through the catalyst, probably poisoning some sites and leaving active some other sites, until a steady situation is reached at which we have the so-called equilibrium catalyst. A similar situation occurs also in the selective oxidation. For instance, in the phenol hydroxylation to catechol and hydroquinone, nonselective sites are forming polyphenols, that are poisoning part of the catalyst until reaching a stationary situation at which we again have the equilibrated catalyst. That is not exactly the same catalyst that we had at the beginning, but is a different one, in which part of the sites have been modified by the deposit in one case of coke, and in the other case of polyphenols.

Christopher Jones: One of the advantages of these materials compared to some of the heterogeneous catalysts we heard yesterday, with regard to identifying active sites and calculating turnover frequencies, is that they are crystalline materials and that we understand the structure well. We can intentionally introduce an active site, for example, by substituting tin for silicon, giving us a Lewis acid center. And, in contrast to a lot of the materials that were discussed yesterday, many of those active sites are defects, defects on well-defined materials that are, in some cases, challenging to identify and in other cases challenging to count. So, even within the realm of heterogeneous catalysis you'll find classes of materials where the identification of the active site for getting site-based rigorous kinetics is relatively straightforward. And there are other cases where, frankly, it's very close to impossible.

John Casci: I'd like to build on the comments made by Giuseppe Belussi. I think the work that has been shown looking at identifying active sites is incredibly important and we need to continue that. But we need to look not only at the fresh catalyst but also at those which have undergone aging. Aging and deactivation occur in many, many ways. Often there is a fast deactivation and a slow deactivation, and understanding what is happening to the active sites under each of these stages and for different reactions is of great commercial and practical benefit, I would say. So, look at the fresh materials by all means. But I think we also need to be looking at aged and de-activated materials.

Enrique Iglesia: So, in every one of the examples that I showed you, the sites have been characterized during the reaction by effectively introducing the titrant during the reaction and counting the number of protons. Undoubtedly, we don't operate for hundreds of hours under commercial conditions, but my point is that if you want to understand the chemistry in one of the zeolites, you need to try to preserve at least some of the dignity of these acid sites, you know, in order to be able to count them and look at the mechanism, and so on. But whatever happened to the sites during reaction has been probed both spectroscopically and by titration.

Mark Davis: Let me further it a little bit. It's just to put it in context, in comparison to homogeneous catalysis. Unfortunately, these crystals are so small that a lot of the single crystal methods that you can do in other systems we can't do here. It's rare when we can get a crystal and get it ordered enough that we can actually use single crystal methods to try to understand structures. So, we rely on other methods. One that, at least for low-temperature liquid-phase reactions, has worked in several cases is to build molecular models. So, for example both Rutger's (van Santen) group and our group, and others in the world have built, basically a cube of silicon dioxide. And you can substitute one of the silicons for titanium, tin, so that now you can start to use the tools of homogeneous catalysis. You can use high-resolution NMR, you can use mass spectrometry. And if those experiments are designed correctly, they do recapitulate the chemistry that's happening in these full extended structures. There are a few cases where we can get a higher resolution of the structural information, when we can build molecular models. It's not the norm, but it can be done for some of the low-temperature liquid-phase reactions that are being conducted. So, there is a lot of analogy that can happen between solution reactions and reactions that can occur inside of these frameworks.

Takashi Tatsumi: Yesterday, Mark Davis pointed out that titanium zeolite has a similarity, true similarity, to homogenous catalysts. Titanium zeolite and titanium environment in the titanium silicate are totally different than, you know, titanium oxide, titania. Titanium is only surrounded by SiO ligands, we can say it is a ligand. In that case, we have heard yesterday, somebody talked about the effect of solvent. Solvent is very important in the liquid-phase oxidation. Originally, methanol was used for this reaction by Enichem. After that, Avelino (Corma) used water not methanol which is more efficient. In that case, we believe water or methanol acts as a ligand. I mean a coordination ligand. Somehow it controls the activity of the oxidation species like titanium–O–OH. I mean, coordination ability of the solvent we have seen that it significantly affects the electrophilicity of the active site. So, titanium silicate is a very close example of resemblance to homogeneous catalysis.

Giuseppe Belussi: I will pick up from the remark of Professor Tatsumi to add some comments. I agree that titanium silicate is something that could be considered as very close to a homogeneous catalyst. As you may know, in the dry catalyst, titanium is in tetrahedral coordination, substituting one of the framework silicon. As it enters in contact with some water, very easily one of the titanium-silicon bond is hydrolyzed giving rise to the formation of one titanol and one silanol. The same occurs with hydrogen peroxide, giving rise to the formation of one silanol and one titanium hydroperoxo species. And the titanium hydroperoxo species has an acidic character, as is well-known. For instance during the propylene epoxidation to propylene oxide, the epoxide ring opening occurs due to some acid sites which are present on these zeolites. Moreover, we have observed that when changing the solvent in which the oxidation (selective oxidation) and epoxidation of propylene occurs, some different reactivity was observed. Particularly when adding traces amount of methanol in a water solution, it was possible to see an increase of the formation of glycols and poly-glycols due to the ring opening of the oxirane-ring. This is probably due, as it has been reported in some papers, by the possible interaction of methanol with the titanium hydroperoxo species by forming a five-membered ring, which enhanced the acidity of the proton associated to the peroxide.

Rutger van Santen: I would like to make a case that there is also something unique to titanium that is part of the zeolite framework. We know very well that when we prepare with titanium alkoxide a monodispersed titanium catalyst on silica, that this catalyst is active for the epoxidation of propylene by hydroperoxide. But, when we use the same catalyst for epoxidation with hydrogen peroxide, it will not work. It will only decompose hydrogen peroxide. So apparently, there's something different with the titanium cation when connected through three oxygen atoms to the silica surface compared to the reactivity of titanium that is four — connected with the zeolite framework. I agree that, yes, when you react hydrogen peroxide with the titanium containing zeolite you will partially open the silica-oxygen bond (as you very elegantly demonstrated also with methanol). Apparently the rigidity of the zeolite framework compared to that of the silica surface maintains the titanium cation that is incorporated in the zeolite framework in a state of different reactivity than present on the silica surface. Second point is, as Mark Davis was mentioning, that titanium incorporated in a silsesquioxane cluster also is only a selective epoxidation catalyst with hydroperoxide and again it will decompose hydrogen peroxide. The titanium cation in the silsesquioxane cluster is three coordinated as the reactive titanium center on the silica surface. Clearly titanium that is part of the zeolite framework has unique reactivity due to its four coordination to the zeolite framework and zeolite framework rigidity.

Giuseppe Belussi: This is really an interesting point. Titanium silicalite is able to work in the presence of water and hydrogen peroxide. The Shell catalyst works only in the presence of alkyl hydroperoxide. It has been reported in several papers that, in the presence of large-pore zeolites or zeolites such as MWW in which titanium is inserted in the framework (so it is an exposed titanium), again, there is a higher reactivity when using alkyl-hydroperoxide with respect to water and hydrogen peroxide. This probably is related to the possibility of titanium not only to break more than one titanium–oxygen bond but also to interact with water and coordinate more water molecules, changing completely the reactivity.

Enrique Iglesia: Since I won't get a chance to ask questions I think, or to make comments if not asked, I wanted to make a point that maybe we can follow up as we go through the open session. If we look at chemical kinetics, irrespective of whether we're working on heterogeneous, homogeneous or enzymatic catalyst, we would all agree that we measure a rate, and that reactivity relates to free-energy differences between a transition state and the relevant precursors. Moreover, if we look at a similar transition state, we can all agree that it is stabilized by ionic or electrostatic or covalent or dispersive interactions. And at that point there is no longer any mix-ups on the language that we use because we actually are measuring a rate, a Gibbs free energy. And then we're looking at it chemically in terms of the interactions that make that transition state more or less stable with respect to whatever it is that is on the reaction coordinate. That kind of language allows me to understand quite a bit of the enzymatic kinetics when I read it, and allows me to look at the homogeneous catalysis when kinetics are the focus of that work. I don't think there is a difference once you get to that level of detail, in spite of the names that we may want to give the things in between, or the mysteries that we may want to create about the magical effects of frameworks or ligands or anything like that. It's all down to what do we measure when we measure a rate, how it relates to transition state theory, and what stabilizes those transition states.

Mark Davis: Let's go to the coffee break. And the other thing I would like you to think about, if you don't have enough to think about at coffee break, is maybe we can also talk about combinations of catalysts when you come back as well.

General discussion

Mark Davis: Welcome back. Now we will open questions to anybody. Anyone would like to start, please? Ah! sorry, go ahead.

JoAnne Stubbe: I think we were talking about stabilization of the transition state and how there were similarities between biological catalysis and near zeolites. For all the enzymes I have worked on, the chemistry is not rate-limiting. It's all conformationally gated. And so, separating out the conformational gating from the chemistry is lots of times really difficult, depending on the catalyst you are looking at. So, I think that's probably distinct from what you are looking at.

Enrique Iglesia: It may not be as distinct. In order to be able to extricate our version of the gating, which is the diffusional problems that we have, you have to conduct your experiments quite carefully. So it is what also in the practice of catalysis gets in the way, or perhaps translating what we get on well behaved systems to the other systems. We also have the kind of gating in zeolites that has to do with conformational flexibility, and that is that the zeolites pay a lattice distortion in order to be able to maximize the van der Waals contacts, so...

JoAnne Stubbe: So, is there a real way that you can measure that experimentally? In other words, that's the key, the key thing, because you know, you have to perturb it, to see it, to remove it, so you can study the chemistry.

Enrique Iglesia: The diffusional problems can be corrected for, so to speak, because we can measure those rates. But also because we can change the number of sites that we have within a given volume, and that creates more or less of a concentration gradient, so when we measure kinetics we try to design materials that are best suited to give us the kinetics. And then we sort of build the complexity of the diffusion by independent measurements that allow to practice, then, the process in the real world.

Avelino Corma: Yesterday we were talking about trying to get connections between heterogeneous, homogeneous and enzymatic catalysis, and I think that after the talk of today we are saying that yes, we have those connections very clear. We are talking about active site identification. You know it in homogenous catalysis because you make it directly. In enzymatic catalysis, you work hard to find it and you know it. And as you see here in the systems that we saw today, we also can identify the active sites or the potential active sites, and we can also titrate them. So, from that point of view, we have it. And then, the second part that we were talking about was the short – medium – long range type interactions that stabilize the free energy of the transition states, and that is common for the three same disciplines that we are working on. We have seen how one can rationalize that through spectroscopic and also through kinetics studies. So, if we apply those concepts and those techniques and we can go through the transition states, and talk about the transition states, how is the interaction that we are having; which reactants with our products with our system that we have for catalysis? I think we are talking then the same language. We can quantify also. And then we can compare on the bases of activation energies, heat of adsorption, and then we can be talking the same language. Regardless, if one is using ligands to achieve the effect that it wants, we can rationalize that later, or be using the type of interactions that we saw here today. So, let's try to go a little bit farther in that direction, please.

Judith Klinman: I just want to say a comment and then a question. The idea of transition state stabilization is really just the definition of catalysis using transition state theory. We run into this in enzymology quite a lot because that is what the textbook says the origin of enzyme catalysis is, whereas in fact that doesn't gives us the physical parameters and properties that are needed to achieve catalysis. So, there is something that I just want to emphasize. The question has to do with how you define sites like acid sites? What kind of tools are used to do that?

Enrique Iglesia: We use titrants that are basic, and we actually can irreversibly titrate the acid sites and count how many titrants we have put in. We can look at the infrared spectrum of the titrants to look at how they may be interacting with the surroundings. We can measure the heat of adsorption of the titrant. We can make titrants of different size so that we can look at different places where the titrants may or may not get in. But it is no different than the way that one would titrate a liquid acid, as long as it is quantitative and irreversible, so that we can do it during reaction and watch the rate go down. We can use titrants that are coordinating and non-coordinating to tell the difference between Lewis acid sites and Brønsted acid sites, and look out which one of the two is involved in the catalysis by whether they decrease the rate or not.

Judith Klinman: OK, you can't really quantitate how many acids would be within a particular cavity or anything like that, so it's...

Enrique Iglesia: You can. If it is restricted in access, then you could. So, for example, in the eight-membered rings in Mordenite we cannot even get pyridine to go in, but ammonia goes in. Whereas in MFI or pentasil, it is very difficult because in order to get to the large places you have to go through the small places. But in many cases, we can.

Judith Klinman: OK, thank you.

Bert Weckhuysen: I would like to ask a question to JoAnne Stubbe, because she mentioned conformational gating, and that you have experimental evidence for this. So, I want to learn from the bio-catalysis field, how do you do that? How can we, for example, try to transfer that knowledge then to heterogeneous catalysis, so how do you know this?

JoAnne Stubbe: Well, I can tell you about the system we worked on. So, everything is conformationally gated. The turnover number is five per second, but if we can figure out how to remove the conformational gate, the rate constants are over 14.000 per second. And the way we perturb the system, this involves transient radical intermediates. We put in unnatural amino acids where we can perturb the pKa and the reduction potentials, and so by several perturbing the system to make these measurements. And therefore we can change the rate-limiting step from conformational gating to chemistry.

Donald Hilvert: I wanted to continue the analogy between biological catalysts and the zeolites. I was particularly intrigued by Mark's (Davis) example of the glucose isomerase. I was wondering how efficient is that reaction, and is epimerization of other sites within the molecule a problem? How general is it with other types of sugar starting materials?

Mark Davis: Actually, the rates are very high. They are almost enzymatic. And, as you know, those reactions are equilibrium-limited. So, one of the nice things about going with the molecular sieve is that we can go over temperature ranges that we can't do with the protein. We can do rates and concentrations like you see with glucose isomerase, 60°C and so forth, but we can also go up to higher temperatures and shift up to very high conversions, because the thermodynamics go that way. It is very general. So, for whole families of sugars, we can do the same reaction on them. And it is very selective to do that reaction only, and not doing both reactions simultaneously. It really is interesting that you can tune the active center to either do the isomerization or the epimerization and have it so that it cleanly goes one way or the other way. Right now, it is a pretty intriguing system to try to figure out, and within the limits of what you can measure with the solid, it does look like the pathway is the same as you are seeing on the protein catalyst. Erick?

Erick Carreira: I am curious, in these confined spaces do solvent effects disappear? Is it irrelevant? Given how much attention has been paid to it in earlier sessions. And I have a follow-up question, but...

Enrique Iglesia: So, to the extent that the solvent is an extended phase that requires reorganization of many many molecules, those effects are no longer present because there is seldom enough room there in order to characterize that as a solvent. But that is not to say that, as you increase the pressure of the reactant for example, you begin to accumulate more and more molecules inside, and those can have an additional solvation effect, but not of the kind of a sort of reorganization that you get in an extended solvent, because of the restrictions that you have...

Erick Carreira: Presumably effects of traffic going in and out, no?

Enrique Iglesia: There is in many cases effect of traffic. There is a higher concentration inside than outside. So, you do get some cooperative behavior. But in general, you don't have the same kinds of solvent effects that you would have in an extended phase.

Erick Carreira: Let me guess, the cage is the solvent of sorts, right?

Enrique Iglesia: To a great extent, I think that the interactions are mostly with the cage at most practical conditions, especially as you go to higher temperatures.

Erick Carreira: My follow-up question, I think it was Mark's (Davis) talk. How do you get these to be more hydrophobic? I think you said something about "there are some hydrophobic zeolites."

Mark Davis: These are basically silicon dioxides. And if you make them defect-free, you have a very perfect surface, or near perfect surface. Those walls will exclude liquid water. So, they become very hydrophobic. When you partition species out of the aqueous solution, when they actually go into the zeolite, you lose their hydration sphere. So, you actually get that entropic driving forces well for going in because they're not hydrated. Actually the molecule goes in and then the oxygen atoms on the framework are substituting for the oxygen atoms on water when it's solvated. But they're just so hydrophobic that it takes thousands of pound's per square inch (psi) of hydrostatic pressure to drive liquid water into the pores of those samples.

Kurt Wüthrich: What is the actual activity of the solvent in these cavities, when compared to the activity of the solvent in the bulk?

Enrique Iglesia: To the extent that there is equilibrium between the outside and the inside, the activity, by definition, would have to be the same. Because that's the definition of chemical equilibrium. So, we generally think of that equilibrium as being established during the reaction, and the chemical reaction being the rate-determining step. So, there is a chemical potential equality between the inside and the outside that is just the defining term for equilibrium.

Christopher Jones: Just following up on Erick's (Carreira) comment: if you read the literature on solvent effects in zeolites, when people talk about a solvent effect it will mostly be manifested in a partitioning. So, we can think of it as two-solvents liquid-liquid extraction and when people talk about solvent effects in literature for zeolites, it's usually changing the base solvent to get a different partitioning of the reactants into the cages.

Dan Herschlag: Why don't we try to take a step back, and maybe pose the hypothetical that maybe the panel won't like to start with, but stay with me for a second. That is: imagine that there was no commercial benefit to zeolites, so you had to do something else with them. What would you do with them? Another way of saying that is: what's the disruptive technology? There is a remarkable usefulness of these now, but is there something? What can one imagine is next? And let me try to seed that with maybe an idea because it seems that one of the limitations that comes up a lot in chemical catalysis is that each time you want something new, you've got to make something new, and that takes time. So now, imagine that there is a way to multiplex and maybe you had a larger cavity. The reason these things work is because the cavities are the right size as we so beautifully showed. But now imagine that you have a larger cavity and you can use a library of sort of plug-ins like Legos that you can put into that cavity to make a lot of different properties. So now you're creating something like the amino-acid side chains of a protein that you're putting into these cavities, but still having enough space to do reactions. I'm just trying to sort of see that there are ways to be able to more systematically look at the properties, look at the potential of what one can do in these sites. And maybe something that's underlying this is (maybe following what Judith (Klinman) said), just looking at kinetics and sort of, you know, products or protons. Probably this is going to be enough to have a full understanding of the systems. So I am trying to see if there is a way to imagine what is this disruptive technology? And is there a way to expand what one can look at in these systems?

Enrique Iglesia: Well, the answer is "I don't have a very good answer for that." But coming from the chemical reaction side, rather than the synthesis side, I think, in my view, the true disruption would be the ability not to have to make a new material every time you want to do a new thing, but to actually be able to take a very large arsenal of materials that we already have and be able to fit them to the right purpose. And to do that by putting the aluminum in the right place, by putting obstacles in the way, by filling half of the cavity so that the cavity is now small. But only through the guidance of some particular requirement that we have identified. So, we're not just trying things to see if they work. But I'm sure from the synthetic side all the people would have better ideas, but that's how I view the evolution of that area into something more intelligent.

Avelino Corma: I think we can bring together the two things: the synthesis and the reactivity. And what we can do is what I was trying to say yesterday, not just to prepare zeolites and then to see what they can be good for or how we can modify them, but to go from the other way. Let's try to make, as a template, the mimic of the transition state. Then, they will leave the site where we want it, and they will optimize the transition state. So, that would be another way of going directly into the zeolite. And you saw several examples already that we made. Another challenge: imagine they would be good for nothing. Well, that is difficult to imagine because at least in the introduction of papers we always say that these will be good for such and such a thing. But, I would like, for instance, to make zeolites where I could have chiral sites, chiral structures. Of course, as Mark (Davis) said, very good, excellent. But if you could have the chiral sites, that would be a big achievement and then we will have chiral solids. Look what we tried to do to get chirality on the solids. It has been done by adsorbing some chiral organic molecules, but at the end they cover your surface. It is very difficult to control. In this case, we can have the chiral sites and all the benefits that we talked before about adsorption and stabilization. So, that would be another thing.

Mark Davis: I'm going to try to answer your question. I'm struggling to. I guess one of the things I've been thinking about for a long time, but don't know how to do it, is: Is there a way to do an analogous situation with these types of solids that you do with the polyketide synthesis? So that you basically have a set of genes, and you can swap those genes around depending upon which molecule you want to make. We've been trying to see if we could do it with just two functions. Could you do a synthesis by design? You would have to go down one trajectory and make it come back another trajectory. We think if you could do some of that, it would be really new and different. So, has anybody done anything like that yet? No. But, that's kind of one that I really look out as a pie in the sky, if you could do something analogous to the polyketide work.

Graham Hutchings: Just following on from Avelino's (Corma) point on the chirality aspect. The chiral pockets that you're going to need are going to be quite small. And the obvious one is the zeolite beta where there is just too large a space. The thing is like a small hand in a large glove. I think the only way you can do it is to have some flexibility built into the structure. And so, I suppose one of the key questions is how flexible are these structures? How flexible can you make these structures at the temperatures at which they are used? Because then I think you're opening up the possibilities. If they are rigid, as I think we started all thinking them as rigid at the start, then that's not going to be really possible because by the time you get the structure for the molecule you want, it is not going to be able to get in there, or get out. It's going to have some flexibility, as enzymes have. They change their conformation. So, can we make, not going to MOFs, but can we make more flexible structures? Is that possible?

Mark Davis: I'll answer that. It depends on the structure. There are some really good examples of how structures can change dramatically. The one I'm thinking about is the one that was published in Nature where the window, actually when you do an adsorption, changes completely from circular to almost American-football like. And it does so in a very dynamic sense as you have adsorption. Kind of like what Enrique (Iglesia) is saying to us that as soon as you have the adsorbates the whole structure changes quite significantly. So, there are some examples of those. There are others that are much more rigid. It really is structure-dependent. Some of them have just massive changes even with simple cation exchange from, say, a lithium to a sodium, the unit cells can change 10% sometimes.

Avelino Corma: But let's think on the possibility also. We are thinking all the time, or I think we are thinking all the time that we have the metal that we have in the framework and we have tetrahedral coordination. Well that is one extreme situation. We can have it bonded through four silicons through the oxygens. We can have it to three or even to two. Now you are getting more and more flexibility, in that case. And then, you are having those attached maybe by two and these opened completely to whatever space you have. Now, we can start to think about that chirality.

Joachim Sauer: I would like to come back to the previous question. Imagine that the zeolite would not have a use, why would they be of interest? I start with the end when you mentioned that we could include some other functionality inside. This is the direction the MOFs have developed. You can imagine you are bringing in as linker molecule something that has a function itself. This is actually done by the people who do catalysis in MOFs. They have a catalytic function connected or attached to the linker. For the other question, why would one be interested in zeolites? I write in my introductions always two things. Apart from their industrial use, they are attractive as model systems. Yesterday, we have seen how difficult it is to know the local structure on a solid oxide catalyst. With zeolites we have a system that, at least in first approximation, can be described by a crystal structure. We can use diffraction methods. And within this frame, zeolites have amazing variability. For a given crystal structure, we can change the silicon-aluminum ratio and tune the acidity. We can, with the same framework structure, prepare a pure silica material or an aluminum-phosphate material. So, we can have the same Brønsted active site in a silica environment or in an aluminum-phosphate environment. And then, for both of these materials, we can vary the framework structure, different cavity sizes with different van der Waals interactions around the site. For somebody who is interested in understanding in detail the structure reactivity relations, such a broad variability is great. Apart from a proton as an active site, you can have a transition metal cation, and this way bring in different transition metal functionality. So, this is such a great system even if the industry would not be interested. You must study this if you would like to understand catalysis. But I would also like to mention that this is the ideal structure that gives rise to a great research program, but, of course, reality is more complicated. And, when we talk for example about Brønsted acidity we know that a much-studied system like H-MFI may have imperfect silicon-oxygen-silicon bridges, so-called internal silanols, which contribute to the acidity. We know that we have some more acidic species where we don't really know in the moment what they are. Whether there is some extra framework aluminum species attached to Brønsted sites, or whether they form internal hydrogen bonds and create bridges across the cavity. So, there is still room for complexity beyond this nice first approximation.

<u>John Casci:</u> I was puzzled by your question because what I thought you said was: what if there were no interest in catalysis? What were the disruptive technologies? So, my mind was going there. So, I'm not sure if you said there is no applications or no catalytic applications. And, regardless of what your question was, let me just remind you of a couple of points that Mark (Davis) made. They are used extensively in ion exchange. And while their use in detergency may not seem very high-tech, they are used in some incredibly high-tech applications for ion exchange. So, a simple thing from radioactive nuclide removal through to, more recently, in purification of groundwater sources which are heavily polluted in parts of the world. So, I would say these are disruptive technologies on a societal level. The adsorption separation has been touched on by people. So yes, they're used in what might be regarded as low-tech applications, in double glazing for example. But you've heard about oxygen enrichment in air, which is used in medical applications and this is used extensively in hospitals now, for example. They are used in a whole variety of adsorption separations, but I would say that one of the disruptive technologies, which we have investigated for some time, is their use as membranes. So, they would perturb chemical processes allowing you to eject species, as I say, to perturb the chemical equilibria. So, there are formidable challenges in terms of flux and fabrication, but these are two not directly catalytic applications which are disruptive I would suggest.

Kurt Wüthrich: In at least one of the talks there was explicit discussion of tetrahedral sites. What about incomplete coordination and open valencies being the actual catalytic sites?

Giuseppe Bellussi: There are different situations depending on the kind of catalytic site considered. In the case of oxidation catalysis with TS-1, the catalytic site, as I told before, is a defective site because it is a site in which hydrogen peroxide is interacting with the tetrahedral titanium, breaking one titanium-oxygen bond and leaving a silanol and a titanium hydroperoxy species. This reaction is reversible, and upon drying very easily the original tetrahedral titanium site is restored. So, this is probably one of the most important properties of titanium silicalite in selective oxidation. This kind of interaction can be monitored very easily by using water labelled with different oxygen isotopes, and monitoring what is going to happen upon catalyst hydration and dehydration with the infrared spectroscopy, for instance. So, I can't exclude that the opened titanium site can have other interactions with molecules that are diffusing through the zeolite channel, but I never saw direct evidences of a different kind of coordination. Someone has reported the possibility that titanium can be coordinated inside the zeolite by other molecules, giving rise to the formation of penta-coordinated species. But in my opinion there are not enough evidences to conclude about that. The situation can be different in the case of the acid sites. In the case of the acid sites, the acid site is generated when aluminum is in tetrahedral coordination surrounded by silicon because of the difference in the valence state of aluminum and silicon. But generating a defect in this area, the Brønsted acidity is lost and silanol groups are generated giving rise to the formation of defective Lewis sites. So, the activity TS-1 can change completely when the original site is modified by the breaking of one of the Ti–O–Si bond.

Mark Davis: I can give you a fun example but it's not catalysis. With these aluminum phosphates, there is now a heat pump that is commercialized in Japan by Mitsubishi. The aluminum actually coordinates two water molecules and goes octahedral. And so, that reversible water off going tetrahedral to octahedral is the origin of the heat pump. Now that's very efficient. That is a very low energy for cooling skyscrapers in Tokyo.

Christopher Jones: In the MOF space, under-coordinated Lewis acidic metal nodes are probably the most common type of active site that's created. And you can increase the number of those for example by introducing defects on purpose. Dirk (De Vos) who's here in the back of the room, has published papers on that. And so, in the MOFs field, dealing with these under-coordinated Lewis acid transition metals is probably the most common type of active site that's been created.

Xinhe Bao: My question is concerning the MOF materials as a potential catalyst. So, as Professor Jones showed, there are two different very important effects in the

MOF materials. One is the order of the pore sizes and the other one is the metal in the framework. I just want to know, in the future, if we focus on such a material as catalyst shall we pay attention to the pore property or to the metal property? If we talk about pore, I think the pore seems a little bit too big.

Christopher Jones: My answer is, yes, you should focus on both. And the beauty of the MOFs is that some of the ones that I showed you were quite big because I was trying to contrast them compared to zeolites. But, there's a huge array of pore structures and pore sizes that you can achieve with MOFs. So, certainly if you want to work with materials that have things that look like eight- and ten-ring zeolites pores, those MOFs most certainly exist. Ultimately though, I think the key thing for MOFs and catalysis is to generate systems that display unique reactivity that you can't get in another way, and display sufficient stability that you can get the number of turnovers you need for your application. If it's a very high-value product that might be a hundred turnovers or a thousand turnovers. If it's a more commodity product, that's going to have to be many millions of turnovers. There's a lot of opportunity, but as of right now I'm not particularly bullish on what I've seen.

Ben Feringa: Continuing with this, we heard this beautiful new world of the MOFs. Actually, I have two questions. One is, can we combine MOFs with all their versatility with the stability of the zeolites in synthetic approaches? For instance using them as template and then introducing functionality? Or is that completely incompatible? And second, besides MOFs, there is also a whole world now emerging of COFs, Covalent Organic Frameworks, and graphene-type materials, etc., which have really amazing properties. And I was wondering what the opportunities are there, in designing these new kinds of heterogeneous structures?

Christopher Jones: In the COFs space, if you create COFs, covalent organic frameworks that have strong bonds, so the irreversible synthesis, then you get very stable materials that can be used in a wide variety of especially liquid-phase applications. As you go to the type of applications that this panel has primarily talked about, which are vapor phase, high-temperature hydrocarbon applications, what do you do when it cokes? Can I burn off the coke? Obviously in a COF you can't do that. So for those types of materials, in my view at least, higher-value low-temperature liquid-phase applications, I think, are the most promising. But, you also have to work with materials that are essentially irreversibly synthesized. Some of the ones that have more dynamic linkages based on imines, that are very common, are not going to be particularly useful under those conditions. Coming back to the MOF case and "can you make MOFs stable in the same way that the zeolites are"? With 5000 structures, I'm sure that there's probably someone who would say "Yes, I've done that", but I'm not familiar with it off the top of my head, and I would say that from a practical perspective I am somewhat doubtful. If you have a MOF that you can show has the type of stability that zeolites have, it's certainly a nice Science or Nature paper. So, I would say anybody who wants to try, go for it.

Henk Lekkerkerker: I don't know whether I'm allowed to switch a little in the subject. We saw some beautiful examples of calculations. I want to ask Rutger van Santen: When you say that the difference in wide and narrow pores zeolites is that the wide pores are dominated by entropy. We can imagine that if you have a bigger volume, the entropy is higher. How many Boltzmann units would that be? Is that a fraction of a Boltzmann unit or is this really a big factor? You gave numbers, I agree, but I just would like to be reminded.

Rutger van Santen: The numbers I showed were energies not entropies.

Henk Lekkerkerker: No, no, but entropy must play a role.

Rutger van Santen: So, the reason that the enthalpies are different, activation enthalpies are different for this wide pore and narrow pore zeolites, is basically the screening of the positive and negative charges in the zeolite, because you have an expense to separate the charges. That's the basic reason. The entropies... My colleague, of course...

Henk Lekkerkerker: It's a question to you both.

Rutger van Santen: The entropies, of course, are extremely important. And maybe you (turning to Iglesia) can also talk about this, but basically what you see is that when the pore is wider then the effect of entropy can be significant and will change the free energies, at least by something like 5 or 6 kJ/mol.

Henk Lekkerkerker: Your neighbor would agree with that?

Enrique Iglesia: Well, I am not sure that I really understood the point he made so I can't comment on Professor van Santen's statement. If I may just comment about argues about enthalpy and entropy: the main difference between a small pore and a large pore is what you give up in entropy in order to get enthalpy gain. We do not think that it has very much to do with the electrostatic component and what it does with it, because the negative conjugate anion in those zeolites is similarly stable in the large and the small pores. It is only the van der Waals interactions where you do the tradeoff between entropy and enthalpy, from what we know for the systems that I have talked about today.

Henk Lekkerkerker: OK, thank you.

Bert Weckhuysen: I would like to switch to Lewis acidity in zeolites and especially also (triggered by the comment of John Casci on binder effects), that the zeolites are compounded in silicas, aluminas, etc. How can we characterize them? What do we know? How are they involved for example in catalytic cracking?

John Casci: The point I was making, I will answer the question, is that when we apply these materials we are not using the simple zeolite. The zeolite is in a complex form where we have binders that provide multiple functions. So, I think I have made the point that they provide mechanical stability, which can be incredibly important in the duty. They provide textural properties, which means the porosity to allow diffusion. And they carry the zeolite. But, you're absolutely right: they are multifunctional so they can provide other features. Some of the other features is that they can be catalytically active in their own right. I guess the most common binder that is used is probably a clay when I come to think of it, probably in volume terms. But gamma alumina is often used, and there we have an amphoteric material, so we do have some acidity associated with that particular alumina, as you well know. So, you have a range of functions there. But, I guess the point I was alluding to is that we have a complex assembly which is different from, I would say, the homogeneous system. But also they provide a number of functions, some of which are catalytic, but some of them are simply adsorption as well. We talked about biomass yesterday, and there is an increasing work where binders are now being used to soak up some of the impurities that we're seeing in biomass feeds for example.

Bert Weckhuysen: Yes, but it alludes also to Bellussi's comment who, like you, stressed the point of studying systems which are not fresh, but has been the ECAT (Equilibrium Catalysis) system that has already been undergoing some regeneration cycles. Then we are sure that already some damage has occurred by having steaming, etc. And then my question is: how sure are we then about the balance between the proton on the Brønsted acid sites and the Lewis acid sites? And have we any insights at what point activity is then dominated by one over the other?

Giuseppe Bellussi: Well, the way to measure the acidity has been already described by Professor Iglesia. The problem of equilibrated catalyst for FCC (Fluid Catalytic Cracking) is not only a problem of which kind of modification has occurred upon steaming. Typically, an FCC catalyst, before to be loaded into a laboratory reactor, is steamed so as to simulate the regeneration procedure. By doing this, you can analyze what kind of differences there can be between a catalyst that undergoes steaming with respect to a fresh catalyst. But the situation of an equilibrated catalyst is even more complex, because in the FCC reactor there is an average catalyst composition, which is a mixture of fresh and coked and steamed catalyst. So, there are different particles of catalyst that have been deactivated in a different way. The equilibrated catalyst is a situation that can be represented by the presence of catalyst particles at a different degree of deactivation. So, it is rather difficult in that case to describe the contribution of the different sites. But, again, there is the possibility to have some reliable measurements of the different kind of acidity by the technique already mentioned by Prof. Iglesia. The real problem with this kind of catalysts is to analyze accurately the diffusion limitation effect, due to the presence of coke, or due to the interaction of the binder with the crystal of zeolite. Usually binder is gamma-alumina that, being rather reactive, can interact with the framework of zeolite, and this can change the local composition. But, again, the technique that is usually utilized to measure the acidity of the pure active phase can be used also to measure the acidity of the real catalyst containing the bound zeolite.

Avelino Corma: I would like to make a comment that connects three points that have been raised here. One by Kurt (Wüthrich), before, where you talked about the defects that you generate, the breaking there. Another, about the polarity that was commented before. And another one about the Lewis acidity. You will see that they can be pretty-well related. First, it was said before, and Mark (Davis) said that when you make a very perfect zeolite, no defects, you have a highly hydrophobic material. Now, when you start to make defects in that zeolite, you start to make it more hydrophilic. So, at the end, that is one way we have to go from more hydrophobic to more hydrophilic. Secondly, when you talk about Lewis sites there (and we have been talking titanium but it can be also tin) titanium was acting for oxidation, but it is a Lewis acid site. But tin also. In a Meerwein–Ponndorf type of reaction, where you have a bimolecular reaction, you have opened and you have generated that defect which is a silanol now. And now you have two sites really, and two sites close-by. One, the silanol that is able to activate one of your molecules. The other, the Lewis that activates the other. And with the other defects you have the transition state. So, in fact, silanols, they can also intervene stabilizing when you have the transition state. So, it is very important to consider those from the three points of view that I said. First the polar, therefore the adsorption properties. And secondly, how they can intervene also in stabilizing your transition state.

Graham Hutchings: I have just a point of clarification to Avelino (Corma) and others. If I understand that you're talking about defects that are being created, either post synthesis and treatment by steaming, or post synthesis by the reaction taking place at the active site. Are there strategies to make inherently defective solids at the start? Because, as we were talking about yesterday afternoon, most of the active centers involve defects. We actually made them at the start, and they were the active catalyst right at the start. So, am I missing something, or is there a strategy that is available?

Avelino Corma: Well, we know that by different synthesis methods, either working on OH media for the synthesis, or in fluoride media, you are going to change the number of defects that you have there. You can make it, if you want, in a clear controlled way by introducing atoms there that you can remove relatively easy after that. OK? And you would generate those defects.

Mark Davis: There are synthetic methods to do just what you said. So, it's not post-synthetic. With some of the zeolites, for example, if you do these in fluoride-based media you can make them defect-free and then if you dial back in sodium ions, every time you put a sodium ion back in you create a defect. So, we showed years ago, where you can just dial right-in as many of these as you want, based on the composition of the synthesis media.

Graham Hutchings: Just to follow up on that, the follow up question is: at the defect sites that I was talking about yesterday, where you can activate oxygen... Now that oxidation today will be involving hydrogen peroxide, which is activated on the defect site that is created by breaking one of these bonds. So, have we got the possibility of creating structures here that will activate dioxygen at reasonable temperatures? I know it's possible at high temperature, because there is literature going back to the 60s and 70s where people have done hydrocarbon oxidation over zeolites at high temperature making lots of CO_2. So, have we got an angle here where we could activate oxygen at reasonable temperatures?

Mark Davis: I don't have a good answer to that one, Graham. Yes, there are actually sites that activate dioxygen. They are really hard to control, and the number densities are quite low. So, you're right, that is a difference. I thought you were talking about tuning the hydrophobicity. That, we know how to do. The oxidation sites, I don't think you can control them so well right now, to be honest with you.

Graham Hutchings: So maybe that's an answer of a disruptive technology, going back to where we were earlier in this discussion, because that is may be something we could focus on. Because if we could do that, we could do some really interesting transformations.

Stephen Buchwald: Mark (Davis), can I ask you a naive question again. With respect to coking, where does it occur in the zeolite? When you say coke, is it one material? Is it a set of materials?

Giuseppe Bellussi: I am using the term coke very often when speaking about coke precursor. In reality, coke precursors can evolve in different ways to form a solid particle of coke. So, you can find on a deactivated cracking catalyst polynuclear species and their evolution towards the solid particle of coke.

Stephen Buchwald: Just some carbonaceous material?

Giuseppe Bellussi: It is a process in which carbon compounds are losing hydrogen, first forming polynuclear aromatics, and then evolving towards even graphitic coke.

John Casci to Stephen Buchwald: I think you asked if it is in the outside or in the inside, and the answer is: it can be both or either. And materials do sometimes de-activate from the outside-in and from the inside-out. The material on the inside, I don't call it coke, it is a deactivating species, and for zeolites, as I think Mark (Davis) showed us, any molecule that is too large to escape from the zeolite. So, it just blocks up the pores. So, even if you cut it in a oligomerization reaction, I think as one of the panel has shown, you can lead to deactivation. If it's on the outside, then it can block port mouths and that can then lead to deactivation. Decoupling the two can be quite tricky but it can be done under strategies to try to remove, for example, a coking or graphitic material on the external surface. Coming back to the binder, however, you can actually start laying down coke on your binding systems. And that can then have like a cascade onto the zeolite. So, the real catalyst systems can be very complicated indeed. And many of us spend a lot of time trying to work out how to minimize that, and then how to develop strategies for regeneration.

Benjamin List: Can you make an enantiopure zeolite by using a chiral organic structure-directing group and then coke it off, as you said? Is that possible? Probably, you've tried it.

Mark Davis: I've been working on that for about 30 years now. So, we do have some evidence now that you can make an enantio-enriched sample.

Benjamin List: But how do you determine the ER (enantiomeric ratio) then?

Mark Davis: It's very hard to do with a powder. What we're doing is actually by direct visualization in TEM. And so, you can only do so many crystals because it's very, very time consuming. Again, you could always argue that you didn't do enough crystals. As far as we know, there's no way to take a powder sample and say you have an enantiopure powder sample, today. So, if we take ten crystals and we get nine out of ten we are pretty happy with it.

Benjamin List: Along those lines, can you catalyze zeolite formation? Maybe enantio-selectively?

Mark Davis: Well, in this case we are using enantiopure organic structure-directing agents. And we can show that if you take one enantiomer you make one particular enantiomer structure. You can take the other enantiomer as the organic, you go to the other enantiomer structure. So, at least you have some rationale to what you're doing. I think it's getting there. There are few people in the world making progress in this right now, but it has been an ongoing project for a long time.

Kyoko Nozaki: Related to this point, I think asymmetric catalysis is just one shot and it's very difficult to get EE's (enantiomeric excess). But if it's chromatographic separation you can get a reasonable separation of the two enantiomers, if you put your zeolite into a column for HPLC separation.

Mark Davis: Agreed.

Mark Davis: OK, thank you, it's been a fun morning. I think we now all have to go to the photo, right?

Kurt Wüthrich: Yes. Next is a photo session in the coffee room.

Mark Davis: Yes, all right. The session is adjourned.

Session 4

Catalysis under Extreme Conditions: Studies at High Pressure and High Temperatures — Relations with Processes in Nature

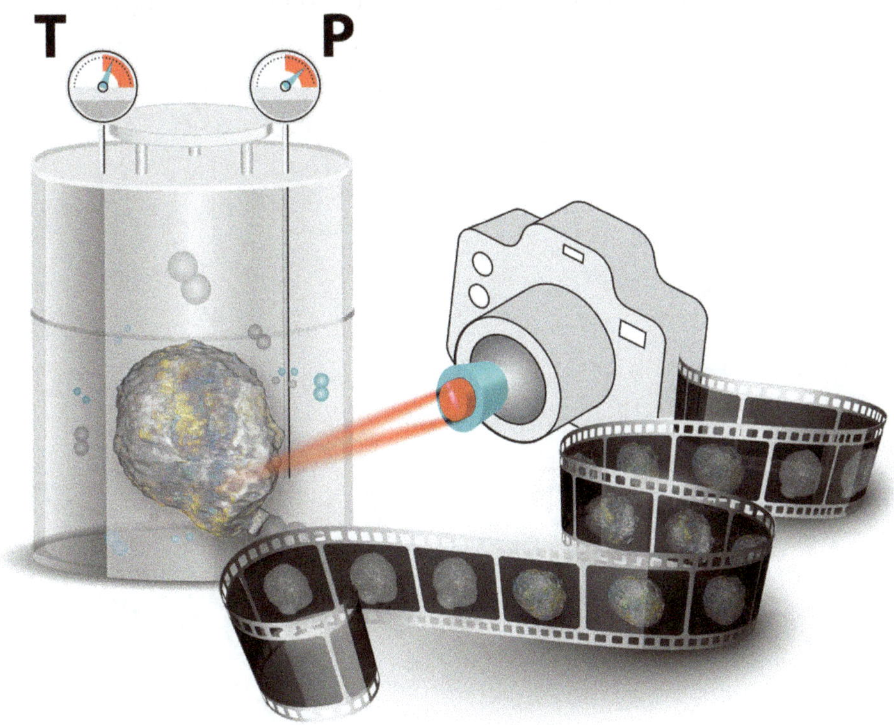

Schematic illustrating the scientific ambition to film real-life catalysts under extreme reaction conditions.
Image by: Bert Weckhuysen, Utrecht University.

CATALYSIS UNDER EXTREME CONDITIONS: STUDIES AT HIGH PRESSURE AND HIGH TEMPERATURES & THEIR RELATIONS WITH PROCESSES IN NATURE

HENK N. W. LEKKERKERKER

Van't Hoff Laboratory for Physical and Colloid Chemistry, Debye Institute for Nanomaterials Science, Utrecht University, 3584 CH Utrecht, The Netherlands

View of the present state of catalysis under extreme conditions

Catalytic solids are important materials in our modern society as they offer ways not only to accelerate, but also direct chemical reactions [1, 2]. Many chemical processes are industrially performed at high temperatures and pressures in the presence of one or more heterogeneous catalysts. A selection of catalytic solids, including their chemical composition and the related reaction conditions are summarized in Table 1. It is evident that many of these systems, including *e.g.* the Haber–Bosch and Fischer–Tropsch synthesis catalysts, require high reaction temperatures and pressures [1]. Clearly, the catalytic solids involved undergo extreme conditions, which imply that materials should maintain their structure and porosity when reactants and reaction products travel to and from the active sites embedded in the catalyst material. Secondly, the solid catalysts themselves are not structurally and chemically inert and will also be susceptible to chemical reactions as they do not exist in their thermodynamically most stable form. This is even enhanced by the fact that the reaction products, which is in the case of *e.g.* Fischer–Tropsch synthesis next to long-hydrocarbons also water, may react themselves with the catalytic material [3]. In other words, the catalytic solids themselves are in a metastable state, and will undergo a kind of corrosion process as the reaction conditions imposed are extremely often severe. Interestingly, in the case of the Phillips catalyst the continuous generation of porosity and related cracks is a necessity to make the continuous growth of polyethylene chains possible [4]. Polymerization catalysis leads to the break up of an active catalyst particle, followed by the creation of new active sites and the further growth and generation of polymer chains. As a result, a sort of cauliflower is formed with a size an order of magnitude larger than the original polymerization catalyst particle.

It is these dynamics in catalyst composition and structure, which makes that there are clear parallels to be made with nature. Well-known are the occurrence of minerals with chemical compositions very close to those used in catalyst compositions. Prototype materials are zeolites, which form the basis of the current technology to crack crude oil fractions into fuel components, such as gasoline, in the so-called fluid catalytic cracking process [5]. Natural zeolites are often found near volcanoes illustrating the requirement of temperatures and pressures to form these microporous solids [6]. However, the analogy with nature is even more evident when comparing the reaction conditions employed in industrial reactors and those found near volcanoes and hydrothermal vents in deep-sea oceans (Table 1). It is important to realize that our current understanding of heterogeneous catalysis under extreme conditions is still very limited, mainly due to the fact that our tools to observe these processes are not yet well developed [7]. Furthermore, the relationships with chemical processes taking place in nature are also very limited. For instance, we do not know so much about chemistry taking place at deep-sea hydrothermal environments. Interestingly, we are aware that nature has developed systems to harvest geothermal and geochemical energy to sustain the action and function of catalytic systems needed for life. On the other hand, how nature makes use of these temperature and chemical gradients induced by *e.g.* volcanoes and hydrothermal vents remains largely unknown. Clearly, both fields of research could inspire each other, and also the exchange of experimental and theoretical tools between these fields may increase our fundamental knowledge on these chemical processes taking place under extreme conditions.

Table 1. Overview of the catalyst composition and reaction conditions of various high pressure and high temperature industrial chemical processes and their comparison with those found in nature.

Catalytic processes	Chemical Composition and Reaction conditions		
Process	Chemical composition	Temperature (°C)	Pressure (atm)
Fischer–Tropsch synthesis	Fe/Si/O	250	30
Phillips catalysis	Cr/Si/O	70–100	30–40
Hydrodesulphurisation	Co/Mo/S/Al/O	300–400	50–150
Haber–Bosch process	Fe/K/O	400–500	200–250
Methanol synthesis	Cu/Zn/O	200–300	50–100
Aqueous Phase Reforming of Biomass	Pt/Al/O	200–300	25–50
Natural processes	Chemical composition	Temperature (°C)	Pressure (atm)
Earth's crust	O/Si/Al/Fe/ Ca/Na/K/Mg	200–600	20–400
Hydrothermal vents	Si/O/Fe/Mn/Ca/S	400	250

Heterogeneous catalysis is not only conducted at the gas/solid interphase, but also at the liquid/solid interface. Examples of the latter include biomass conversion reactions in the presence of *e.g.* water or ethanol/methanol [8]; but also more recently electrocatalysis is a growing field of research where the stability of catalyst materials is of great importance [9]. People have noted that catalyst compositions, designed for the processing of *e.g.* apolar molecules, such as those typically found in crude oil fractions, are not fully compatible with those used in catalytic biomass conversion. Here it is noted that hydrothermal conditions make that zeolite-based catalysts have to be stabilized. Similarly, for electrocatalysis the materials are put at rather extreme pH and temperatures. Also here, the analogy can be made between heterogeneous catalysis and nature. More specifically, one can think of hydrothermal vents near underwater volcanic mountains. Here, the conditions can be extreme as at 2500 m below the sea pressures of 250 atm and temperatures up to 400°C can be reached. This implies that minerals in the presence of organic matter may also start to catalyze chemical processes leading to the formation of chemicals, including light gases. In addition, the solids may undergo structural and chemical changes.

Contributions in this session to catalysis under extreme conditions

The presentations in this session have been centered around the following three topics and the invited speakers have been asked to deliver their viewpoints on the most recent developments in these fields of research. Their respective contributions are published in the same Proceedings of this Solvay Conference on Chemistry.

Watching catalysts in action under extreme conditions

This topic has been discussed in detail in the contributions of Dr. Stig Helveg (Haldor Topsoe, Denmark) and Prof. Bert Weckhuysen (Utrecht University, the Netherlands). Dr. Helveg, a senior research scientist at Haldor Topsøe, is specialized in the use of advanced transmission electron microscopy (TEM) techniques to problems related to energy and the environment with special emphasis on catalytic solids. Helveg has outlined in his contribution the progress in transmission electron microscopy, the development of proper *in situ* reaction cells, which offers new possibilities to address the role of gas surface interactions on catalysts in the working state. Several showcases have been discussed including one in which Electron Energy Loss Spectroscopy (EELS) has been combined with TEM. Prof. Weckhuysen, professor of inorganic chemistry and catalysis at the Faculty of Science at Utrecht University, has been pioneering *in situ* and operando spectroscopy and microscopy methods for investigating catalytic solids under high temperature and pressure, revealing detailed insight in their working principles and deactivation. The focus of his talk was on visualizing both the organic and inorganic worlds of the concept of chemical reactivity. More specifically, how can we understand the complexity and dynamics of a heterogeneous catalyst under extreme reaction conditions? To achieve such insights it is necessary to develop the tools for recording a real-life molecular movie of a catalytic solid at work. The challenges and future prospects have been discussed.

Why high activation temperatures and why high temperature processes in numerous heterogeneous catalysts?

This topic will be discussed by Prof. Xinhe Bao (Dalian Institute for Chemical Physics, China) and Prof. Christophe Copéret (ETH Zurich, Switzerland). Prof. Bao has devoted his research to the surface chemistry and dynamics of catalytic solids, with the emphasis on the development and employment of nanotechnologies and *in situ* analytical techniques in order to understand the nature of catalysis. Xinhe Bao's recent work has largely contributed to single atom catalysis, which has lead among other discoveries to new high-temperature processes and related catalytic solids, based on *e.g.* isolated iron in an oxide, for methane activation, which has no or a limited amount of carbon deposits. The scientific interest of Prof. Copéret lies at the frontiers of molecular, material and surface chemistry, with the aim to design functional materials with potential applications in the field of heterogeneous catalysis. The lecture of Prof. Copéret focused on how we can get ultimate control over the active sites of a catalytic solid by fundamentally understanding how the local environment of a surface influences the molecular structure and reactivity of an active species. For this purpose, he has made use of *e.g.* olefin metathesis and polymerization catalysts.

The extreme conditions of electrocatalysis and electrocatalysis under extreme conditions

Two contributions have been focused on this topic; the first lecture was given by Prof. Marc Koper (Leiden University, The Netherlands), while the second one was given by Dr. Ryuhei Nakamura (RIKEN, Japan). Prof. Koper, professor at the faculty of science of the Leiden Institute of Chemistry of Leiden University, studies electrochemical interfaces using a surface science approach, combining electrochemical techniques, *in situ* spectroscopy and imaging, theory, and computational methods, with an emphasis on their electro-catalytic properties. Electrochemical reactions of interest include those involved in the hydrogen-oxygen cycle, the carbon cycle (such as carbon dioxide reduction) and the nitrogen cycle. One specific aspect of electrochemistry that makes it "extreme" compared to other fields of catalysis is the convenience with which highly reducing or oxidizing conditions can be imposed. Take, for example, a hydrogen-oxygen electrochemical cell, in which water is converted to hydrogen and oxygen gas. Under equilibrium conditions, the cell potential is given by the Nernst equation:

$$E_{cell} = \Delta E^0 + \frac{RT}{4F} \ln \left(p_{H_2}^2 p_{O_2} \right)$$

where $\Delta E^0 = 1.23$ V, and the other symbols have their usual meaning. This simple equation indicates that every 30 mV change in cell potential corresponds to one order of magnitude change in (corresponding) hydrogen pressure. Conversely, changing the cell potential by a mere 300 mV would correspond to ten orders of magnitude in hydrogen pressure. The invited lecture by Prof. Koper provided some examples of the surprising chemistry that can happen under these "extreme" conditions. The final contribution of this session was given by Dr. Nakamura, a research scientist at the RIKEN Center for Sustainable Resource Science in Japan. He is making major achievements in the development of biologically inspired catalysts and their application to energy conversion and production systems. Specifically, his group has attempted to exploit the nature's ingenuities of multi-electron catalytic reactions, electron/proton transport, metabolic regulation, flexible response to external stimuli, and the robust energy management in a deep sea environment. It was this topic he explicitly addressed in his invited lecture, as also compiled in the Proceedings of this Solvay Conference on Chemistry.

Outlook to future developments of research on catalysis under extreme conditions

Research in this field should be directed in the following directions:

1. Can we find inspiration from the high-temperature and –pressure chemical processes taking place in geological sediments and near deep sea hydrothermal vents to design new catalyst materials?
2. How has nature developed its strategies to make new porous materials under these extreme conditions of high temperatures and pressures in *e.g.* the presence of water?
3. Is catalytic chemistry in large-scale industrial reactor vessels similar to the chemistry observed in these natural environments?
4. Can we adopt experimental and theoretical methodologies from the field of geochemistry to accelerate discoveries in the field of heterogeneous catalysis and vice versa?
5. Can we make and test porous catalytic solids with a chemical composition and stability similar to those found in nature? Are these materials better or worse than those currently under development for *e.g.* biomass conversion processes and electrocatalysis?

This session could be a start to stimulate discussions to gather inspiration from nature to advance the field of catalysis under extreme conditions.

Acknowledgments

The author thanks Prof. B. M. Weckhuysen for numerous illuminating discussions and his help with writing this note.

References

[1] (a) G. Ertl, H. Knözinger, F. Schüth, J. Weitkamp (eds.), *Handbook of Heterogeneous Catalysis*, 2nd Edition (Wiley-VCH, 2008); (b) J. Hagen, *Industrial Catalysis: A*

Practical Approach, 2nd Edition (Wiley-VCH, 2006); (c) M. Beller, A. Renken, R. A. van Santen (eds.), *Catalysis: From Principles to Applications* (Wiley-VCH, 2012).

[2] (a) G. Ertl, *Angew. Chem. Int. Ed.* **47**, 3524 (2008); (b) G. A. Somorjai, Y. Li, *Introduction to Surface Chemistry and Catalysis*, 2nd Edition (Wiley, 2010).

[3] (a) A. Y. Khodakov, W. Chu, P. Fongarland, *Chem. Rev.* **107**, 1692 (2007); (b) E. de Smit, B. M. Weckhuysen, *Chem. Soc. Rev.* **37**, 2758 (2008).

[4] M. P. McDaniel, *Advances in Catalysis*, Vol. 53, B. C. Gates, H. Knözinger (eds.), 123 (Academic Press, 2010).

[5] (a) E. T. C. Vogt, B. M. Weckhuysen, *Chem. Soc. Rev.* **44**, 7342 (2015); (b) E. T. C. Vogt, G. T. Whiting, A. Dutta Chowdhury, B. M. Weckhuysen, *Advances in Catalysis*, Vol. 58, F. C. Jentoft (ed.), 143 (Academic Press, 2015).

[6] G. Gottardi, E. Galli, *Natural Zeolites* (Springer, 1985).

[7] B. M. Weckhuysen, *Chem. Soc. Rev.* **39**, 4557 (2010).

[8] A. Hornung (ed.), *Transformation of Biomass: Theory to Practice* (Wiley, 2014).

[9] C. F. Zinola (ed.), *Electrocatalysis: Computational, Experimental and Industrial Aspects* (CRC Press, 2010).

ELECTRON MICROSCOPY IN HETEROGENEOUS CATALYSIS

STIG HELVEG
Haldor Topsoe A/S, DK-2800 Kgs. Lyngby, Denmark

My view of the present state of research on characterization in catalysis

The introduction of *in situ* and *operando* characterization techniques has had a vast impact on the understanding and developments in heterogeneous catalysis. Such techniques have shown that heterogeneous catalysts can undergo dynamic transformations upon changes in the reaction environment and that the alterations can have a pronounced effect on the catalytic performances. This realization implies that the active state of a catalyst may only exist during the catalytic process and therefore emphasizes the need for *in situ* and *operando* studies to obtain relevant structural and chemical information [1]. So far, *in situ* studies have mainly employed photon-based techniques that operate in *e.g.* the infrared, visible, X-ray and γ-ray region and that average information over a macroscopic scale being far larger than the extension of the catalytic active phases. Complementary observations of the catalysts at atomic resolution and sensitivity have not been available under similar reaction conditions, and this limited our understanding and challenged our imagination of the structure and dynamics of the catalyst surface. As a consequence, it has been difficult to directly elucidate the role of the gas-surface interactions on the working catalysts and to directly relate information derived from simpler and well-defined model systems to heterogeneous catalysis — a challenge which has been referred to as the pressure and materials gaps in catalysis and surface science.

My recent research contributions to electron microscopy in catalysis

In recent years, new possibilities for addressing this fundamental challenge in catalysis research have opened up by the substantial progress made in transmission electron microscopy (TEM). Advancements in electron optics and detection methods have made TEM techniques capable of delivering images and spectra of matter, ultimately, at atomic resolution and sensitivity. Hereby, TEM enables detailed visualizations of small clusters, surfaces and, in some cases, even molecular species. Figure 1 demonstrates this capability by observations of a hydrotreating catalyst consisting of Co-promoted MoS_2 nanocrystals dispersed on a carbon support. The images and spectra reveal directly the atomic structure and stoichiometry of a

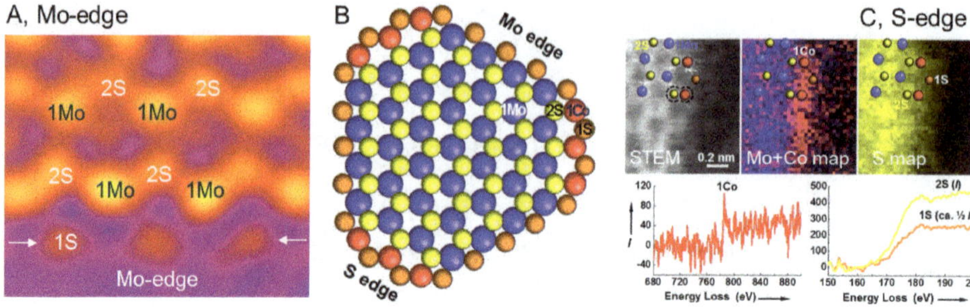

Fig. 1. Edge structures of graphite-supported Co-Mo-S nanocrystals. (A) High-resolution scanning-TEM image of a Mo-edge of a single-layer MoS_2 nanocrystal, oriented with the [001] direction along the electron beam direction. The image is adapted from Ref. [3] and processed as described therein. (B) Ball-model of a Co-Mo-S nanocrystal derived from a quantitative intensity analysis of the images and element maps in (A, C). (C) High-resolution scanning-TEM image of the S-edge of another single-layer MoS_2 nanocrystal and the corresponding Mo, Co and S element map derived from electron energy loss (EEL) spectroscopy. The EEL spectra show an integrated Co $L_{2,3}$ ionization edge, representing the sum of six 1Co positions along the S-edge, and the S $L_{2,3}$ ionization edges of a 2S and 1S atomic column. The data are adapted from Ref. [4] and processed as described therein.

single-layer MoS_2 nanocrystal and provide a profile-view of the surfaces (edges) that are commonly considered as catalytic active sites [2–5]. In fact, interplay with previous *in situ* X-ray absorption spectroscopy data and an extended series of model studies confirmed the observations as representative for the catalytic important Co-Mo-S phase in hydrotreating catalysis [4, 5]. The example therefore demonstrates that complex nanostructures and surfaces, even at atomic dispersions, may now be examined on an atom-by-atom basis.

While electron microscopy is mainly conducted under the high vacuum conditions, instrumental advancements have also focused on functionalizing electron microscopes for *in situ* and *operando* studies. The introduction of reactive gas environments can be accomplished by means of differentially pumped vacuum systems and closed, electron-transparent Si-based devices. Such miniaturized reactors have opened up for TEM examinations of catalysts during exposure to reactive gas environments at pressures of up to the atmospheric level and temperatures of up to several hundred degrees Celsius, even while maintaining atomic-resolution imaging capabilities [6–8]. Under these reaction conditions, time-resolved TEM observations have led to surprisingly new insight into many dynamic phenomena in heterogeneous catalysis and shown that the catalyst surface generally is far more dynamic than hitherto anticipated [9]. In fact, the atomic-resolution TEM images of a catalyst may be correlated with concurrent measurements of its functionality, as shown recently for the CO oxidation over Pt nanocrystals [10]. Figure 2 displays such a correlation as the reaction undergoes oscillatory behavior and shows that a reversible shape transformation of the Pt nanocrystals is synchronized with a periodic variation in their catalytic activity. Such observations were beneficially combined

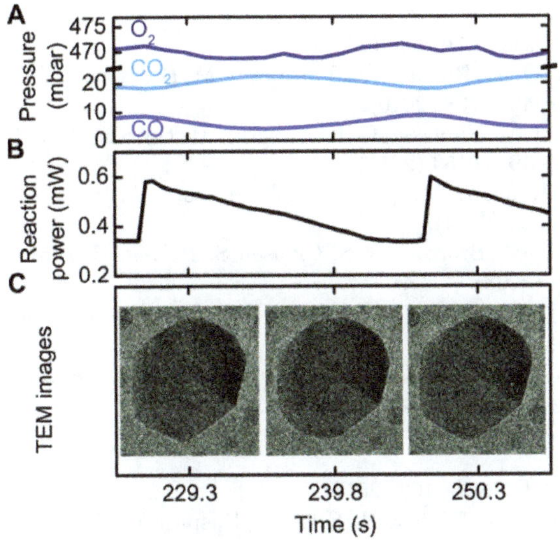

Fig. 2. Correlation of time-resolved reaction data for the oscillatory CO oxidation with simultaneous TEM images of a Pt nanocrystal obtained using a Si-based nanoreactor. (A) Mass spectrometry of the CO, O_2 and CO_2 pressures, (B) reaction power data, and (C) TEM images of a Pt nanocrystal as a function of time. The gas entering the reaction zone is 1.0 bar of CO:O_2:He = 3%:42%:55% and the nanoreactor temperature is 659 K. The data are adapted from Ref. [10] and processed as described therein.

with reactivity data, calculated from simpler model systems, and a mass-transport description of the reactor to demonstrate the refacetting of the nanocrystals as a mechanism that causes reaction oscillations. Thus, the example demonstrates that electron microscopy provides dynamic information about catalyst surfaces and enables the molecular-level description of dynamic properties and functions in catalysis to be extended with information about the exposed surface sites.

Outlook for future developments of electron microscopy research in catalysis

The advances in electron microscopy instrumentation and methodologies have opened up new opportunities for "*live*" observations of catalysts that are beneficial for the understanding of their dynamic properties and functioning at the atomic-level. However, electron microscopy is by no means trivial due to the strong interaction of the electron-beam with matter. It is therefore crucial to exercise control over the electron illumination, *e.g.*, by operating at variable electron energies and dose-rates, in order to suppress beam-induced alterations of the catalyst and reaction environment and to ensure that chemical relevant insight is derived [8, 9]. Further advancements of electron illumination and detection schemes may, in the future, expand the electron microscopy capabilities for visualizing molecular species and their turn-overs at individual sites.

References

[1] H. Topsøe, *J. Catal.* **216**, 155 (2003).
[2] C. Kisielowski, Q. M. Ramasse, L. P. Hansen, M. Brorson, A. Carlsson *et al.*, *Angew. Chem. Int. Ed.* **49**, 2708 (2010).
[3] L. P. Hansen, Q. M. Ramasse, C. Kisielowski, M. Brorson, E. Johnson *et al.*, *Angew. Chem. Int. Ed.* **50**, 10153 (2011).
[4] Y. Zhu, Q. M. Ramasse, M. Brorson, P. G. Moses, L. P. Hansen *et al.*, *Angew. Chem. Int. Ed.* **53**, 10723 (2014).
[5] F. Besenbacher, M. Brorson, B. S. Clausen S. Helveg, B. Hinnemann *et al.*, *Catal. Today* **130**, 86 (2008).
[6] S. Helveg, C. López-Cartes, J. Sehested, P. L. Hansen, B. S. Clausen *et al.*, *Nature* **427**, 426 (2004).
[7] J. F. Creemer, S. Helveg, G. H. Hoveling, S. Ullmann, A. M. Molenbroek *et al.*, *Ultramicroscopy* **108**, 993 (2008).
[8] S. Helveg, C. F. Kisielowski, J. R. Jinschek, P. Specht, G. Yuan *et al.*, *Micron* **68**, 176 (2014).
[9] S. Helveg, *J. Catal.* **328**, 102 (2015).
[10] S. B. Vendelbo, C. F. Elkjær, H. Falsig, I. Puspitasari, P. Dona *et al.*, *Nat. Mater.* **13**, 884 (2014).

OPERANDO SPECTROSCOPY OF A CATALYTIC SOLID: TOWARDS A MOLECULAR MOVIE

BERT M. WECKHUYSEN

Debye Institute for Nanomaterials Science, Utrecht University, 3584 CG Utrecht, The Netherlands

My view of the present state of research on catalyst characterization

Catalysts are the workhorses of chemical industry: more than 80% of chemicals have come into contact with at least one catalyst material during their manufacturing process [1]. Catalysts are also key to develop a more sustainable society; for example chemical processes should become less dependent on fossil resources, such as crude oil [2].

Catalytic solids, especially those used in practical applications, can also be very complex: solids with high surface areas, in combination with promotor elements and binder materials, for example, possess many different potential active sites in their structure, making it in most cases difficult to unambiguously identify the active site that actually does the job [1]. Studies have also shown that catalytic solids are highly dynamic and act almost like chameleons, changing their surface structure under reaction conditions as a function of both space and time [3–6].

Figure 1(a) illustrates the behavior and dynamics of a catalytic solid. The schematic highlights the different physicochemical processes taking place, while *e.g.* an Fe-based Fischer–Tropsch synthesis catalyst is placed in a reactor vessel in the presence of CO and H_2 at high pressures and elevated temperatures (*e.g.* 300°C and 20 bar) producing long-chain hydrocarbons. To illustrate that this picture can also be captured by modern characterization methods, a 3D chemical image of a real-life Fischer–Tropsch catalyst particle of the dimensions of $\sim$20–30 μm is shown in Fig. 1(b) [4]. Scientists want to understand the intricate details of the gas-solid-liquid processes, leading to the formation of the wanted end products, which are in the case of Fischer–Tropsch catalysis either diesel-type molecules or light olefins. However, also important surface and solid-state processes are taking place in the catalytic solid, leading to metal-support and promoter/poison effects, often directly influencing activity, selectivity, and stability of the catalyst. It is this type of chemistry that after more than one century of detailed spectroscopic and microscopic investigations still provides inspiration to many academic and industrial scientists in the field of heterogeneous catalysis. One of the dreams is to make a molecular movie of catalytic solids down to the level of a single particle with nanometer

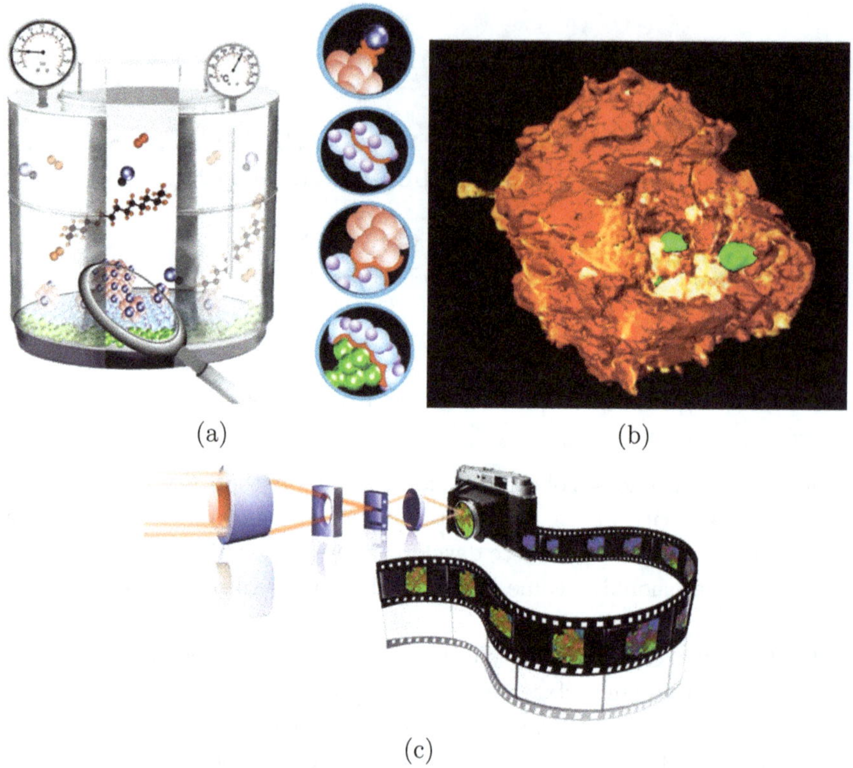

Fig. 1. (a) Physicochemical processes taking place in a catalytic solid. Showcase is a Fischer–Tropsch synthesis catalyst particle, of which a 3D picture has been measured with X-ray nano-tomography (b) red, Fe; green, Zn promotor; and Ti promotor, yellow). Gas-liquid-solid surface processes are happening, but also solid-solid interactions, which often determine the overall selectivity and stability of a catalyst. To answer questions, such as "What is the precise role of a support oxide as well as additional oxide phases, which may act either as a promoter or a poison in catalysis?" we need analytical tools (c) capable of recording a molecular movie of catalytic solids down to the level of a single particle with nanometer resolution. (b is reproduced from Ref. [4]).

resolution and single molecule sensitivity under real-life reaction conditions (i.e., elevated temperatures and pressures) [5]. This ambition is graphically illustrated in Fig. 1(c), in which snapshots of a working catalytic solid are taken by a transmission X-ray microscope equipped with optics focusing the hard X-rays down to the nanometer level.

This invited article discusses the recent strides and future perspectives of the field of making snapshot pictures of catalysts while they are working. This perspective aims to highlight some of the recent developments in this field of research and provides some outlook of what may become feasible in the foreseeable future. By no means this article intends to be a comprehensive review and consequently several of the described trends are only highlighted making use of examples from my own research group.

Catalyst dynamics and operando spectroscopy

From the above considerations it is evident that determining the active site in real-life heterogeneous catalysts and elucidating their reaction mechanism remains an intellectual challenge. It is of paramount importance for new roads towards the rational design of catalysts [3–6]. This road will offer clear prospects for improved formulations of existing catalyst materials, but also the possibility to create more effective and selective catalytic solids from scratch. It is important to realize that the latter will be the most intellectually rewarding. Unfortunately, in most cases rational catalyst design remains a pipe dream: as of today the experimental tools available for monitoring catalysts as they work are still, in the main, too rudimentary [7].

About 15 years ago Professor Eric Gaigneaux (Belgium), Dr. Gerhard Mestl (Germany), Professor Miguel Banares (Spain) and myself have launched the "operando spectroscopy" methodology for doing catalysis research [8–10]. We felt that spectroscopic as well as microscopic approaches should be (further) developed so that we could really tell how an operating catalyst material was working. This required not only to perform spectroscopic measurements under realistic reaction conditions, but also that the catalyst performance of the material was unambiguously determined by *e.g.* on-line gas chromatography or mass spectrometry. Although the name of operando spectroscopy was new in the years 2002–2003, the conceptual ideas behind this experimental approach were already explored by some research groups in the years before. Examples are the development of infrared [11] and UV-Vis [12] spectroscopy in combination with on-line gas product analysis, the latter being used to determine the active site for the dehydrogenation of light alkanes over supported metal oxide catalysts and explaining why alumina-supported systems are more active than silica-supported ones [12]. In the years after its introduction, many research groups have explored the capabilities of the operando spectroscopy approach, leading to various powerful experimental methods, often in a combined fashion, to investigate catalytic solids at work. A testimony of all past and more recent research activities on operando spectroscopy can be found in the proceedings of the past five editions of the *International Conference on Operando Spectroscopy*, held subsequently in Lunteren (The Netherlands, 2003), Toledo (Spain, 2006), Rostock-Warnemünde (Germany, 2009), Brookhaven (USA, 2012) and Deauville (France, 2015).

Clearly, the experimental approaches developed in the past 15 years are now becoming more or less standard characterization tools, often first constructed in beamline hutches of synchrotron radiation sources [13, 14], but nowadays also based in a normal academic laboratory [15, 16]. One of these powerful set-ups is the combination of X-ray diffraction with optical methods, such as Raman spectroscopy, with on-line performance measurements [15]. Figure 2 illustrates this set-up for the case of a 15 wt% Co/TiO_2 Fischer–Tropsch Synthesis catalyst operating at 10 bar and 250°C. A 3D plot of operando Raman spectra of the catalyst, as a function

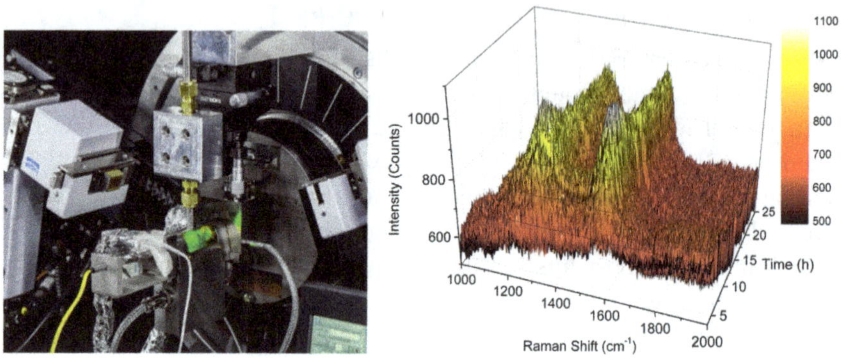

(a) (b)

Fig. 2. (a) Photograph of the laboratory-based combined operando X-ray diffraction/Raman spectroscopy/on-line GC set-up for monitoring the phase changes in a catalytic solid at elevated temperatures and pressures, including activity measurements providing direct structure-activity relationships. Crux in the set-up is the use of a relatively high-energy MoKα radiation source. (b) Showcase example is a 15 wt% Co/TiO_2 Fischer–Tropsch Synthesis (FTS) catalyst at 10 bar and 250°C, of which a 3D plot of operando Raman spectra as a function of time during the first 25 h of FTS reaction is included in the figure. The spectra are dominated by the so-called D and G bands, located at ~1350 and 1600 cm^{-1}, due to hydrocarbon deposits formed during reaction. Reproduced from Ref. [15].

of time during the first 25 h of Fischer–Tropsch synthesis reaction, is included in Fig. 2. The advantage of a laboratory-based set-up is that very long experiments can be conducted making it possible to investigate the activation and deactivation of catalytic materials. Such operando studies elegantly demonstrate how the catalyst structure changes when altering the reaction temperature and pressure, as well as the gas environment (*e.g.* CO/H_2 ratio).

Although in the past decade many experimental set-ups for high-pressure and high-temperature characterization of catalytic solids have been developed for gas-phase reactions, the number of set-ups and related studies for liquid-phase processes are much more limited [17]. One of the reasons is that investigations in the liquid-phase are more complicated because of the interference with solvents, making it difficult if not simply impossible to observe the chemistry, taking place at the catalyst surface. Many efforts are currently under way to make sure that techniques become more sensitive to the chemical processes taking place at catalyst surfaces. Figure 3 gives an illustration of two set-ups able to capture the catalytic chemistry of a liquid-phase process converting lignin, one of the key building blocks of biomass, into valuable chemical building blocs, namely aromatics [18].

Control over catalyst heterogeneities

In the past years it has been shown that by using complementary spectroscopy and microscopy methods valuable structure-activity data can be obtained at the

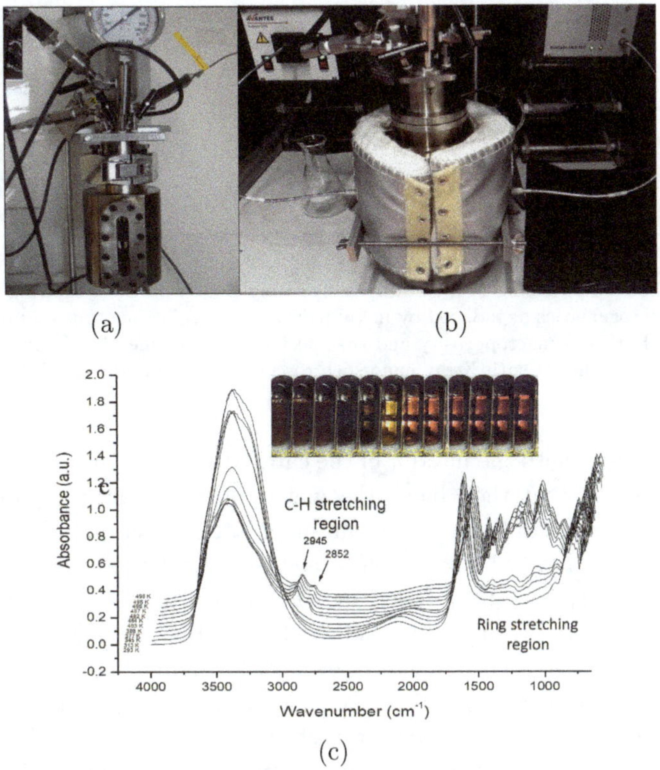

Fig. 3. (a) Photograph of a high-pressure liquid phase reactor for performing catalytic experiments on *e.g.* biomass conversion processes; equipped with an optical window, which allows both visual inspection as spectroscopic analysis of what is happening during catalysis at elevated temperatures and pressures. (b) Photography of the same reactor set-up but now connected to an optical fiber UV-Vis spectrometer. (c) Operando IR spectra measured with an Attenuated Total Reflection Infrared (ATR-IR) reactor set-up in a high-pressure liquid phase reactor for monitoring the catalytic decomposition of lignin in aromatic hydrocarbons and hydrogen in a so-called aqueous phase reforming (APR) process. The catalyst is a Pt/Al_2O_3 system, while the reaction temperature and pressure are respectively 29 atm and 225°C. Reproduced from Ref. [18].

level of a catalytic reactor, even up to the level of a pilot-scale system, mm-sized catalyst extrudates, micron-sized zeolite grains and nanometer-sized metal particles, often even under realistic reaction conditions. Given the pace analytical methods develop a lot of progress can be expected in the years to come. Interestingly, many of these studies have clearly provided evidence for the existence of heterogeneities in and between catalyst particles down to the level of single molecules and atoms [5]. Figure 4 illustrates as a showcase the existence of these intra- and inter-particle heterogeneities as observed for an ageing Fluid Catalytic Cracking (FCC) catalyst [19]. X-ray nano-tomography has been able to demonstrate that the metal poisons, such as Fe and Ni, are non-homogeneously demonstrated in the FCC particle and are preferentially located at the outer surface of the particle. Upon ageing the metal

Fig. 4. Snapshots of an ageing Fluid Catalytic Cracking (FCC) particle, showing the intra- and interparticle heterogeneities as induced by metal poisons present in the crude oil processed by the FCC particles. Both the macroporosity and the speciation of two metal poisons, namely Fe and Ni, have been probed in 3D with X-ray nano-spectroscopy. Reproduced from Ref. [19].

poisons move deeper into the interior of the catalyst particle, and one of their detrimental functions is that they block the macropores preventing hydrocarbons to enter the catalyst material. They also can induce agglutination of FCC particles.

The question is now whether heterogeneities are (always) disadvantageous to heterogeneous catalysis or not; and if so, can we get full control over the heterogeneities found in *e.g.* a catalyst bed and catalyst particle to make an overall better catalytic process or design a completely new one. Such important question can only be addressed by performing a multi-scale characterization approach, which is outlined for a working Co/TiO_2 Fischer–Tropsch synthesis catalyst in Fig. 5 [20]. Here it remains unanswered whether effects on the level of atoms and molecules, i.e. at the nanoscale, in a catalyst particle can be translated to differences observed in the reactor bed, i.e. at the macroscale. Ultimately these observations at all relevant lengths scales need to be linked to catalyst performance, including stability. Unfortunately, there is no experimental approach yet available which can couple all this information together in one set-up; and therefore we cannot yet reliable link all the data obtained and develop space-time resolved structure-performance relationships. Furthermore, we often have methods, which are sensitive towards the organic and the inorganic part of a catalytic process; but it is essential to acquire both "sides" of the "same" catalytic event to make proper — preferably quantitative — relationships between structure and function possible.

Indeed, we wish not only to know how a catalyst particle looks like when residing in a reactor bed, including its 3D chemical composition, structure and porosity, but we would also like to see how a feedstock molecule enters the pores of a catalyst particle. Moreover, we want to understand how this molecule further interacts with the catalyst surface, which obviously depends on pressure and temperature, but also on pore size and shape. In other words, we want to obtain a "true" molecular movie of a molecule: how it travels through a catalyst particle until it reacts with the active site, then follow the reaction products that travel further towards the outside of the catalyst particle, and finally how they pass through the catalyst bed. Such a molecular movie is still a dream for scientists in our

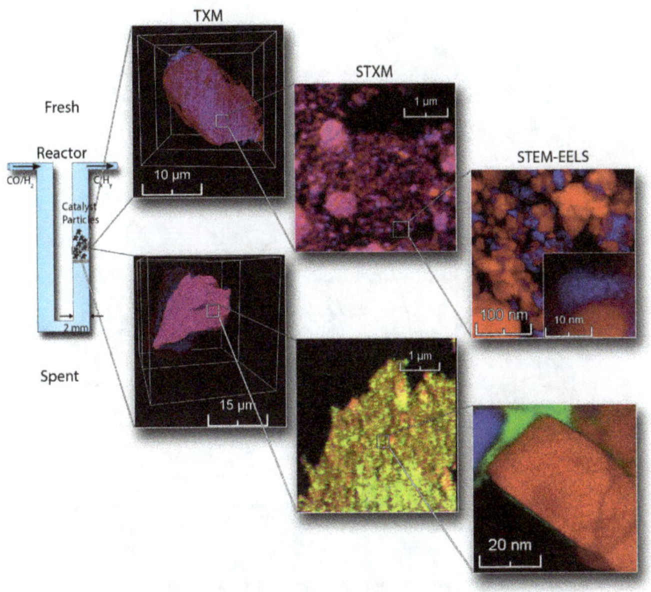

Fig. 5. Schematic of the multi-scale chemical imaging approach in the field of heterogeneous catalysis, taking Co-based Fischer–Tropsch Synthesis (FTS) catalysis as an example. (top) Fresh and spent Co/TiO_2 FTS catalysts were imaged by 3D Transmission X-ray Microscopy (TXM, voxel size: ~50 nm), Scanning Transmission X-ray Microscopy (STXM, spatial resolution: 30 nm), and Scanning Transmission Electron Microscopy–Electron Energy Loss Spectroscopy (STEM–EELS, spatial resolution: 0.5 nm). In TXM, TiO_2 is shown in red, Co is shown in blue. In STXM, TiO_2 is shown in red, Co_3O_4 is shown in blue and Co^{2+} is shown in green. In STEM-EELS, TiO_2 is shown in red, Co is shown in blue and carbon is shown in green. Reproduced from Ref. [20].

field. Certainly obtaining such a motion picture requires cooperation with scientists from other fields, including people active in mass transport simulations and fluid dynamics. Furthermore, when one wishes to make such movies we will have to have access to reliable micro-reactor systems, which allow performing for example single molecule-single particle-single droplet experiments. This micro-reactor technology platform also provides us with the possibility to investigate in detail interparticle heterogeneities and offers means to evaluate whether heterogeneities are essential or detrimental to catalyst performance.

My recent research contributions to catalyst characterization

X-ray nano-tomography of a single catalyst particle

We have developed in recent years a correlative approach combining 3D X-ray imaging techniques at different length scales for the analysis of metal poisoning of an individual catalyst particle [21].

The correlative nature of the characterization data obtained allowed establishing a macro-pore network model that interprets the accumulations of the metal poisons

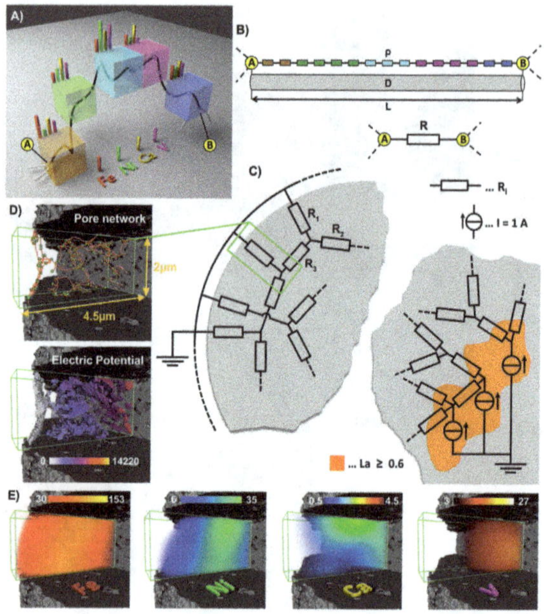

Fig. 6. Establishing the resistor network from data acquired with 3D μ-XRF and nano-TXM tomography. (A) Schematic of a pore channel (black) connecting nodes A and B that represents the pore space passing through 5 volume units with distinct elemental concentrations. (B) In analogy to a conducting wire this connection can be represented by its length L, the average pore channel cross section D, and a specific resistivity ρ, defined as one plus the summed average concentrations of poisoning metals along the pore channel. (C) Every pore channel of the studied FCC particle can be expressed by a resistor with a resistivity defined as $R = \rho \times L / D$. Then nodes of the pore network located in regions of elevated La concentrations were defined as current sources of 1 A, while entry nodes at the particle surface were set to ground potential (0 V). (D) An arbitrarily selected sub-volume located near the surface of the particle and the corresponding pseudo-electric potential (in Volts) as calculated using nodal analysis of the established virtual resistor network. (E) The poisoning metals' μ-XRF signal over the same sub-volume, which were used to determine the specific resistance as schematically depicted in (A) and (B) (color scales report the μ-XRF counts). Reproduced from Ref. [21].

as a resistance to mass transport and can, by tuning the effect of metal poison deposition, simulate the response of the network to a virtual ageing of the catalyst particle. Figure 6 illustrates the developed approach for an individual FCC particle, which has been (partially) deactivated by the metal poisons Fe, Ni, V and Ca [21]. The combination of data obtained from nano-TXM and μ-XRF (X-ray fluorescence) tomography on the same catalyst particle allowed an assessment of the degree of pore blockage or "resistance" that the metal poisons pose to mass transport through the pore network. First, the pore space was skeletonized, resulting in a set of points and connecting cylinders with diameters that correspond to the determined pore diameters. Then, using La as a marker for the embedded zeolite domains, regions of elevated La XRF intensities were defined as the active zones that should be connected to the exterior of the particle. Therefore, all nodes of the pore network located within these zeolite domains were labeled as "source nodes," while all nodes at the surface of the FCC particle were labeled 'sink nodes'. In order to assess how easily molecules can travel between source and sink nodes we used the analogy of an electrical resistor network. Figure 6 shows schematically how this pseudo-resistor network was established based on the measured macro-pore network and the elemental concentration distributions of La, Fe, Ni, Ca, and V.

Single molecule fluorescence microscopy of a single catalyst particle

If we wish to record a "true" molecular movie of a catalytic process taking place at high temperatures and pressures, as outlined above, it will be necessary to include a characterization method which also has the sensitivity to visualize single molecules, while they react. One of these potential methods, next to *e.g.* Surface-Enhanced Raman Spectroscopy (SERS) [22], is fluorescence microscopy [23], provided either the reactant or reaction product, or potential reaction intermediates have the required optical properties, making single molecule imaging possible.

In the case of Brønsted acid sites residing within *e.g.* zeolite crystals two reactions have been explored in the past years [24, 25]. The first one, illustrated in Fig. 7 (top), is the oligomerization of furfuryl alcohol, a molecule which is non-fluorescent (OFF) in its monomeric state, while its oligomeric form has a conjugated structure, leading to photon absorption when excited with a visible laser (ON). As a result, the position and reactivity of Brønsted acid sites can be visualized in 3D with nanometer spatial resolution within a zeolite crystal. As a result, it is possible to make 3D surface reactivity maps of *e.g.* a zeolite crystal as a function of the z-direction. This is illustrated in Fig. 7 (bottom) for a zeolite crystal upon steaming. Clearly, there are large heterogeneities in the nanoscale chemical reactivity of Brønsted acid sites present in zeolite ZSM-5 crystals. This indicates that the Al^{3+} sites, responsible for creating Brønsted acid sites in the zeolite framework structure, are non-homogeneously distributed within the zeolite crystals. An observation we have been able to confirm making use of atom probe tomography on specimen isolated from the large ZSM-5 crystals [26]. Furthermore, steaming of the ZSM-5 crystals, especially under mild conditions, lead to an even more inhomogeneous distribution of Brønsted acid sites, explaining the 2D reactivity zoning at *e.g.* the zeolite surface (Fig. 7, bottom, sample H-ZSM-5 MT).

In a follow-up study [25], this work on large zeolite ZSM-5 crystals has been extended to investigate proton-transfer processes at the nanoscale. For this purpose, we have turned our attention to another Brønsted acid-catalyzed reaction, namely the oligomerization of styrene derivatives. The advantage of this probe reaction, as illustrated in Fig. 8 (left), is that distinct dimeric and trimeric fluorescent carbocations can be formed. Both species are characterized by their different photostability and as a result they can be discriminated with single molecule fluorescence microscopy. It was found that the kinetics of the oligomerization of styrene

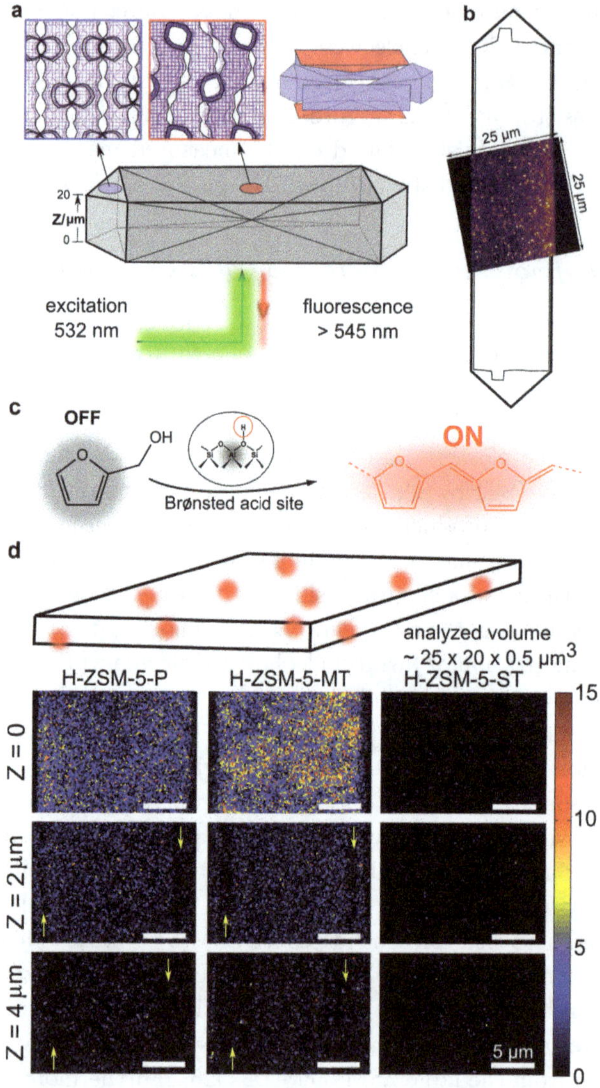

Fig. 7. (Top) Schematic of the single molecule fluorescence microscopy approach used to map in 3D the reactivity of a single ZSM-5 crystal. (a) The intergrowth structure of a large zeolite ZSM-5 crystal indicating the direction of straight and sinusoidal pores in different subunits (color-coded). (b) Accumulated image of individual fluorescent products depicted with respect to the size of the zeolite crystal. (c) The formation of fluorescent products (red) upon the protonation of furfuryl alcohol (black) on a Brønsted acid site. (d) An estimate of the analyzed crystalline volume depicting the 3D distribution of fluorescent molecules (red). (Bottom) Single molecule reactivity maps for a parent crystal (H-ZSM-5-P), a mildly steamed (H-ZSM-5-MT) crystal and severely steamed (H-ZSM-5-ST) crystal, recorded at three different focal depths (Z = 0 (surface), Z = 2 μm and Z = 4 μm). The yellow arrows indicate the regions with lower reactivity due to a different crystallographic orientation of the subunits. Reproduced from Ref. [24].

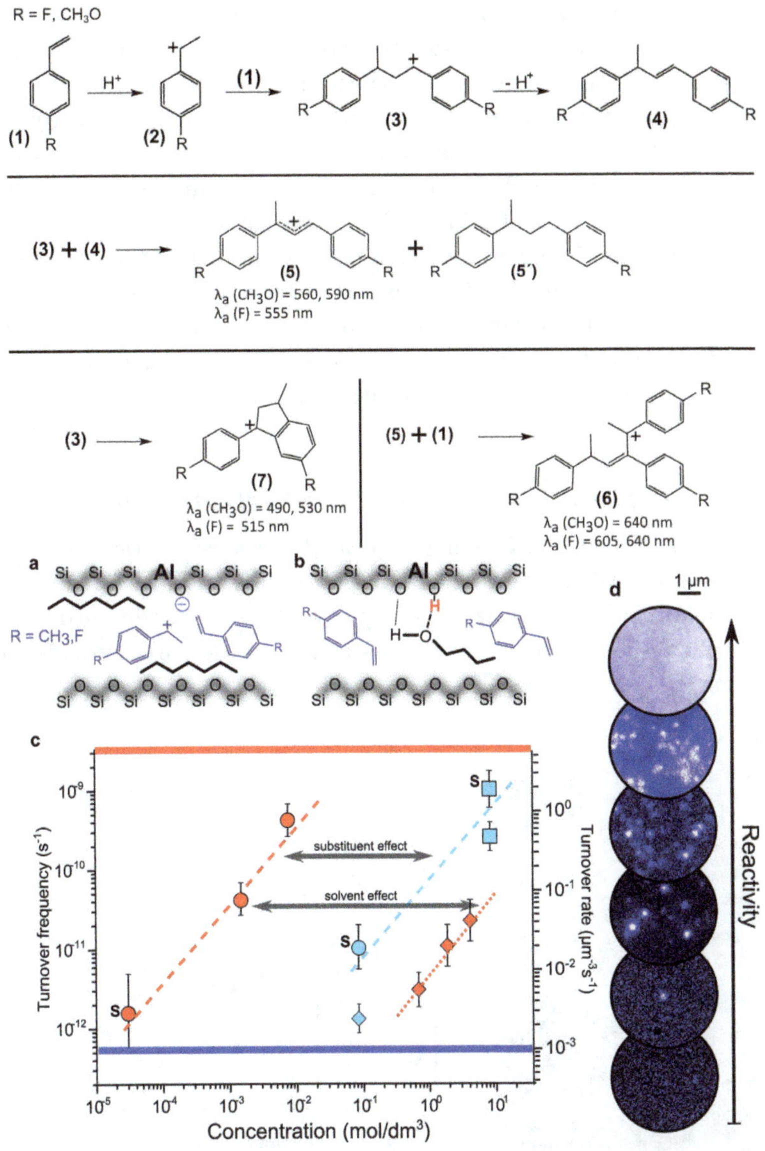

Fig. 8. (Left) Reaction mechanism for the Brønsted acid-catalyzed oligomerization of styrene derivatives, leading to dimeric and trimeric carbocations, which are photo-active, and can be visualized down to the single molecule level with fluorescence microscopy. (Right: a, b) Schematic of the styrene oligomerization reaction in (a) n-heptane, and (b) 1-butanol as solvent. (c) Reactivity of 4-methoxystyrene (in red) and 4- fluorostyrene (in blue) as probe molecules for single molecule chemical imaging in n-heptane (circles) and 1-butanol (rhombi). The squares represent 99% pure 4-fluorostyrene. Averaged turnover frequencies of the parent and steamed zeolite ZSM-5 crystals (denoted as "S") were recorded close to the outer surface of the single crystals. The concentration axis denotes the concentration of the probe molecules in the solvents. The blue and red lines indicate the lowest and highest turnover rates that can be imaged by the presented single molecule approach. (d) Frames indicating the density of the individual emitters and an approximate level of reactivity for turnover frequencies. Reproduced from Ref. [25].

derivatives is very sensitive to the reaction conditions and the presence of the local structural defects in the large zeolite H-ZSM-5 crystals. The remarkably photostable trimeric carbocations were found to be formed predominantly near defect-rich crystalline regions in the zeolite ZSM-5 crystals. Even more interestingly, as shown in Fig. 8 (right), replacing *n*-heptane with 1-butanol as a solvent led to a reactivity decrease of several orders and shorter survival times of fluorescent products due to the strong chemisorption of 1-butanol onto the Brønsted acid sites. A similar effect was achieved by changing the electrophilic character of the *para*-substituent of the styrene moiety. Both investigations clearly show the power of single molecule fluorescence spectroscopy to investigate local reactivity differences and solvent effects in zeolite-based catalytic reactions. These studies once again demonstrate the existence of large heterogeneities within catalytic solids.

Operando liquid-phase spectroscopy of a catalytic reactor

All the above, i.e., chemical imaging of a *single* catalyst particle by *e.g.* X-ray microscopy, and identifying *single* molecules during reaction with fluorescence microscopy, should then be integrated in a *single* reactor set-up, equipped with proper windows to allow such detailed and first-of-its-kind single particle-single molecule-single reactor studies.

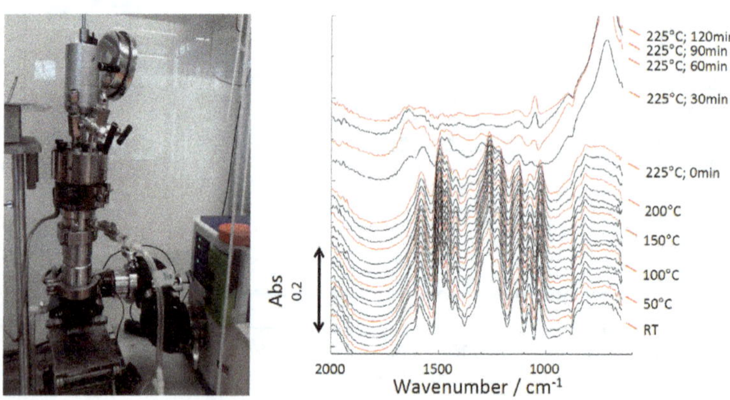

(a) (b)

Fig. 9. (a) Photograph of a high-pressure liquid phase reactor for performing catalytic experiments on *e.g.* biomass conversion processes; equipped with an Attenuated Total Reflection Infrared (ATR-IR) system for monitoring the catalytic decomposition of lignin in aromatic hydrocarbons and hydrogen in a so-called liquid phase reforming (LPR) process. The catalyst is a Pt/Al_2O_3 system, while the temperature and pressure are respectively 58 atm and 225°C. (b) Corresponding ATR-IR spectra taken during the heating up process of Kraft lignin in a water/ethanol mixture and further monitoring the reaction process as a function of time. The characteristic C–O stretching frequency at $\sim$1230 cm^{-1}, also found in β-O-4 model compounds, decrease in intensity upon lignin depolymerization.

Clearly, we are not yet there, but in what follows I summarize some of our efforts to measure in the liquid-phase catalytic processes with optical spectroscopy. Indeed, in the past years our group has been working on the catalytic valorization of lignin, one of the most recalcitrant bio-polymers on earth, which constitutes about 20–30% of plant biomass keeping the cellulose and hemicellulose fibers together [27]. Lignin is very rich in aromatic rings, and could serve as a renewable resource for the production of aromatics, key-building blocks in our current chemical industry. Our group has developed a process, which is called liquid-phase reforming (LPR), in which lignin is gradually converted into aromatics [28]. The crux is the addition of ethanol, which leads to the partial formation of ethoxylated products. In the case of only water a lot of very condensed products are formed. This observation, and the related chemical analysis of the formed reaction products, is indicative for the formation of reactive intermediates, which are prevented to condense in the presence of ethanol. Figures 3 and 9 show an experimental strategy to capture the details of this reaction process making use of operando ATR-IR and UV-Vis spectroscopy. While the distinct color changes (see *e.g.* the pictures shown in the insert of Fig. 3c) and related UV-Vis absorption bands can be used to follow the dissolution of *e.g.* Kraft lignin, originating from pulp industries, into aromatics, ATR-IR in combination with proper reference compounds (*e.g.* with characteristic features representing the β-O-4 linkages) allows to monitor the vibrational features taking place during this catalytic process at a temperature of 225°C and a pressure of 58 atm. Such studies lead to reaction process schemes, as schematically outlined in Fig. 10.

Outlook to future Developments of research on catalyst characterization

One of the scientific ambitions in this field of research is: "How can we push the spatial and temporal resolution of operando spectroscopy so it becomes possible to make detailed molecular movies of a catalytic solid at relevant reaction temperatures and pressures?" Here the focus should not only be on the organic or inorganic part of the catalytic process; but on both worlds, as it is the interplay between both which leads to the concept of chemical reactivity. Although understanding the molecular structure of the active sites of catalytic solids and their relation to reactivity is now possible in specific cases, the complexity and dynamics of heterogeneous catalysts under extreme conditions, i.e., at high pressures and temperatures, remains a clear challenge. We do not fully understand yet what the roles are of the support oxide, binder material, promotor elements as well as poisons on the genesis and stability of the active phase.

A dream I wish to realize with my team is to make am operando set-up combining synchrotron and optical microscopy methods, which provide 3D spectroscopic information down to a spatial resolution of 1 nm and with a time resolution of 1 ms or faster. Such a novel set-up, operating at high reaction temperatures and

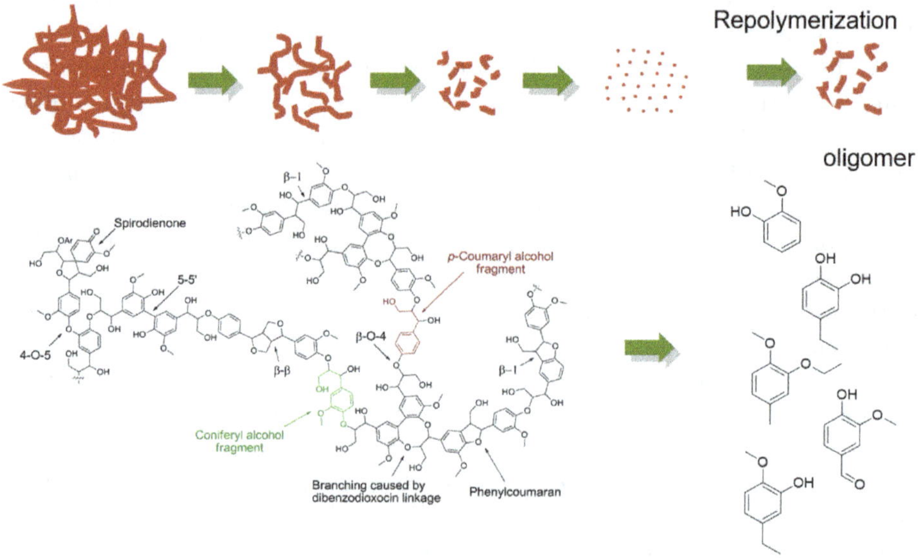

Fig. 10. Schematic of the lignin depolymerization process in the presence of a heterogeneous catalyst, such as Pt/Al_2O_3, in the so-called liquid-phase reforming (LPR) process at high pressures (58 bar) and temperatures (225°C). The reactive intermediates, formed by breaking the chemical bonds in lignin, can readily repolymerize; and *e.g.* ethanol in the mixture prevents this recondenzation reaction by ethoxylating the aromatic intermediates formed. In this manner, the yield of mono-aromatic compounds can be substantially increased.

pressures, hopefully will provide the ultimate resolution to perform pulse-type experiments in which the catalyst dynamics at the level of a single molecule and a single nanoparticle can be fully assessed. The question is if the reactivity and kinetics of one single catalyst particle can be explained in terms of its 3D structure, chemical composition and porosity.

Acknowledgments

Utrecht University (in the frame of the Strategic Theme Sustainability), the Netherlands Organization for Scientific Research (NWO, in the frame of their Gravitation program MCEC, Multiscale Catalytic Energy Conversion) and the European Research Council (ERC, Advanced Grant) are gratefully acknowledged for their financial support. B. M. W. is also grateful to all his PhD students and postdoctoral fellows who have contributed over the years to the advancement of the field of operando spectroscopy and microscopy. Special thanks are directed towards current and former staff scientists, Dr. Pieter Bruijnincx (Utrecht University, UU), Dr. Florian Meirer (UU), Dr. Rosa Bulo (UU), Dr. Gareth Whiting (UU), Dr. Javier Ruiz Martinez (currently working at AkzoNobel) and Dr. Andy Beale (currently professor at University College London, UCL).

References

[1] (a) G. Ertl, H. Knozinger, F. Schuth, J. Weitkamp (eds.), *Handbook of Heterogeneous Catalysis*, 2nd Edition, (Wiley-VCH, 2008); (b) J. Hagen, *Industrial Catalysis: A Practical Approach*, (Wiley-VCH, 1999).

[2] (a) G. W. Huber, S. Iborra, A. Corma, *Chem. Rev.* **106**, 4044 (2006); (b) A. Corma, S. Iborra, A. Velty, *Chem. Rev.* **107**, 2411 (2007).

[3] (a) G. Ertl, *Angew. Chem. Int. Ed.* **47**, 3524 (2008); (b) G. A. Somorjai, Y. Li, *Introduction to Surface Chemistry and Catalysis*, 2nd Edition, (Wiley, 2010).

[4] I. D. Gonzalez-Jimenez, K. Cats, T. Davidian, M. Ruitenbeek, F. Meirer *et al.*, *Angew. Chem. Int. Ed.* **51**, 11986 (2012).

[5] (a) I. L. C. Buurmans, B. M. Weckhuysen, *Nat. Chem.* **4**, 873 (2012); (b) B. M. Weckhuysen, *Angew. Chem. Int. Ed.* **48**, 4910 (2009); (c) B. M. Weckhuysen, *Natl. Sci. Rev.* **2**, 147 (2015).

[6] (a) U. Bentrup, *Chem. Soc. Rev.* **39**, 4718 (2010); (b) J. D. Grunwaldt, C. G. Schroer, *Chem. Soc. Rev.* **39**, 4741 (2010); (c) I. E. Wachs, C. A. Roberts, *Chem. Soc. Rev.* **39**, 5002 (2010).

[7] B. M. Weckhuysen, *Nature* **439**, 548 (2006).

[8] (a) B. M. Weckhuysen, *Chem. Commun.* 97 (2002); (b) B. M. Weckhuysen, *Phys. Chem. Chem. Phys.* **5**, 4351 (2003).

[9] (a) M. O. Guerrero-Perez, M. A. Banares, *Chem. Commun.* 1292 (2002); (b) M. A. Banares, M. O. Guerrero-Perez, J. L. G. Fierro, G. G. Cortez, *J. Mater. Chem.* **12**, 3337 (2002); (c) M. A. Banares, *Catal. Today* **100**, 71 (2005).

[10] A. M. Beale, M. G. O' Brien, B. M. Weckhuysen, *Characterization of Solid Materials and Heterogeneous Catalysts: From Structure to Surface Reactivity*, Vols. 1 & 2, M. Che, J. C. Vedrine (eds.), 1075 (Wiley-VCH, 2012).

[11] (a) N. Y. Topsoe, *Science* **265**, 1217 (1994); (b) N. Y. Topsoe, H. Topsoe, J. A. Dumesic, *J. Catal.* **151**, 226 (1995); (c) N. Y. Topsoe, J. A. Dumesic, H. Topsoe, *J. Catal.* **151**, 241 (1995).

[12] (a) B. M. Weckhuysen, A. Bensalem, R. A. Schoonheydt, *J. Chem. Soc. Faraday Trans.* **94**, 2011 (1998); (b) B. M. Weckhuysen, R. A. Schoonheydt, *Catal. Today* **49**, 441 (1999); (c) B. M. Weckhuysen, A. A. Verberckmoes, J. Debaere, K. Ooms, I. Langhans *et al.*, *J. Mol. Catal. A: Chem.* **151**, 115 (2000).

[13] M. Rønning, N. E. Tsakoumis, A. Voronov, R. E. Johnsen, P. Norby *et al.*, *Catal. Today* **155**, 289 (2010).

[14] E. de Smit, F. Cinquini, A. M. Beale, O. V. Safonova, W. van Beek *et al.*, *J. Am. Chem. Soc.* **132**, 14928 (2010).

[15] K. H. Cats, B. M. Weckhuysen, *ChemCatChem* **8**, 1531 (2016).

[16] N. Fischer, B. Clapham, T. Feltes, M. Claeys, *ACS Catal.* **5**, 113 (2015).

[17] S. J. Tinnemans, J. G. Mesu, K. Kervinen, T. Visser, T. A. Nijhuis *et al.*, *Catal. Today* **113**, 3 (2006).

[18] J. Zakzeski, B. M. Weckhuysen, *ChemSusChem* **4**, 369 (2011).

[19] (a) F. Meirer, S. Kalirai, D. Morris, S. Soparawalla, Y. Liu *et al.*, *Sci. Adv.* **1**, e1400199 (2015); (b) F. Meirer, D. T. Morris, S. Kalirai, Y. Liu, J. C. Andrews *et al.*, *J. Am. Chem. Soc.* **137**, 102 (2015); (c) F. Meirer, S. Kalirai, J. Nelson Weker, Y. Liu, J. C. Andrews *et al.*, *Chem. Commun.* **51**, 8097 (2015).

[20] K. H. Cats, J. C. Andrews, O. Stéphan, K. March, C. Karunakaran *et al.*, *Catal. Sci. Technol.* **6**, 4438 (2016).

[21] Y. Liu, F. Meirer, C. M. Krest, S. Webb, B. M. Weckhuysen, *Nature Comm.* **7,** 12634 (2016).

[22] T. Hartman, C. S. Wondergem, N. Kumar, A. van den Berg, B. M. Weckhuysen, *J. Phys. Chem. Lett.* **7**, 1570 (2016).
[23] (a) G. De Cremer, B. F. Sels, D. E. De Vos, J. Hofkens, M. B. J. Roeffaers, *Chem. Soc. Rev.* **39**, 4703 (2010); (b) K. P. F. Janssen, J. Van Loon, J. A. Martens, D. E. De Vos, M. B. J. Roeffaers *et al.*, *Chem. Soc. Rev.* **43**, 990 (2014).
[24] Z. Ristanović, J. P. Hofmann, G. De Cremer, A. V. Kubarev, M. Rohnke *et al.*, *J. Am. Chem. Soc.* **137**, 6559 (2015).
[25] Z. Ristanović, A. V. Kubarev, J. Hofkens, M. B. J. Roeffaers, B. M. Weckhuysen, *J. Am. Chem. Soc.* **138**, 13586 (2016).
[26] (a) D. E. Perea, I. Arslan, J. Liu, Z. Ristanović, L. Kovarik *et al.*, *Nature Comm.* **6**, 7589 (2015); (b) J. E. Schmidt, J. D. Poplawsky, B. Mazumder, O. Attila, D. Fu *et al.*, *Angew. Chem. Int. Ed.* **55**, 11173 (2016).
[27] R. Rinaldi, R. Jastrzebski, M. T. Clough, J. Ralph, M. Kennema *et al.*, *Angew. Chem. Int. Ed.* **55**, 8164 (2016).
[28] J. Zakzeski, A. L. Jongerius, P. C. A. Bruijnincx, B. M. Weckhuysen, *ChemSusChem* **5**, 1602 (2012).

NOVEL CONCEPTS IN C1 CHEMISTRY

XINHE BAO
State Key Laboratory of Catalysis, Dalian Institute of Chemical Physics, CAS, Dalian, 116023, China
Department of Chemistry, Fudan University, Shanghai, 200433, China

The technology converting CO molecules into liquid fuels via hydrogenation is known as Fischer–Tropsch synthesis (FTS). It was first developed by two German scientists, Fischer and Tropsch, in the 1920s and has now become the core technology for Gas-to-Liquid (GTL) and Coal-to-Liquid (CTL) in industry. Chemicals produced via FTS have also gone beyond liquid fuels, including other high-value hydrocarbons, such as light olefins, wax and oxygenates. Despite significant advances in both fundamental understandings and commercial applications, there are two major drawbacks in FTS technology: (1) a wide distribution of hydrocarbons with different chain lengths, i.e. a poor product selectivity; (2) the large consumption of H_2, which requires the energy intensive Water-Gas-Shift (WGS) process to generate more H_2 to replenish the coal-derived syngas. Recently, we developed a technology, completely different from conventional FTS, based on a nanocomposite catalyst composed of metal oxide and zeolite (OX-ZEO). Such a composite enables the direct conversion of coal-based syngas (with H_2/CO ratio at ~0.5–0.8) to light olefins, which thus does not need the additional WGS process. The selectivity of light hydrocarbons containing two to four C atoms (C_2–C_4) is over 94% and that of light olefins ($C_2^=$-$C_4^=$) reaches 80%. The CO activation and C–C coupling are separated onto two different types of active sites with complementary properties. Particularly, the selective chain growth via C-C coupling is controlled within the confined environment of zeolite pores. This OX-ZEO differs from conventional FTS in the design concept, catalyst and reaction mechanisms.

The state of the art of C1 chemistry and the challenges for the direct synthesis of light olefins from syngas

Since its invention in the 1920s, FTS has become the core technology for converting CO into hydrocarbons with more than one C atom. The great success in commercialization and fundamental studies has led to the unshakeable standing of FTS in the field of C1 chemistry [1]. FTS involves a collection of reactions, which can be simplified and described by the equation:

$$nCO + 2nH_2 \rightarrow (-CH_2-)n + nH_2O$$
$$\Delta H = -158 \text{ kJ/mol } CH_2(250°C).$$

FTS was usually catalyzed by metal or metal carbides such as iron, cobalt, nickel and ruthenium. Extensive efforts have been made to tune the product selectivity. By controlling the catalyst composition, promoters (*e.g.* K, Mn and Zn) and reaction conditions (*e.g.* temperature, pressure and reactants), the reaction could be tuned to favor the production of methane, light olefins or liquid fuels, such as synthetic oil, and wax. It is generally accepted that FTS proceeds via surface polymerization of CH_x ($x = 1$, 2 or 3), i.e. addition of CH_x monomer units to the adsorbed alkyl species on the open metal surfaces. Briefly, the reaction involves: (1) the adsorption of CO over metal or metal carbide surfaces; (2) the activation and dissociation of CO and the formation of surface carbides; (3) the reaction of surface hydrogen atoms with oxygen from dissociated CO forming H_2O (consumption of hydrogen) and hydrogen with surface carbon species forming CH_x intermediates (generally considered as CH_2); (4) surface polymerization of CH_x and the desorption of hydrocarbon products. The above pathway requires the surface to be active for the dissociative adsorption and activation of both CO and H_2, whilst facilitate the polymerization of intermediates such as CH_2. These requirements can be fulfilled by metal and metal carbides, particularly transition metals such as Fe, Co, Ni and noble metal such as Ru, which are shown to be the best FTS catalysts.

The above reaction mechanism leads to a few intrinsic drawbacks of FTS. The first one is the large consumption of hydrogen. To convert CO to olefins, oxygen has to be removed from CO, which can only be realized either by hydrogenation to form H_2O or reacting with CO to form CO_2. Since metal and metal carbides exhibit high activity in the dissociative adsorption of CO and H_2, the surface hydrogen atoms and oxygen atoms readily react with each other, via the L-H mechanism, forming H_2O. However, the result of this is the necessity of the energy-intensive WGS reaction ($CO + H_2O \rightarrow H_2 + CO_2$) to generate more H_2 when using coal-derived syngas. That is, one H_2 molecule, produced from one CO molecule and one H_2O molecule, is accompanied by emission of one CO_2 molecule. Therefore, FTS is an energy-intensive, water-consuming process accompanied by generation of waste water and emission of CO_2. The second drawback is the poor selectivity of desired products. The reaction involves polymerization of CH_x intermediates over an open metal surface with no confinement and control. This polymerization is driven by the entropy of products with the carbon number ranging from 1 up to 100, and even higher. The statistical distribution of products can be described by the Anderson–Schultz–Flory (ASF) model. The weight percentage of a hydrocarbon containing i C atoms can be correlated with the chain growth probability α:

$$\lg(\mathrm{Mi/I}) = \lg(\ln 2\alpha) + \lg\alpha.$$

One can see that higher hydrocarbons are produced at a higher chain growth probability. If lower hydrocarbons are desired, α has to be reduced. However, reducing α also favors methane formation. By varying catalysts (active components or additives) and reaction conditions, α can be altered. However, the studies to date indicate that to go beyond the ASF limit remains a challenge.

Taking the synthesis of light olefins for example, the highest selectivity of C_2–C_4 (including parrafins and olefins) is 56.7% at $\alpha = 0.4$–0.5 and meanwhile, methane selectivity reaches 29.2%. To improve the selectivity to light olefins, modified FTS for olefin synthesis has been studied extensively for more than 50 years. de Jong and coworkers in Netherland synthesized iron-based catalyst with uniform nanoparticle size (~6 nm) [2]. A light olefin selectivity as high as 61% was achieved at a CO conversion of $< 1\%$. When the CO conversion was increased to over 70%, the light olefin selectivity dropped to 30–50%. More recently, Sun and coworkers in China used Mn-promoted cobalt carbide quadrangular nanoprisms with preferentially exposed {101} and {020} facets and achieved a light olefin selectivity of 61% at a CO conversion of 30% [3]. However, this superior performance strongly relies on the operation conditions and may deteriorate significantly upon changing conditions. Therefore the challenge of light olefin synthesis from syngas lies in a reduced water consumption and the improved selectivity.

A new concept in C1 chemistry and OX-ZEO process for highly efficient synthesis of light olefins directly from syngas

The hydrogen consumption and poor product selectivity can be attributed to the reactivity of surface hydrogen with oxygen forming H_2O, and the lack of structural confinement to limit chain growth. Therefore, the development of new catalysts and reactions can be directed to eliminate water formation and generate a confined environment to control the selectivity. One solution is to use CO to remove surface oxygen although it is still in debate whether this is more preferable than using H_2. The overall CO_2 emission is comparable for both routes. If H_2 is used to remove the CO dissociated oxygen, WGS is needed to generate more H_2, which is accompanied by CO_2 emission. If CO is used, then the emission of CO_2 takes place during the synthesis of light olefins. In the former case, H_2O is recycled in the whole process, feeding H_2O for WGS and producing H_2O during olefin synthesis. On the contrary, WGS is not needed if the surface oxygen is removed by CO, although some energy input is necessary for the separation of products. The economic assessment shows that in fixed bed reactors concerning conversion of coal-derived syngas, it is more economic to use CO to remove oxygen, particularly in the water-deficient regions of China.

To this end, we have developed a composite catalyst for direct conversion of syngas to light olefins. One component of the catalyst is metal oxides with a spinel structure such as Zn/Cr, rather than metal and metal carbides used in the conventional FTS [4]. *In situ* reduction leads to formation of oxygen vacancies on the surface, i.e. coordinatively unsaturated sites. Preliminary results indicate that Zn is the main active component, while Cr or Mn serves as a promoter, which modulates the redox properties, the formation of oxygen vacancies and the stabilization of active sites. Temperature-programmed desorption and ambient-pressure XPS, as well as recent high-pressure STM studies show the disproportionation reaction of

CO over partially reduced oxides at 400°C:

$$CO_{ad} + CO_{ad} \rightarrow C^*_{ad} + CO_2^{\uparrow}$$

The results show that H_2 does not readily adsorb and dissociate over oxide surfaces we studied and hence the reaction of hydrogen with oxygen is avoided. On the other hand, surface C^*_{ad} species are highly active and react facilely with H_2 via the E-R mechanism, forming CH_2 or CH_x intermediates.

$$C^*_{ad} + H_2(g) \rightarrow CH_2^{\uparrow}$$

In conventional FTS, the surface adsorbed CH_x intermediates undergo surface polymerization whereas the metal oxide surface does not exhibit strong adsorption capability towards CH_x intermediates, which therefore desorb into gas phase. Consequently, the surface polymerization is avoided. However, the structure of surface carbon species remains to be elucidated.

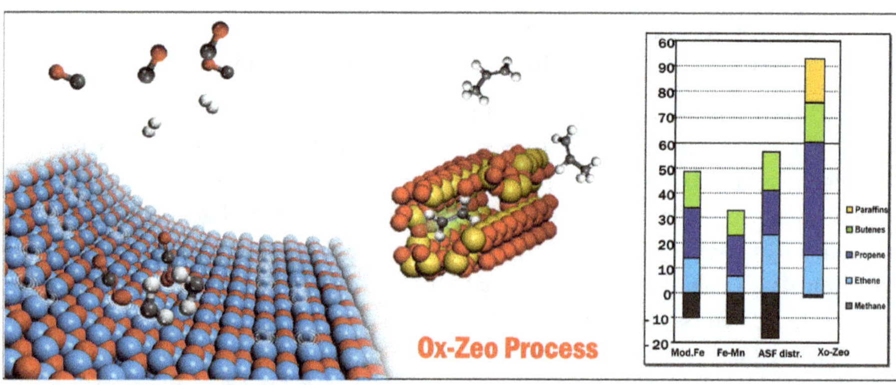

Fig. 1. The schematic mechanism and the product distribution of the new process based on a nano-composite catalyst composed of metal oxide and zeolite (OX-ZEO). Such a composite enables the direct conversion of coal-based syngas (with H_2/CO ratio at ~0.5–0.8) to light olefins. The selectivity of light hydrocarbons containing two to four C atoms (C_2–C_4) is over 94% and that of light olefins ($C_2^=$-$C_4^=$) reaches 80%. The CO activation and C–C coupling are separated onto two different types of active sites with complementary properties. Particularly, the selective chain growth via C–C coupling is controlled within the confined environment of zeolite pores.

The other component of the composite catalyst is a zeolite. The desorbed CH_x intermediates diffuse to the pores of zeolites, where the shape selectivity is well known. Thus, selective chain-growth may be controlled within such a confined environment. The experiments validate this assumption. By varying the acidity and size of pores, the product selectivity could be tuned. In our published work, mesoporous SAPO-34 was used with the micropore size of 0.34 nm, which leads to light olefins $C_2^=$–$C_4^=$ as the main products. Among that, propylene is around 50%.

Though some understanding has been obtained about the reaction, we still do not know three key steps involved in the reaction: the desorption of reaction intermediates; their migration into the zeolite pores; and their conversion to final products.

First of all, the reaction intermediates and their structures need further confirmation. Our data indicate that CH_2 free radicals are likely the intermediates, which are extremely active and readily transform into stable products such as methane in the presence of hydrogen. We employed synchrotron-based vacuum ultraviolet photoionization time-of-flight mass spectrometry to detect intermediates at reaction conditions. At a photon energy of 9.72 eV, a signal of mass/charge ratio of 42.01 corresponding to ketene (CH_2CO) was detected in the effluent in addition to stable products. CH_2CO is the reaction product of CH_2 radical with CO, which is relatively more stable than CH_2. Further experiments indicate that CH_2CO can diffuse by several micrometers and even over ten millimeters distance at reaction conditions (around 1 atm). Although a lot of experimental data show that ketene is likely the reaction intermediate, more sophisticated experimental and theoretical studies are needed to pin down the exact intermediates and their structure.

Secondly, the reaction pathways of intermediates in the zeolite pores, i.e. the mechanism of C–C coupling is not known yet. The mechanism of the first C–C bond formation is a long-pursued question. For methanol-to-olefins (MTO) reaction, the methoxyl group is generally accepted as the reaction intermediate and the reaction follows the mechanism of carbon pool. Briefly, methanol dehydrates or DME decomposes forming methoxy (CH_3O), which goes through hydrodeoxygenation within the zeolite pores forming aromatics with branched chains. Upon the catalysis by acid sites of zeolites, these aromatics crack and form a variety of lower hydrocarbons (including olefins and paraffins). The chain length and saturation of carbon bonds depend on the physiochemical properties of zeolites. During reaction, these aromatic molecules may go through further coupling, forming polycyclic aromatics and even coke, which could block the pores and eventually leads to the deactivation of catalyst. Initially, we tended to borrow the carbon pool mechanism for the OX-ZEO reaction because our NMR experiments also showed the presence of similar aromatics with branched chains and polycyclic aromatics. However, further experiments revealed the absence of oxygenates such as methanol, DME and very little H_2O. More importantly, the catalyst is very stable and no obvious deactivation has been observed within a 3000 h test. Under comparable conditions of MTO, the catalyst deactivates quickly within hours. In addition, the acidity of the zeolite pores modulates mainly the ratio of olefins/paraffins but has little effect on the formation rate of hydrocarbons. These characteristics imply that the reaction mechanism of OX-ZEO may differ significantly from that of MTO. The elucidation of reaction mechanism will rely on *in situ* studies, using, for example, high pressure NMR and IR as well as theoretical calculations.

Thirdly, it is intriguing that the presence of hydrogen under reaction conditions does not cause deep hydrogenation of the unsaturated intermediates and a high olefin selectivity is obtained. In fact, analogous composites with metal oxides and zeolites had been applied for syngas conversion previously. Although high CO conversion could be obtained, the final product was mainly paraffins with little olefins. The presence of paraffins was also found in our early research. Upon optimization of the catalysts, particularly the structure of metal oxides and zeolites, as well as their coupling modes, a product distribution with a high content of olefins is obtained over a composite catalyst with an appropriate granule size. In contrast, if metal oxide is assembled as the core and zeolite as the shell, or metal oxide is highly dispersed on zeolite via *e.g.* impregnation method, formation of paraffins and methane is much more favored. Experimental data suggest that it is extremely important to couple the two types of active sites for CO activation and C–C coupling at a proper distance. The preliminary study indicates that this might be related to the stability of two types of free radicals produced in the reaction. As stated above, the relatively stable CH_2CO can diffuse intact as far as several micrometers and even millimeter scale. At the same time, there likely exists H free radical generated by surface catalysis, which is highly active and cannot transport very far. It is expected that it may react with CH_2 or CH_2CO intermediates or olefins before the intermediates can reach the active sites of zeolites. When the two types of active sites are close enough, < 1 nm for instance, hydrogenation readily happens, which will lead to formation of a large amount of paraffins. The preliminary study on the transport of the free radicals CH_2CO and H under reaction atmosphere supports our assumption.

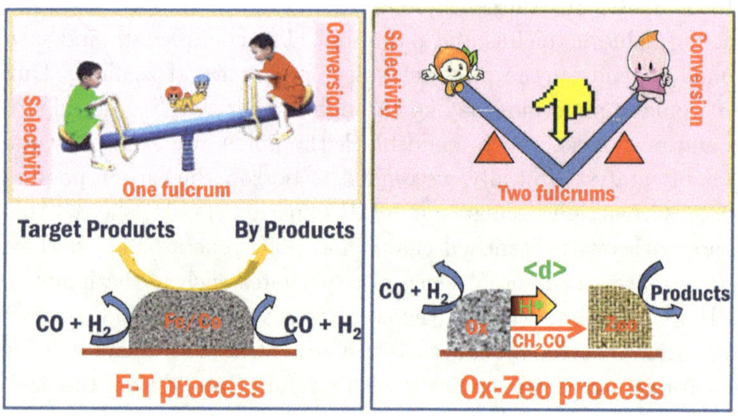

Fig. 2. The formation of CH_x intermediates and chain growth via C–C coupling are separated onto two types of active sites in Xo-Zeo process, which allows tuning of both activity and selectivity simultaneously.

Our study demonstrates that using the nanocomposite with a mixed metal oxide and zeolite, the hydrogen consumption to remove surface oxygen and the random surface polymerization of CH_x are avoided. More importantly, the formation of CH_x intermediates and chain growth via C–C coupling are separated onto two types of active sites. This is expected to allow tuning of both activity and selectivity simultaneously, which is usually difficult in catalysis and often behaves like a teeter totter. This strategy may provide a general guidance for the rational design of highly efficient catalysts for similar reactions.

A new horizon of syngas conversion, a solution to the production of chemicals for the post-petroleum era

Currently, more than 90% chemicals are produced from petroleum. With the diminishing oil reserve and its wildly fluctuating prices, it is extremely urgent to develop technologies from alternative resources during the post-petroleum era. The discovery of large reserves of natural gas and its applications, the recognition of coal resources, reducing cost of biomass utilization, as well as low-cost hydrogen from future renewable resources, provide a variety of resources for future production of chemicals and fuels. Conversion of these resources into syngas, which is further converted to chemicals and materials via catalysis, attracts increasing attention. The sustainable development of the society requires higher efficiency, smarter and cleaner solution.

Taking synthesis of basic chemicals such as light olefins from coal-derived syngas as an example, there are two approaches available. One is the indirect approach involving syngas conversion to methanol, and further dehydration to form olefins. The second is the direct approach based on FTS. The common feature of these two approaches require relatively high H_2/CO ratio. However, the coal-derived syngas contains H_2/CO ratio of 0.5–0.8. Therefore, water-gas-shift reaction (WGS) is conventionally required to generate more hydrogen to raise the H_2/CO ratio to around 2.0. In addition, how to improve olefin selectivity remains a great challenge. These approaches exhibit drawbacks either a long operational path, and/or high water and energy consumption, and/or emission of waste water. In contrast, the newly developed OX-ZEO process enables the direct conversion of syngas to light olefins with a high selectivity, which does not need extra hydrogen to remove oxygen. In principle, the molar ratio of H_2/CO at 0.5 is sufficient for the direct conversion. Thus, WGS reaction can be discarded with no waste water produced from the reaction itself.

Most recently, we optimized the metal oxide with a spinel structure, combined with mesoporous SAPO-34, which delivers a single pass CO conversion close to 35%, C_2–C_4 hydrocarbon selectivity 95%, while $C_2^=$–$C_4^=$ selectivity over 80% with a low methane selectivity ($< 2.5\%$). The catalyst has gone through a life test for 3600 h (equivalent to 150 days) in laboratory. Furthermore, the product distribution can be tuned by varying the acidity and pore size of zeolites. The preliminary results show that other high-value added chemicals such as liquid fuels (C_{6+} and aromatics) can also be obtained by OX-ZEO process.

In summary, the OX-ZEO technology based on the nanocomposite catalysts of oxide and zeolite enables the direct and selective conversion of syngas. This opens up a new horizon for C1 chemistry, providing a perfect solution to diversify carbon resources for chemical production during the post-petroleum era.

Acknowledgments

This work was financially supported by the National Natural Science Foundation of China (grant nos. 21425312, 21321002, and 91545204), the Ministry of Science and Technology of China (no. 2013CB933100), the "Strategic Priority Research Program" of the Chinese Academy of Sciences (grant XDA09030101).

References

[1] H. M. T. Galvis, K. P. de Jong, *ACS Catal.* **3**, 2130 (2013).
[2] H. M. T. Galvis, J. H. Bitter, C. B. Khare, M. Ruitenbeek, A. Iulian Dugulan *et al.*, *Science* **335**, 835 (2012).
[3] L. Zhong, F. Yu, Y. An, Y. Zhao, Y. Sun *et al.*, *Nature* **538**, 84 (2016).
[4] F. Jiao, J. Li, X. Pan, J. Xiao, H. Li *et al.*, *Science* **351**, 1065 (2016).

ELECTROCHEMISTRY FOR THE PRODUCTION OF FUELS, CHEMICALS AND MATERIALS

MARC T. M. KOPER

Leiden Institute of Chemistry, Leiden University, 2300 RA Leiden, The Netherlands

My view of the present state of research on electrocatalysis

Renewable electricity is generated by intermittent sources such as solar and wind. Storage of electricity will therefore become an increasingly important issue as the importance of renewable electricity increases. While electricity storage in batteries is a good option for large-scale energy storage, the electrochemical production of high-energy density fuels ("electrofuels") in principle allows for higher energy density and greater flexibility. Hydrogen is certainly the most likely "electrofuel." The conversion of carbon dioxide, water and solar (electricity) into a carbon-based fuel remains another exciting but challenging option. In addition, the further integration of renewable electricity into chemical industry and the importance of electrocatalytic conversions have brought the area of electrosynthesis back into the spotlight. One further specific interest in electrochemistry and electrocatalysis is the ability to perform chemistry at extreme values of electrochemical potential.

My recent research contributions to electrocatalysis

Multiple proton-electron transfer in electrocatalysis

My work in electrocatalysis aims at obtaining a fundamental understanding of basic electrocatalytic reactions on well-defined electrode surfaces using a combination of electrochemical experiments, *in situ* spectroscopy, theory and computational chemistry (mainly Density Functional Theory). Recently, focus in my group has been on proton-coupled electron transfer (PCET) reactions in electrocatalysis, specifically on the importance of pH as a descriptor in optimizing the rate of (catalytic) pathways [1, 2]. Multiple PCET reactions follow pathways such as indicated schematically in Fig. 1.

Catalytic and electrocatalytic reactions involving multiple intermediates are limited by so-called energetic scaling relations between intermediates [3, 4]. In case of a single intermediate, typically a two-electron transfer reaction such as hydrogen evolution, optimization of the binding energy of the catalytic intermediate is possible such that an optimal reversible catalyst with near-zero overpotential is obtained. Platinum is a good example of an optimal reversible catalyst for hydrogen evolution

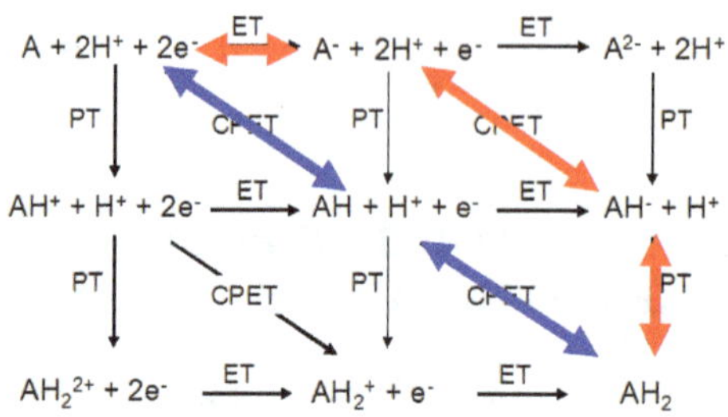

Fig. 1. Square scheme for proton-coupled electron transfer reactions.

and hydrogen oxidation, because it binds the hydrogen intermediate with just the right binding energy. However, for reactions involving the transfer of more than two electrons, say four, six or eight, more than one catalytic intermediate exists. The relative binding energies of these intermediates are typically fixed by scaling relationships, leading to fundamental limitations in the extent to which such reactions can be catalyzed, leading to non-zero overpotentials [3, 4]. In addition, many multiple electron reactions, such as those occurring in the water-oxygen-hydrogen cycle, the carbon cycle and the nitrogen cycle, also involve proton transfer. Such proton-coupled electron transfer reactions can feature pathways in which proton and electron transfer are decoupled, such as illustrated by the red arrows in Fig. 1. These pathways involve charged or dipolar dissolved or surface-adsorbed intermediates that interact strongly with the electrolyte double layer existing at the electrocatalyst-electrolyte interface. Such pathways are important in many electrocatalytic reactions, such as carbon dioxide reduction [5], and they manifest by a strong pH dependence, which is not expected for reactions featuring only concerted proton-electron transfer reactions (blue arrows in Fig. 1). Carbon dioxide reduction by an immobilized cobalt-porphyrin [6] or by a gold electrode [7], involving a catalyst-bound CO_2^-, is a good example of such a pathway (see Fig. 2).

Electrochemistry at extremely oxidizing and reducing conditions

Electrochemistry offers the opportunity to perform chemistry under extremely oxidizing or reducing conditions. On a suitably inert electrode such as boron-doped diamond, hydroxyl radical can be generated at highly anodic potentials, which may help in the breakdown and mineralization of organic pollutants [8]. For instance, we have shown that cellulose can be converted into formic acid by such a treatment [9]. More surprisingly, strongly reducing conditions can lead to apparently novel chemistry involving the breakdown of metals, a phenomenon called cathodic corrosion. Metals polarized at negative potentials generate nanoparticles, and this process is mediated/accompanied by a striking anisotropic etching of the electrode surface [10].

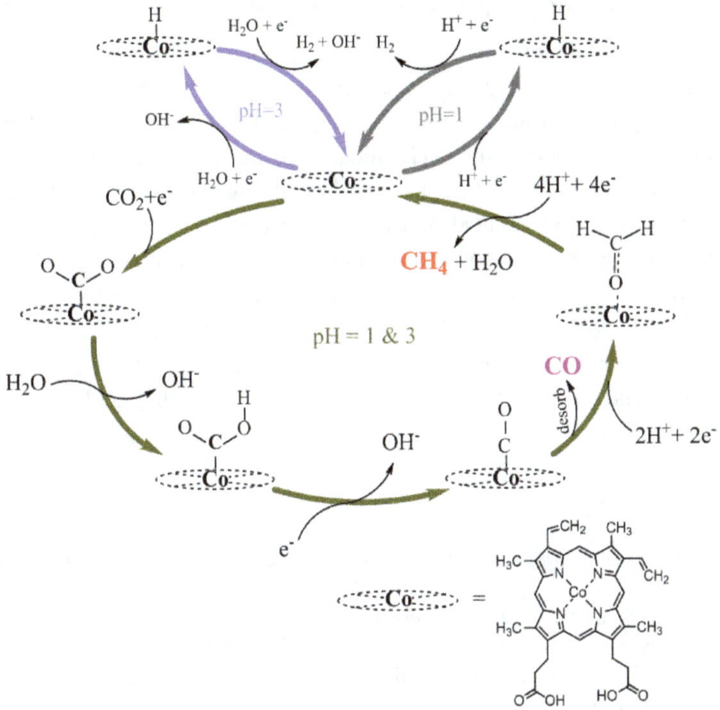

Fig. 2. Catalytic cycle for CO_2 reduction to CO and CH_4 on a cobalt-porphyrin electrode, showing the importance of the CO_2^- intermediate [6].

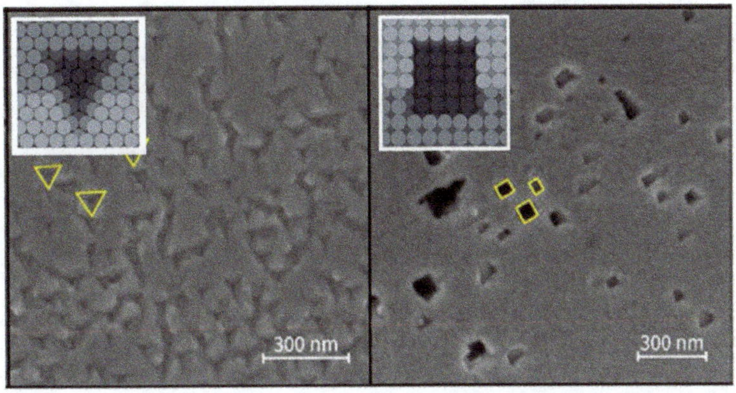

Fig. 3. Anisotropic etching of a platinum electrode and the formation on triangular and rectangular etch pits at negative potentials. Left panel shows a (111) facet etched at −0.6 V (versus RHE) in 10 M NaOH; right panel shows a (100) facet etched at −0.8 V (versus RHE) in 10 M NaOH [10].

Outlook to future developments of research on electrocatalysis

With the increasing generation and growing importance of renewable electricity, it is to be expected that electrocatalysis and electrochemistry will start playing an increasingly important role in energy storage and in the chemical industry. Using electrochemistry for making fuels, chemicals and materials is highly promising with many opportunities still left largely unexplored. A better and more detailed understanding of the fundamental aspects of the electrode-electrolyte interface will require the development of new experimental and computational methods to study the solid-liquid interface, and the further development of "electrochemical surface science" is highly desirable. In addition, electrochemistry should explore the ability to perform chemistry at more extreme conditions, not only in terms of extreme electrode potentials, but also in terms of studying electrochemical processes at high temperature and pressure.

References

[1] M. T. M. Koper, *Phys. Chem. Chem. Phys.* **15**, 1399 (2013); *Chem. Sci.* **4**, 2710 (2013).

[2] S. Hammes-Schiffer, A. A. Stukhebrukhov, *Chem. Rev.* **110**, 6939 (2010).

[3] M. T. M. Koper, *J. Electroanal. Chem.* **660**, 254 (2011).

[4] I. C. Man, H. Y. Su, F. Calle-Vallejo, H. A. Hansen, J. I. Martinez *et al.*, *ChemCatChem* **3**, 1159 (2011).

[5] R. Kortlever, J. Shen, K. J. P. Schouten, F. Calle-Vallejo, M. T. M. Koper, *J. Phys. Chem. Lett.* **6**, 4073 (2015).

[6] J. Shen, R. Kortlever, R. Kas, Y. Birdja, O. Diaz-Morales *et al.*, *Nature Comm.* **6**, 8177 (2015).

[7] A. Wuttig, M. Yaguchi, K. Motobayashi, M. Osawa, Y. Surendranath, *Proc. Natl. Acad. Sci. USA* **113**, E4585 (2016).

[8] A. Kapalka, G. Foti, C. Comninellis, *Electrochemistry for the Environment*, C. Comninellis, G. Chen (eds.), 1 (Springer, 2010).

[9] Y. Kwon, S. E. F. Kleijn, K. J. P. Schouten, M. T. M. Koper, *ChemSusChem* **5**, 1935 (2012).

[10] T. J. P. Hersbach, A. I. Yanson, M. T. M. Koper, *Nature Comm.* **6**, 12653 (2016).

ELECTROCATALYSIS AT DEEP-SEA HYDROTHERMAL VENTS

RYUHEI NAKAMURA

Biofunctional Catalyst Research Team, RIKEN Center for Sustainable Resource Science, Wako, Saitama 351-0198, Japan

My view of the present state of research on catalysis in extreme conditions

From my view, our understanding of catalysis under extreme conditions and its relation with processes in nature are very limited. For instance, we do not know so much about chemistry taking place at deep-sea hydrothermal environments. We know that nature has developed the systems to harvest geothermal and geochemical energy to sustain the action and function of catalysts needed for life. Yet, how nature utilizes temperature and chemical gradient fostered by geochemical settings remains largely unknown.

My recent research contributions to catalysis in extreme conditions

For the research field of catalysts under extreme conditions, our main contribution is discovery of the multi-functionality of earth crust minerals that interconvert chemical-thermal-electrical energy. We have fabricated the series of mineral electrodes and found that the deep-sea hydrothermal vent minerals, composed of Fe/Cu/S, are capable of converting geochemical gradient into electricity with excellent electron conduction and electrocatalysis [1, 2]. In addition, we found that this extremely abundant minerals can function as thermoelectric materials, which can covert temperature gradient into electricity. ZT value, which is the index of thermoelectric performance of the materials, is almost identical with the laboratory-synthesized materials, even after intensive optimization [3]. We identified that one of the reasons for thermoelectric effects of the natural minerals is the ability to de-couple electron and heat transfer, due to the existence of nano/micro-sized surface structuring.

Based on the findings of those multi-functionality of earth crust minerals, we proposed the new model: de-coupled electron and heat transfer at deep-sea hydrothermal vents (Fig. 1) [4, 5]. The point of the proposed model is temperature term of Nernstian equation, which becomes important for catalysis under high temperature conditions. As predicted by Nernstian equation, the 3D columnar structure of hydrothermal vents can generate the highly reductive electron inside hot hydrothermal fluid. Of note here is that this high energy electron can be propagated to a cold ocean/vent interface and trigger CO_2 reduction and fixation, if the hydrothermal vent wall can effectively de-couple heat and electron transfer (i.e. maintaining temperature gradient across the wall of hydrothermal vent). As a consequence of de-coupling, the system can generate high-energy electrons almost comparable to those generated by solar radiation in photosynthesis.

This is a simple, but potentially to be general idea to understand the interconversion of chemical-thermal-electrical energy under the conditions of high temperature and high pressure, and offers the insight into the nature's system to utilize not only ΔpH, but also ΔT to feed the energy for CO_2 fixation without the use of solar energy.

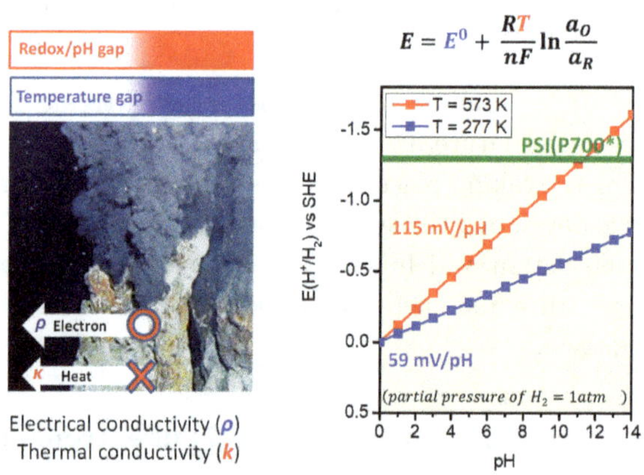

Fig. 1. Temperature and pH dependence of the standard redox potential of H^+/H_2 couple. The pH and temperature gap between hydrothermal fluid and sea water can generate the reduction power as high as that of photosynthetic CO_2 fixation center (PSI).

Outlook to future developments of research on catalysis in extreme conditions

Chemical reactions at deep-sea hydrothermal vents is one of the most outstanding chemistry under high temperature and high pressure conditions. Catalysis at the deep-sea vent is a heart of solar-independent modern biospheres, and may tightly link with the origin and evolution of life. Thus, the intriguing perspective of "*Catalysis under Extreme Conditions*" is to predict the first engine that drives the life emergence and fosters life complexity [6]. In parallel veins, utilizing the temperature term of Nernstian equation is a state-of-art challenge to develop electrochemical devices that can harvest not only redox gradient, but also temperature gradient.

Acknowledgments

The author thanks Drs. K. Takai, M. Yamamoto for Marine-Earth Science and Technology (JAMSTEC) for discussions about microbiological and biogeochemical aspects of deep-sea hydrothermal ecosystems. The author also thanks Dr. T. Mori for National Institute for Material Science (NIMS) for analyzing thermoelectric properties of natural minerals. This work was financially supported by a Grant-in-Aid for Specially Promoted Research from the Japan Society for Promotion of Science (JSPS) KAKENHI Grant Number 24000010, and The Canon Foundation.

References

[1] R. Nakamura, T. Takashima, S. Kato, K. Takai, M. Yamamoto *et al.*, *Angew. Chem. Int. Ed.* **49**, 7692 (2010).
[2] M. Yamamoto, R. Nakamura, K. Oguri, S. Kawagucci, K. Suzuki *et al.*, *Angew. Chem. Int. Ed.* **52**, 10758 (2013).
[3] R. Ang, A. U. Khan, N. Tsujii, K. Takai, R. Nakamura *et al.*, *Angew. Chem. Int. Ed.* **54**, 12909 (2015).
[4] A. Yamaguchi, M. Yamamoto, K. Takai, T. Ishii, K. Hashimoto *et al.*, *Electrochim. Acta* **141**, 311 (2014).
[5] A. Yamaguchi, Y. Li, T. Takashima, K. Hashimoto, R. Nakamura, *Solar to Chemical Energy Conversion*, M. Sugiyama, K. Fujii, S. Nakamura (eds.), 213 (Springer, 2016).
[6] M. J. Russell, L. M. Barge, R. Bhartia, D. Bocanegra, P. J. Bracher *et al.*, *Astrobiology* **14**, 308 (2014).

CONTROLLED FUNCTIONALISATION AND UNDERSTANDING OF SURFACES TOWARDS SINGLE SITE CATALYSTS AND BEYOND

CHRISTOPHE COPÉRET

Department of Chemistry and Applied Biosciences, ETH Zürich, CH-8093 Zürich, Switzerland

My view of the present state of research on single-site catalysis

Industrial processes mainly rely on heterogeneous catalysts because they are more suitable for continuous efficient processes (easier recycling, separation and regeneration combined with less operating units). However, industrial heterogeneous catalysts are very complex and can thus be mainly developed via empirical approaches. In contrast molecular chemistry has made tremendous progresses in the past 40 years, enabling the development of homogenous processes via more rational (structure–activity) approaches. Striking examples can be found across many types of reactions: hydroformylation, methanol carbonylation, asymmetric hydrogenation, alkene metathesis, oligomerisation or polymerisation to name but a few. With the power of molecular principle, it has been proposed as early as in the 70's to design and develop heterogeneous catalysts via a molecular approach [1]. However, it is only recently that such step is fully possible thanks to the emergence and the development of advanced characterisation techniques for solids.

Here, we will first discuss how molecular approaches towards the design and the development of well-defined supported catalysts can help understanding industrial catalysts; we will discuss two illustrative examples: olefin metathesis and polymerisation. In the second part, we will discuss how solid-state NMR spectroscopy can provide a unique approach to obtain unprecedented information about surface site structures. We will finally examine where the field will likely be going in the coming years [2].

My recent research contributions to single-site catalysis

Bridging the gap between well-defined and industrial catalysts

Olefin metathesis is a prototypical example as some of the earliest catalysts were based on supported group 6 and 7 metal oxides and are still industrially used today despite many revolutions on the molecular side. Indeed, while Chauvin proposed a mechanism involving metallocarbene and metallacyclobutane in 1971 [3], there is no spectroscopic evidence for the formation of these intermediates in the aforementioned industrial catalysts. In sharp contrast, organometallic chemists have developed alkylidene complexes as isolable and fully characterised compounds, which display today very high catalytic efficiency (activity, selectivity and stability). In fact, homogeneous catalysts have now increased their impact in industrial settings. One might wonder whether a molecular approach could not help developing or at least understanding heterogeneous metathesis catalysts.

Such approach became successful in 2001, with the development of the first well-defined and highly active supported catalyst based on a silica-supported Re-alkylidene [4]. Such species displayed much higher activity than both its molecular and heterogeneous analogues. Thanks to its known structure, it was possible to interrogate the origin of its efficiency through a computational approach and propose a model to develop better metathesis catalysts. Indeed, it was shown that these well-defined alkylidene complexes had an ideal set of ligands: the metal center comprised a d^0 pseudotetrahedral site with strong and weak σ-donor ligands [5]. This first hit and model made possible the development of improved catalysts as novel molecular complexes become available (Fig. 1). The latest generation of catalyst is based on well-defined silica-supported cationic W oxo alkylidene species, which can yield more than 1'200'000 turnover number (TON) in contrast to the 6000 TON obtained with the original Re-based catalysts [6].

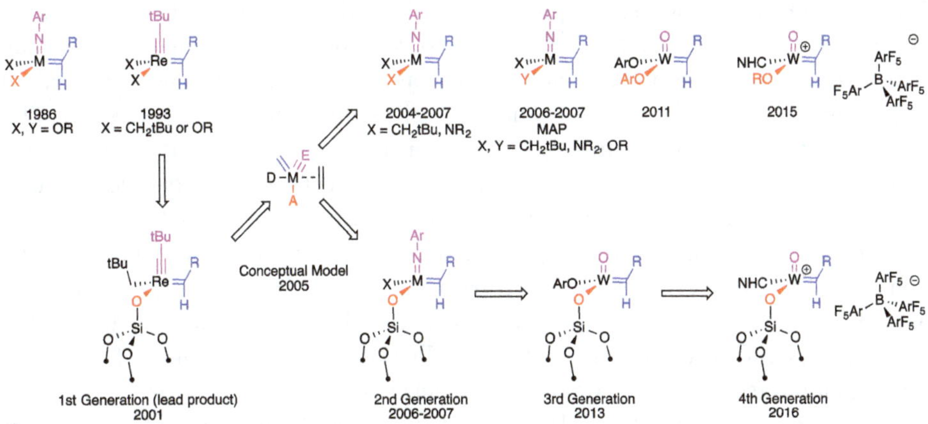

Fig. 1. Evolution of molecular and surface catalyst structure following the advancement of molecular chemistry and the increase understanding of rules for catalyst design.

Worthy of note silica-supported well-defined W oxo alkylidene catalysts display several orders of magnitude higher activity than the corresponding industrial silica-supported W based catalysts, with turnover frequencies reaching $> 200\ \text{min}^{-1}$ at 30°C for the former versus $\ll 0.1\ \text{min}^{-1}$ at 400°C for the latter [6, 7]. In fact, it has been shown that well-defined d^0 W oxo species do not yield active catalysts at low temperatures. This indicates that formation of the alkylidene from an oxo species is a critical (slow) step. However, the most recent studies have shown that organosilicon reducing agent can switch on the activity of W(VI) oxo species, by first reducing them into W(IV) species under mild conditions (70°C) as they can react with olefins to generate metallacyclopentane, metallacyclobutane and alkylidene intermediates (Fig. 2) [8]. This approach opens new avenues to activate classical heterogeneous metathesis catalysts, which currently operate at 400°C.

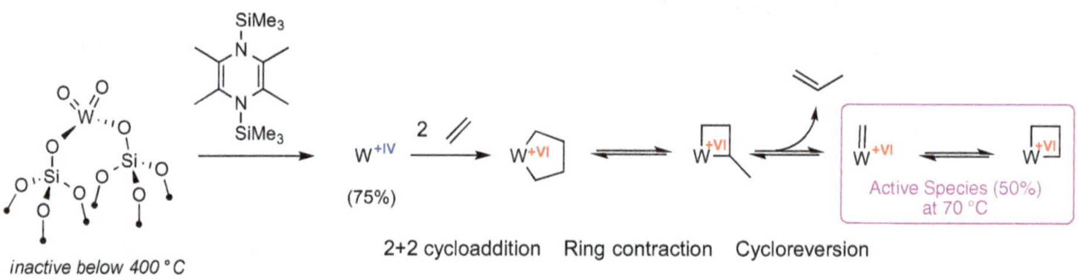

Fig. 2. Formation of active species from inactive well-defined W oxo surface species using an organosilicon reducing agent.

Another interesting and recent study has been the investigation of Phillips Cr-based ethylene polymerisation catalysts (CrO_3/SiO_2), which is responsible for ca. 50% of the world production of high density polyethylene [9]. Since its discovery, the Phillips catalyst has been a mystery and led to many debates regarding its active site structure and the corresponding polymerisation mechanism. It has been recognised very early that under the reaction conditions, the surface chromate species (Cr(VI)) evolves into reduced species, mainly Cr(II), which lead to the formation of the active species. Using a thermolytic precursor approach, it has thus been possible to generate well-defined Cr(II) sites. Such species are inactive in polymerisation. However, such species can be turned into highly active polymerisation catalysts upon activation with N_2O, which yields Cr(III) sites (Fig. 2) [10]. In fact, Cr(III) molecular precursors can also be used to generate isolated Cr(III) sites, which are also highly active in ethylene polymerisation without induction period and change of oxidation state. Computational studies on cluster models have indicated that initiation could take place via an heterolytic C–H bond activation or the insertion of ethylene on Cr(III),O bonds and that polymerisation occurs via the classical Cossee–Arlman mechanism. Worthy of note polyethylene is obtained with a relatively high polydispersity (8–12), which speaks for the presence of a variety of active sites. One can propose that the large distribution of sites probably arise from the amorphous nature of silica, which will lead to many different local environments around Cr. In fact, recent computational studies on amorphous periodic models show that the local environment can lead to significant differences for initiation, from endoergic to highly exoergic C–H bond activation step. In addition, it has also been shown that alternative initiation mechanism could also be competitive and involve ethylene insertion in the Cr,O bond. Overall, this study has shown that the active sites in the

Phillips catalysts is likely a Cr(III), that initiation can take place via C–H bond activation and/or ethylene insertion in Cr,O bonds and that the amorphous character of silica is probably the origin of the high polydispersity of the polymer [11].

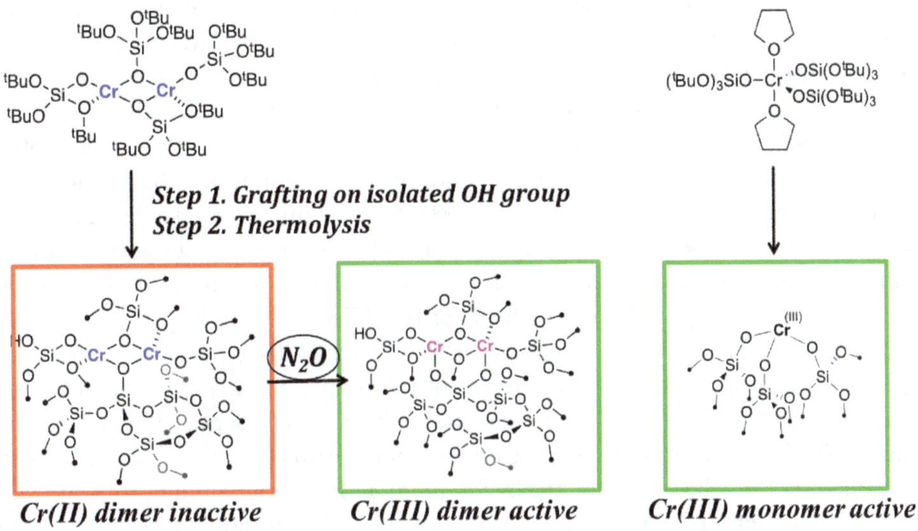

Fig. 3. Molecular approach to construct silica-supported well-defined Cr surface sites: control of oxidation state and nuclearity.

Expedient and Detailed Characterisation by DNP SENS

The development of well-defined supported catalysts relies on detailed spectroscopic data, an essential step in obtaining an active site structure. Of the various spectroscopic methods, solid-state nuclear magnetic resonance spectroscopy (NMR) is probably one of the most information rich because it is highly sensitive to the local environment and to dynamics. On the other hand, NMR suffers from a low sensitivity, which is worsened by the fact that active nuclei often correspond to low abundance isotopes and that only a small fraction of the sample is of interest in the studied material: the surface sites often constitute only few percent. It is therefore typically required to use labeled compounds and long acquisition times, thus preventing most advanced pulse sequences and powerful 2D NMR to be used. Magic Angle Spinning (MAS) and cross-polarization (CP) have been essential steps in obtaining high quality NMR spectra. In particular, the latter exploits the greater polarisation of protons to increase the signal of less sensitive nuclei such as carbon-13 or nitrogen-15. Very early on, Overhauser and Slichter have proposed and shown that electrons could provide alternative ways to polarise nuclear spins (so-called dynamic nuclear polarisation — DNP), thus enabling even greater polarisation transfer [12]. With the development of high field NMR spectrometer and gyrotron technology, Griffin and coll. have developed DNP NMR into a very powerful method to increase of the sensitivity [13]. While originally developed mainly for the structural determination of biomolecules, it has been recently recognised as a very powerful tool to characterise materials, in particular the surface of materials, and is often coined DNP surface enhanced NMR (DNP SENS) [14]. In this case, the sample is contacted with a minimum amount of a solution containing an organic radical (the source of electrons). Under microwave irradiation and MAS at 100 K, polarisation transfer is induced from the unpaired electrons to the proton baths of the solvent and the molecules, and under CP conditions this polarisation is further transfer to the nuclei under investigation. This polarisation transfer can be carried out in a variety of solvents and radicals, tuned to maximise polarisation transfer and to minimise the interaction of the radical with the samples. The radical of choice are typically dinitroxyl radicals of the TEKPOL family [14]. The polarisation transfer dramatically enhances the NMR signals by a factor 100–250, thus decreasing acquisition time by the square of this value and enabling the detection of surface species in an expedient time (hours in place of years). This method is thus applicable to a broad range of samples enabling to establish the connectivity of a monolayer of silica at the surface of alumina in industrial alumino-silicates [15] or the location of an Sn atom in a zeolitic structure [16], to observe the surface Cd atoms in nanocrystal quantum dots [17] or the reaction intermediates of metathesis catalysts [18], and to obtain bond distances and 3D structures of surface sites [19].

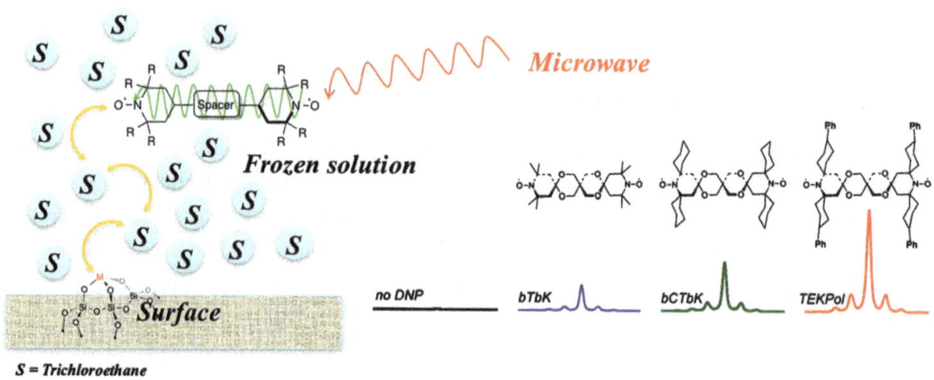

Fig. 4. Schematic representation of polarisation transfer in DNP SENS and a selected family of radicals.

Outlook to future developments of research on single-site catalysis

Understanding the structure of active sites in single-site supported catalysts and their relation to reactivity is now possible and enables more rational development of catalysts. It also provides unprecedented information that can be transferred to existing industrial catalysts, for which the small amount of active sites and the ill-defined nature prevent such study at the current stage. This has been made possible thanks to the development of molecular chemistry, advanced spectroscopic and computational methods.

One may question whether or not well-defined supported catalysts can be used in industrial settings in place of the current ill-defined systems. Indeed, their higher activity and selectivity would lead to much more efficient operational units. A clear caveat towards this direction is their current price and the risk associated with changing industrial processes.

Molecular approaches have changed our view of supported so-called single-site catalysts, but they have not yet penetrated a larger class of catalysts, based on supported nanoparticles. Supported nanoparticles are obviously much more complex as they are constituted of a metal particle with different sites, a support, and an interface with the support, which may be the place where chemistry is taking place. One may hope that it will be possible using molecular approaches to interrogate these much more complex systems and to exploit the cooperativity between the nanoparticles and the support to discover more efficient processes [20].

One may expect major breakthroughs in operando spectroscopy and microscopy, which will enable to directly observe the active sites under reaction conditions and to distinguish active from dormant or inactive species (see contribution of Prof. B. Weckhuysen). Concerning solid-state NMR spectroscopy specifically, one may foresee that DNP SENS will enable 3D structural determination of surface sites, similarly to what was done in molecular biology using solution NMR spectroscopy. Further increase of signal intensity by the improvement of DNP SENS will also enable to compare perfectly defined surface species and materials derived from surface science (see contribution of Prof. H. J. Freund) with these of industrial catalysts, thus bridging the gap between surface science and industrial catalysis. It is clear that further advances in analytical and computational methods including artificial intelligence as well as in molecular chemistry will help design of better catalysts. Acceleration of discovery and transfer of technology is also a critical aspect of modern research. High-throughput experimentation combined with the conceptual approach described here can certainly help reaching that goal.

The discovery and the development of more efficient (environmentally friendly sustainable) process will not be possible without a more global approach to solving problem through transdisciplinary research and collaborations. Time will tell.

Acknowledgments

ETH Zürich and SNF are gratefully acknowledged for their continuous financial support. CC is also grateful to his long standing collaborators, in particular Prof. Eisenstein, Emsley and Schrock, and the students and post-doctoral fellowships who have contributed over the years to the advancement of the field.

References

[1] D. G. H. Ballard, *Advances in Catalysis* vol. 23, D. D. Eley, H. Pines, P. B. Weisz (eds.), 263 (Academic Press, 1973).

[2] C. Copéret, A. Comas-Vives, M. P. Conley, D. Estes, A. Fedorov *et al.*, *Chem. Rev.* **16**, 323 (2016).

[3] (a) J. L. Hérisson, Y. Chauvin, *Makromol. Chem.* **141**, 161 (1971); (b) Y. Chauvin, *Angew. Chem. Int. Ed.* **45**, 3741 (2006).

[4] M. Chabanas, A. Baudouin, C. Copéret, J.-M. Basset, *J. Am. Chem. Soc.* **123**, 2062 (2001).

[5] (a) X. Solans-Monfort, E. Clot, C. Copéret, O. Eisenstein, *J. Am. Chem. Soc.* **127**, 14015 (2005); (b) A. Poater, X. Solans-Monfort, E. Clot, C. Copéret, O. Eisenstein, *J. Am. Chem. Soc.* **129**, 8207 (2007).

[6] M. Pucino, V. Mougel, R. Schowner, A. Fedorov, M. R. Buchmeiser *et al.*, *Angew Chem. Int. Ed.* **55**, 4300 (2016).

[7] M. P. Conley, V. Mougel, D. V. Peryshkov, D. Gajan, A. Lesage *et al.*, *J. Am. Chem. Soc.* **135**, 19068 (2913).

[8] V. Mougel, K.-W. Chan, G. Siddiqi, K. Kawakita, H. Nagae *et al.*, *ACS Central Sci.* **2**, 569 (2016).

[9] M. P. McDaniel, *Advances in Catalysis* vol. 53, B. C. Gates, H. Knözinger (eds.), 123 (Academic Press, 2010).

[10] E. Groppo, A. Damin, C. O. Arean, A. Zecchina, *Chem. Eur. J.* **17**, 11110 (2011).

[11] (a) M. P. Conley, M. F. Delley, G. Siddiqui, G. Lapadula, S. Norsic *et al.*, *Angew. Chem. Int. Ed.* **53**, 1872 (2014); (b) M. F. Delley, F. Núñez-Zarur, M. P. Conley, A. Comas-Vives, G. Siddiqi *et al.*, *Proc. Natl. Acad. Sci. USA* **111**, 11624 (2014).

[12] (a) A. W. Overhauser, *Phys. Rev.* **92**, 411 (1953); (b) T. R. Carver, C. P. Slichter, *Phys. Rev.* **92**, 212 (1953).

[13] (a) L. R. Becerra, G. J. Gerfen, R. J. Temkin, D. J. Singel, R. G. Griffin, *Phys. Rev. Lett.* **71**, 3561 (1993); (b) D. A. Hall, D. C. Maus, G. J. Gerfen, S. J. Inati, L. R. Becerra *et al.*, *Science* **276**, 930 (1997); (c) Q. Z. Ni, E. Daviso, T. V. Can, E. Markhasin, S. K. Jawla *et al.*, *Acc. Chem. Res.* **46**, 1933 (2013).

[14] (a) A. Lesage, M. Lelli, D. Gajan, M. A. Caporini, V. Vitzthum *et al.*, *J. Am. Chem. Soc.* **132**, 15459 (2010); (b) A. J. Rossini, A. Zagdoun, M. Lelli, A. Lesage, C. Copéret *et al.*, *Acc. Chem. Res.* **46**, 1942 (2013).

[15] (a) A. Zagdoun, G. Casano, O. Ouari, G. Lapadula, A. Rossini *et al.*, *J. Am. Chem. Soc.* **134**, 2284 (2012); (b) D. J. Kubicki, G. Casano, M. Schwarzwälder, S. Abel, C. Sauvée *et al.*, *Chem. Sci.* **7**, 550 (2016).

[16] (a) P. Wolf, M. Valla, A. J. Rossini, A. Comas-Vives, F. Núñez-Zarur *et al.*, *Angew. Chem. Int. Ed.* **53**, 10179 (2014); (b) P. Wolf, M. Valla, F. Núñez-Zarur, A. Comas-Vives, A. J. Rossini *et al.*, *ACS Catal.* **6**, 4047 (2016).

[17] L. Piveteau, T.-C. Ong, A. J. Rossini, L. Emsley, C. Copéret *et al.*, *J. Am. Chem. Soc.* **137**, 13964 (2015).

[18] T.-C. Ong, W.-C. Liao, V. Mougel, D. Gajan, A. Lesage *et al.*, *Angew Chem. Int. Ed.* **55**, 4743 (2016).

[19] P. Berruyer, M. Lelli, M. P. Conley, D. L. Silverio, C. M. Widdifield *et al.*, *J. Am. Chem. Soc.* **139**, 849 (2017).

[20] K. Larmier, W.-C. Liao, S. Tada, E. Lam, R. Vérel *et al.*, *Angew. Chem. Int. Ed.* **56**, 2318 (2017).

SESSION 4: CATALYSIS UNDER EXTREME CONDITIONS: STUDIES UNDER HIGH PRESSURE AND HIGH TEMPERATURES — RELATIONS WITH PROCESSES IN NATURE

CHAIR: HENK N. W. LEKKERKERKER
AUDITORS: F. DE PROFT[1], T. VISART DE BOCARMÉ[2,3]

[1] *Vrije Universiteit Brussel, Research Group of General Chemistry (ALGC) Pleinlaan 2, 1050 Brussels, Belgium*
[2] *Université libre de Bruxelles, Chimie Physique des Matériaux et Catalyse, Faculté des Sciences, Campus Plaine CP243, 1050 Brussels, Belgium*
[3] *Université libre de Bruxelles, Interdisciplinary Center for Nonlinear Phenomena and Complex Systems, CP231, 1050 Brussels, Belgium*

Discussion among the panel members

Henk Lekkerkerker: This is quite a variation of topics, I must say it was extreme. But just like a meal with many courses, I hope there was something to your liking in everything. And now I would like to start off the discussion among the panelists. Who would like to start first? Bert Weckhuysen.

Bert Weckhuysen: Here is a question for Christophe Copéret. You showed these defect sites and I would like to come back to my earlier question this morning on Lewis acidity and Lewis acid sites on alumina and how would that affect catalysis. If you think about the comment that was made yesterday by Jens Nørskov when he stated that the exponential is very important and then you showed that you have defect sites at very low amount, so what is then their precise role?

Christophe Copéret: to answer Bert, I think that we see that on oxides: we can easily generate defects by applying temperature. And defects are not weird species, what I call undercoordinated species, and we have shown recently that these defects can activate the C–H bond quite easily at low temperature and that, when you have something like methanol or DME (dimethyl ether), you can easily make the C–C bond at low temperature. And the question is: "is there a relation with alumina and zeolites?" This is a question I cannot answer yet, I can only propose that we know that zeolites contain aluminum and that under extreme conditions which are typically the reaction of MTO i.e. the methanol to olefin process, you can imagine a reorganization. We know there are binders, we know that there is alumina in zeolites as they are made in industry, so the question that I shall pose to all my colleagues and to you as well, Bert, because you can actually image much better than we can do today: are they all related? Is alumina released to activate the C–H bond and then to make a new coupling or is it unrelated because it is possible that the zeolite by itself can have its own life? So I don't know. I don't know, Bert, but maybe you can tell me, since you do imaging. Can we image today defect sites by this new technology that you have developed?

Bert Weckhuysen: Two things. First is, when we think about catalytic cracking in such a real ECAT (equilibrium fluid cracking catalyst), sample of this 50 or 100 micrometres size where alumina is a binder, then one of the issues in the mechanism is that an olefin has somewhere to be made. Ok, you can have protolytic cracking but also you have this carbenium type mechanism but the question is: what is the relative contribution of thermal cracking versus catalytic cracking and where is taking place what; in other words, what is the role of the alumina binder in the generation of active intermediates? When I see your methane activation work, I see an alternative route to make it: not thermal cracking but catalytic. Your question about imaging Lewis acid sites: I can tell you that we are currently working, already for some years, on finding (and we have already some) probes, nanoscale probes to probe Lewis acidity. It is not yet published. It is actually not so easy, because the point is that often these molecules that you want to have visualized, they have an aromatic ring and you do not always know what you make, so there is still an issue on what we are really probing. But we hope at some point, that's my wish, that we could make next to what is currently done in industry and academic labs where pyridine and ammonia-TPD (temperature programmed desorption) are routinely used for acidity probing, we will have tools in our hands to probe at the nanoscale both Brønsted and Lewis acid sites. That would be maybe a time where we just say we have one catalyst particle or we have multiple particles and we can just say this particle has this number of Brønsted acid sites and this number of Lewis acid sites. So it is a matter of time. It is also I think a matter of convincing people that it is worthwhile to further pursue this approach of nanoscale chemical imaging, but that is somewhat a trend.

Christophe Copéret: I want to continue the discussion because I didn't answer all the questions of Bert. Coming back to Lewis acidities, so we know obviously from N_2 coordination that the sites are extremely Lewis acidic otherwise they would not bind N_2 because N_2 is not directly the most Lewis base compound that we know. Clearly this Lewis acidity is very important for the binding, but I would say that in oxides we should be very careful when we use the term Lewis acidity because the close-by oxygen is very important so we always have to deal with something that is not a pure Lewis acid, there is always a Lewis acid associated with a Lewis base. And this is a big difference between molecular chemistry, I would say, and surface chemistry: we can isolate sites and generate sites which are very difficult to make in real molecular chemistry because you can have at the same time, a bit like enzymes in a way, you can have a base and an acid in, say, less than a nm apart. So it would be very important to be able to image both the aluminum site, the oxygen site, because there is a Lewis acid and a Lewis base pair. I think this would be a big challenge that we will have to face in the coming years but you will have to solve it.

Henk Lekkerkerker: Any of the panellists would like to follow up on this part of the discussion? Marc Koper.

Marc Koper: I have a question for Ruyhei Nakamura, because I am intrigued by this idea of working under a temperature gradient and maybe to have a fuel cell working under a temperature gradient, but then you would need an electrolyte that would transfer the ions but would not transfer the heat and typically also the electrolytes in these devices are extremely thin. So, how would you imagine, I understand it is just an hypothesis, but how would you imagine that something like this could work?

Ruyhei Nakamura: Are you talking about a natural system or about technology?

Marc Koper: About the technology.

Ruyhei Nakamura: Nature is an open system, the convection-diffusion decreases in the electrolyte, so we don't have to care about the ion conduction so much, but in the practical application if we can make such a flow system, it is possible.

Marc Koper: I guess in the practical system you would want to minimize the Ohmic loss, so the electrode should be extremely thin, ideally.

Ruyhei Nakamura: In other words, we have to develop new electrolytes which can transfer ions but not heat. So we need new materials.

Marc Koper: Would you have any idea what that material would look like?

Ruyhei Nakamura: No. I don't have.

Christophe Copéret: I have a follow-up question to my colleague Nakamura. More philosophical maybe. We can do electrochemistry in extreme conditions, so how do you see a relationship basically with the origin of life? Obviously, when you can do electrochemistry and we know that the potential can drive a reaction to any direction including reduction of CO_2, can you imagine actually that hydrocarbons and other compounds could be made in the deep sea?

Ruyhei Nakamura: Currently, the origin of life triggered by electrochemistry is an emerging new field. Not many people are working on this: in Japan, there are three groups, in US one group and in the UK there are two groups. Molecular hydrogen is a primary energy source but as you know the reduction of CO_2 by hydrogen is a kinetically and thermodynamically tough reaction so nobody demonstrated experimentally CO_2 reduction by hydrogen under hydrothermal environment. But today, I demonstrated one possibility. If we have some membrane and if it can maintain the pH gradients and temperature gradients we can generate high energy electrons which can actively electrochemically reduce CO_2. Experimentally, we already demonstrated CO_2 reduction to methane, but the performance is very very low.

Christophe Copéret: Do you see also sulphide forming as well because when you see metal sulphide and you come from the petrochemical point of view, metal sulphides are very good to do hydrodesulphurization. You can remove sulphur from basically any aromatic compound, thiophenes, so you can imagine to do the reverse in nature. Have you tried to look for any metal organic sulphur containing compound in the system as well?

Ruyhei Nakamura: Unfortunately, I haven't considered this kind of possibility.

Bert Weckhuysen: A question to Stig Helveg. First, fantastic work and also a nice movie. I have two questions. The first is on beam damage, actually, you alluded to it but I want to hear it a bit more clear because it is a very important point and I also want to indicate that such things may also happen with X-ray microscopy. We also know that you have to be very careful and that even in the beam you can just have reduction. We have seen for example that, if we are not careful, *e.g.* Mn can be just reduced in the X-ray beam. And especially when you try to go to liquids, then the X-rays are doing a lot with water. So the beam damage, what do you do to avoid it? And what would it imply if you would move to liquid phase catalysis?

Stig Helveg: I also had sort of the same question for you. First of all, if you look at the current state of the techniques in electron microscopy and you look at what amount of energy we deposit in the system with an electron beam of the incident energy we have here, then you will very fast learn that the energy we deposit compared to the dose we need to create a signal and maintain resolution is just like living on the edge of a knife. So you deposit as much energy as in chemical bonds or more with the dose you need to get a detectable signal. And that is why we need to make an optimization of our detection techniques. What saves us is that much of the energy we deposit goes into the electron or phonon degrees of freedom, which means that the rate of delivery of the electrons is something you can beneficially tune when excitations decay reversibly. We employ tools and techniques that we could call a "divide and conquer" concept to acquire data as complementary to the "hit and destroy" approach you have with free electron lasers. But it is an empirical approach to do that, to quantify those effects and to make sure you are in a window where you can, in a quantifiable way, say you are on a safe ground. This is how we attack it, and I think each and every study needs that because we are so close in signal and damage that it depends on the catalyst system that you look at where you have the damage onset. My question to you is of course the other way around that if you look at how much energy you then deposit per elastic scattering event, with an X-ray beam, I know that that is around 100 to 1000 times more. It relates to the fact that you scatter only on electrons and not the nuclei which is what we do with the electron beam, right, we scatter mostly on nuclei, so I guess that the theme is even more important for X-rays and I don't know if it would be limiting, but that is something you can maybe comment on.

Bert Weckhuysen: I already made the comment that we sometimes see this. And you can ask yourself how do we know that then? We published in the Journal of Physical Chemistry a paper where we measured with one technique, UV-VIS in the same *in situ* cell, and we did an X-ray beam off/beam experiment and then we were visualizing what was happening. We actually even had in a later study an optical camera on it. And the chairman even made colloidal particles from a Cu catalyst with water around, it was liquid phase catalysis. With my group I am since 2000–2001 in the field of this X-ray spectroscopy of catalytic solids. Liquid phase catalysis is, (if you see how much we have published so far, although we did a lot), a lot is beam damage or not true what it is. So, biomass catalysis we tried, it is still an issue. Secondly, we found out that the X-ray energy is very crucial. Surprisingly, very hard X-rays, *e.g.* 80 keV, where you can shoot through big reactors, centimetres, there is almost no damage. But when you go to low energy, in the soft X-ray regime, actually it sticks on, and that is also the reason why we are moving to the tender X-rays and the hard X-rays, because there we see less beam damage. But it is all empirical, I must admit, so it is more experience and the way we developed it. I tell it here, nobody even wants a paper to be accepted where you report on beam damage, we had a hard time to get the paper accepted.

Henk Lekkerkerker: So X-ray and electron microscopy are complementary, each having their strength and perhaps weaknesses. I was wondering: are there questions for Xinhe Bao among the panellists?

Xinhe Bao: The first question I want to ask Bert Weckhuysen: the time resolution will be very interesting. I remember the time when I was in the Fritz Haber institute, Prof. Gerhard Ertl discovered such a structured change for the carbon monoxide adsorption. Sometimes time resolution is part of the second scale. So now, of course it is not enough, so my question is: in your opinion, in the future, which time resolution is available for the catalyst research? In other words, in your opinion, what will be the good method for such things?

Bert Weckhuysen: First of all, the movie which I showed for example, every 20 minutes an hour, you can make a decent 3D image in such time period. If you focus on one voxel and you are only taking some energies out of it, you already actually know how the spectrum looks like. So if you say you are only going to measure 3 or 4 X-ray energies then you can really bring the measurement time down. Your synchrotron is actually a pulse source anyway then that is your limitation, and the detector. So then you are in the milli, micro, even further, I have not seen so many results so far with the current detectors, milliseconds, but I think we can go further or faster. Then, free electron lasers are always proposed. We are trying to get beam time or time for that but there we still struggle that you have to have a repetitive process. For photocatalysis, I am optimistic, we can do that, but you have to have a system that you can trigger and in photocatalysis you can do this, then you can go very fast. I think we will have to have multiple methods, electron microscopy, X-ray microscopy, tip enhanced Raman, etc. So these things where you have the molecular organic part (TERS) Raman spectroscopy single molecule, and on the other hand where you can go for oxidation state, metal states, etc.

Xinhe Bao: Another question for the electrochemists. I think now is really the time for us to revisit the electrochemistry. Because nowadays, we have a lot of renewable energy with electricity. A big amount of the electricity cannot be used so good, then a lot of people are thinking to use such energy for a catalytic reaction. In this case, I think we are facing two problems. The first one, if you want to do electrocatalysis, normally you have to use a noble metal. This is again the noble metal problem. Another one is for the electrocatalytic reactor, the scale up has some limitation. If you want to do this for the big systems, you have to do the scale up.

Henk Lekkerkerker: Who wants to answer? Marc Koper.

Marc Koper: To start with the last part of your question, big scale electrochemistry set-ups already exist. So chlorine evolution is already scaled up, a very large scaled up electrochemical process that also works under oxidative conditions and there is a stable catalyst, the only problem is indeed that it makes use of Ru, so it uses a noble catalyst, but in principle, it can be done, of course, it is a big capital investment. For the scale up of electrochemical principles, it is known how to do that. You can also do large scale hydrogen evolution. I think the stability issue is of course something that happens as soon as you work in a liquid. And when you work under oxidative conditions, you need to worry about corrosion and again part of the solution is working in the right electrolytes, *e.g.* by going to alkaline electrolytes, you can make use of oxides and stable oxides that do not dissolve and *e.g.* alkaline electrolyzers make use of Ni electrodes. So I think there is room there to play but obviously this is one of the reasons why fuel cells are still only going slowly in the market and one reason for this is the stability issues. But I think that is something that would be addressed as these things become scaled up but a very big problem there of integrating this into our society is just the capital investment that you would need to do in doing very large scale electrochemistry.

Ruyhei Nakamura: About the second comment, you talk about the electrocatalysts still relying on the noble metal, but a recent development of the theory of electrocatalysis is amazing, like Professor Nørskov described yesterday, because now we have the time to catch up with descriptors for activity, and that depends on the catalyst. Now we are screening the descriptors and then rational design becomes possible.

Henk Lekkerkerker: It is time for the coffee break.

General discussion

Henk Lekkerkerker: I realized that the last session was a mixture of things, I hope it was sufficiently extreme for the organizers. I think it is perhaps a good idea to structure the discussion around the three main topics we addressed and then eventually go to more free style things. I would first of all like to solicit questions, comments, or otherwise about the techniques to see the catalyst in action. Who would like to be the first to open the discussion? If not, the chairman has to do more talking but I see already that Graham Hutchings from Cardiff raises his hand.

Graham Hutchings: On the microscopy, we have seen fantastic developments in the last few years. What is the next big thing we are going to see in microscopy that is really going to help us in understanding the catalysts and looking at them? I mean environmental transmission microscopy (ETEM) is here, we can do it, but not under all the environments that some of us would like to use. What's the next big thing? We have single aberration corrected microscopy, double aberration corrected microscopy, now we can do it *in situ*, is this it? Or, for somebody who uses this technique, do you know if there is something else coming?

Henk Lekkerkerker: I will first ask Bert Weckhuysen to comment on this question.

Bert Weckhuysen: Future predicting is always difficult. There are a few things on microscopy and imaging catalysts at work on the nanoscale. Can we connect kinetics information which is going on in a single catalyst particle or a single metal particle and really connect that to each other? So, can we have surface reconstruction or the dynamics or the heterogeneity of these dynamics, can we correlate that with a kinetic description, can we connect that with each other? That is what I call the marriage between this organic and inorganic chemistry of catalysis and that is a challenge although we had nice movies, etc. We already discussed beam damage as a possibility. So, at the moment organics are around with maybe the exception of CO and NO but at the moment some molecules start to be complex and we cannot ignore that measuring is perturbing and the question is to what extent? Then the molecular movie has not only to have these organic/inorganic aspects but it also will have to include the dynamics in the sense that you would have to push the time resolution. There was already a question about that. For example, I am also working together intensively with Stanford (SLAC) University where we are trying to see how far these synchrotron radiation sources could evolve and in the coming decade I think, we would see that it would really be pushed to the 1 nm regime. We will never beat at this moment electron microscopy with this but I am quite sure we will obtain interesting information. There are already lensless systems. There are ways, for example X-ray ptychography is around, so it is a matter of moving into a regime where we are getting closer and closer to what electron microscopy could offer. If that is the case then we can also bring in spectroscopy, so my question for electron microscopy — but that I have to pass to Stig — if you think about EELS detection and spectroscopy bringing in the electron microscopy, how far can that be brought and how sensitive can it be made and where do they then at some point meet each other as nicely shown in one of the last slides of Stig where he was showing electron microscopy as a bridge between the different worlds, i.e. the real world and also the surface science world? That is where — I hope — this will go.

Graham Hutchings: One of the complexities with the microscopy and what you are talking about is that the microscopy has to be totally vibration-free to get to the level of resolution that we need. I personally think it would be difficult to stick this inside a synchrotron environment or the bits of metal that you have to bring around the microscope; I thought about this. Maybe I'm not thinking about it in the right way. What is your opinion on this?

Bert Weckhuysen: When I entered this field in 2000, then people also stated that certain things were not possible, and I must say that we have shown that things can be progressed a lot. Then I come back to Hajo Freund's statement yesterday, a very important message: he stated that when we have to develop new tools, new methodologies, we also have to make sure that we are really persuasive and that we stick to our ambition. We will have to make compromises in that adventure. Some of the things I showed today originate from concepts developed in 2005, but I have published them only this year or last year. That takes a while, not only to convince synchrotrons to go and invest, etc. It often takes a while before you are where you want to be with such type of experiments. It is often not simple to do such experiments. Questions than pop up. For example, how do you measure the temperature of the catalyst particle? What about the flow through such a cell? We did not discuss the dynamics, the glass flow dynamics through this. So a lot of these things are hidden. We have been, and I am quite sure that Stig will confirm that, thinking and at least checking that.

Kurt Wüthrich: In structural biology, cryo electron microscopy (EM) is considered to extend single crystal X-ray crystallography. Then we have the electron-free laser in development, big moneys are spent on these techniques. What do these techniques promise for heterogeneous catalysis projects?

Stig Helveg: There are a lot of topics on my list, I will try to limit my answers. Concerning cryo EM, it is true that it has evolved tremendously in the last few years. Now we begin to see biological structures with a few Angström resolution. That has come because of optical advancements in optics, cryo-stages and data processing. One thing that you gain there certainly is you get the fixation of the molecules in the ice. There are also some considerations that the ice helps protecting the molecules from beam damage. But it also illustrates another thing, which is that if you can fixate molecules you may get a picture of them. There are also other configurations where people have immobilized molecular structures like DNA, and bundles and so forth, and the first atomically resolved images have been published on that.

We can discuss how much of the internal structure you learn from that or not because the big question here is actually hydrogen: how well can we keep hydrogen in place without shooting it off? That is a big problem in EM, or challenge I would say at this point in time. That brings us back to the question about beam damage. I still think it is so that if you go back to one of Henderson's old papers, it says that if you go through the estimations you will reserve one Angström X-ray beam compared to a 100 kV electron beam such that, per elastic scattering event, you will deposit 100 times more energy in X-rays than in the electrons which gives you a favor for the electrons to the X-rays. We can then discuss these calculations, that's another thing, but I think that this idea reflects the scene that we have today in terms of resolution in the different instruments. So, to come back to you Bert, on where do you see the overlap between the tools. X-rays and electron microscopy are certainly powerful when used back-to-back. With electron microscopy, it is also possible to combine imaging with electron loss spectroscopy, which is particular sensitive to light atoms. This offers means to measure gases at high spatial resolution, which have been used to measure a local temperature or composition of a gas environment confined in the electron microscopes. So that is certainly an advantage of spectroscopy with electrons to address some of your questions. And if I may use two seconds more, to maybe come back to Graham Hutchings: what would be the next big thing in electron microscopy? It is true that within the last 10 years what has happened in EM is simply tremendous, you have resolution well down to sub-Ångströms, you have time resolution on the femto- to picosecond time-scale and with the ability to introduce reactive environments. So that is sort of the scene you have today. I think you don't need to have been many days to school to say "could all this be used in one instrument?" That would be a big thing but of course you need to sit down and think about what would be the challenges ahead of us. One challenge is certainly to find strategies to cope with beam damage and find strategies to work around that. My view on that is we are on a track at this moment to figure out how to handle it and how to describe it. So if you give it a few more years then we will probably say if this microscope may come at some point. But right now, it looks a little bit science fiction. I also see the challenge in the detection principles we have.

You will be limited by noise, noise is the killing factor today. If you see a crappy atom-resolved image from EM it's most likely relevant within chemical terms. It may be a bold statement. But if you look at one that looks good it is important to ask at how it was obtained. There are of course images that can be trusted out there, so the point is to be critical. That would be the next big blow and you can say what can you then gain from it from the catalytic point of view. I mean, if you were free to think, that is how I understood your question. If money and time was not a limit, we have come to the state where you can begin to look at a catalyst at atomic resolution under relevant reaction conditions. Why not dream of having the possibility to look at the adsorbates at a single site? I don't know if it is realistic, but you have the question: if you were free to think, could you do that? And if you come down and have resolution down to even light atoms, like carbon in *e.g.* DNA, graphene, what limits there is fixation or time resolution because if the molecules are not stable enough before you have acquired your signal, then you have lost the game, right? But that is a free thought and maybe we will never get there.

Henk Lekkerkerker: Professor Gerhard Ertl.

Gerhard Ertl: I understand that you are interested to look at the processes taking place at the atomic scale during catalytic reactions. This concerns small particles and these small particles expose different crystal phases so if you use from the beginning a flat surface you can apply all these scanning probe techniques, which are very powerful. They can be used from lower pressure up to atmospheric pressures even in liquids in electrochemical environments, there is no beam damage and they can be very very rapid, you can follow processes on the millisecond time scale. So, I don't see any strong advantages of the electronic microscopy techniques in answering these kinds of questions.

Stig Helveg: To an extent I will agree with you, but Jens Nørskov showed yesterday that to break scaling relations and come into play with new types of chemistry, we should also look for new types of sites, which may not be so easily modelled by a planar surface. And that is where the electron microscope has an advantage compared to the scanning probe, because the scanning probe is good as long as the surface you look to is reasonably flat. I remember some beautiful things that Hajo Freund has been involved in on clusters showing that as long as you have something that begins to step out in a few layers then scanning probe microscopes begin to be challenged in terms of maintaining atomic resolution.

Henk Lekkerkerker: Does this answer satisfy you, Professor Ertl? Yes. I have seen Hajo Freund.

Hans-Joachim Freund: I was about to say the same thing as Gerhard Ertl was just saying. The argument that you have to use a flat sample is true but you don't even have to do STM, scanning tunnelling microscopy. Scanning force microscopy, atomic force microscopy is getting to a stage where you can get as good atomic resolution as you used to get only with STM. You don't even have a current flowing but you only have a contact. Now what you do have to have is a flat surface. You want to understand catalytic phenomena. If you work on the model, there is no limit to complexity, I think one also has to work on establishing the model properly for the material you want to look at. Once you have that, I think an atom probe method is the ideal solution, but you have to work on making these materials or samples that you can look at, right, but you don't have all these problems that you have with electron microscopy. So, I think if both of these strategies go together that would be the best solution to the problem.

Stig Helveg: I tend to agree with you of course, I have a surface science background myself. So I very much like that, but I also know how difficult it can be to establish a good model. The way I understand you is that the approaches can come together because you also need to know what to put into that model from the beginning and that is probably where the electron microscope is coming in as a good partner for the scanning probe microscopes. So, I think we pretty much agree.

Bert Weckhuysen: I am not in STM, but we are currently working and have also worked on AFM-like approaches including spectroscopy. Currently, we are measuring infrared, we also have been working on Raman. One of the first papers even was by Gerhard Ertl, so we are also trying to find model systems for zeolites or for porous materials and we are trying to make thin films which we grow on substrates. Yes, we would like to have this flat surface approach. But also you would like to have a third dimension in it, you want to go from a 2D model to a 3D model. And then the question will be how complex will that 3D model be, how many layers you will have to have where you still can do the nice spectroscopy, but also will be able to see somehow a bit deeper into the 3D material. That compromise, finding a good model system, how it looks like and then combining that with scanning probe methodologies, so I am not an STM specialist at all I am more in this AFM type approach I am now learning my way. There, I see that making a thin film of a zeolite. So, the materials challenge of making a good model system is already a challenge on its own, in its own right. So back to Graham, it would be advantageous if we could translate for example two real systems, for example we take a cracking or a hydrocracking catalyst and we say what do we now have as elementary pieces and how we can then make it to a model system. And the other way, we could think about how to move to another system. How could we then make model systems that we can put in the electron microscope which we grow on these silicon nitride windows? It should be possible. The same with AFM or STM-type methodologies or with X-ray, etc. or single molecule. That could be maybe the marriage where we could go and have then real detailed insight.

Henk Lekkerkerker: Thank you Bert, unless there are burning questions about the future of all these techniques to visualize, I would now like to turn to the provocative statements of Christophe Copéret and Xinhe Bao. Who has questions for these two gentlemen? Kurt Wüthrich.

Kurt Wüthrich: The question has to do with the NMR experiments that Christophe Copéret used together with colleagues in France. The major hurdles that are being overcome in structural biology with new techniques, I mean electron free laser and cryo EM, is that ever smaller crystals can be used for structure determinations. I do not know what corresponds to this when studying surfaces but there is one number I would like to present to the group. When we run an NMR experiment we would typically have 10^{16} or thereabout spins in the sample. That is about the detection limit. Now I am interested to hear what the number of spins is in the samples that are used for these surface area studies with NMR.

Henk Lekkerkerker: That is a very precise question and I expect a very precise answer.

Christophe Copéret: Typically we deal with a material of 100–200 m^2/g. This is very important because if you go to lower surface area materials, if you want to bridge the gap to what we do and surface science, there are many more gaps. So 200 m^2/g typically. If I look at OH groups, we can control basically one OH per nm^2, this is something we know how to do. So, if you look at the number of spins, it is actually not too bad, something like 10^{18}, so it looks much better than proteins. However, because we have a solid, we have several other problems. We have line broadening effects because we not only have one site but we have a distribution of sites. Then we have broadening because we have chemical shift anisotropy and then we have broadening because we have residual dipolar coupling. So what looks to be less challenging than proteins turns out to be much more challenging because the signal at the end of the day is very broad. So there are many techniques that have been developed over 40 years like Magic Angle Spinning. Today we can spin the sample very fast. My colleague Bert Meijer, that you know very well, spins samples up to above 100 kHz and then we can get rid of most of the dipolar coupling in our sample. But if we are talking about a well-defined site on the surface, now if we look at the defect, then you are again two orders of magnitude lower then the defect of ^{27}Al which means that now you have a quadupolar coupling constant, which is basically of the order of 40 MHz. So it is impossible to average. You can only decrease the signal or the broadening but you cannot get the average. To answer your question: basically for a well-defined solid catalyst, it is a challenge without labelling, and for a defect today, especially aluminum defects, it is impossible because it turns out that the aluminum in the bulk has a very low quadupolar coupling constant while the surface defect has a very broad quadupolar coupling constant so the only thing you see is the bulk, whatever you do. Actually we wrote a paper two years ago to tell to people "be careful what you measure, because very likely what you measure has nothing to do with the surface". And today, we don't know how to measure this, we have been discussing with people in France, Massiot in particular, who are experts in quadupolar nuclei. Today, we have no technique available to measure a defect which has a very large quadupolar coupling constant.

Henk Lekkerkerker: I am asking Kurt Wüthrich: does this answer satisfy you?

Kurt Wüthrich: Thank you.

Xinhe Bao: I just want to answer this. We have also done some solid-state NMR for catalysis but not for the surface. You have also visited our laboratory. The main problem is that an efficient catalyst normally is iron or nickel on a surface. The magnetic problem is the problem we are facing if we want to do something for this. Another thing is that, for the solid-state, we have to use magic angle spinning (MAS), also in a very fast manner. By using such a technique, it is quite difficult for us to connect it to the reaction system. So we cannot do it under *operando* or *in situ* conditions, just under *ex situ*. There is another thing for our reactions: sometimes they are at high pressure and also at high temperature, this is also some problem. We are now also working at high pressure but it is difficult to do the *operando in situ*. Now a lot of people are thinking about DMP, DMP is a technique to combine with solid-state NMR of the Fe/Ni-based systems, which will certainly be interesting.

Kurt Wüthrich: Are you actually doing high pressure work with magic angle spinning?

Xinhe Bao: We have, but this is not connected with a gas. For example, for the synthesis of zeolites, there we use also high pressure, sometimes 200 atmospheres. But this is a closed system, and there we also use the magic angle spinning, and this is not connected with another preparation system; this is isolated. We can also get results. But for us, if we want to get operando spectroscopy information, we have to connect it with the reaction. This is a problem for us. So I have a very interesting idea. Now the load is spinning very fast. Later if the magnetic field can also load very fast then it could help. I don't know if this is possible or not. If I can do such a system, then probably your problem would be solved.

Henk Lekkerkerker: Thank you. I would like to move to Hajo Freund from the Fritz–Haber Institute.

Hans-Joachim Freund: I have a general comment. We have been doing EPR, electron spin resonance, on single crystal surfaces where you have of course a sensitivity of something like 10^{-11} or even less spins which is a hundredth of a monolayer. But the problem that arises is exactly the problem that Christophe Copéret addressed: if you don't pay attention, you only see the bulk. And of course you can all do this under high pressure. It's tricky with the cavities you have to use but in principle you can do this. So, in principle, on model systems if you have a good control of your surface, you can make sure that even in an EPR experiment, all your signal comes from the surface, and you would have the sensitivity of say, a hundredth of a monolayer. So we can see defects in an MgO crystal. If we make the complex as thin films, which you can prepare very precisely (you can never do that with a bulk single crystal; you would only see the defects in the bulk) then you can see the surfaces exclusively. And you know exactly that your signal comes from the surface because you can titrate these things. The signal disappears by titrating it with a probe that quenches the signal. So in principle, magnetic resonance could be done, pretty tricky. But sensitivity always would be there.

Henk Lekkerkerker: I think Christophe Copéret wants to react on this.

Christophe Copéret: Hajo Freund actually points very well to the problem. For DNP today, our current record is that we can detect a phosphorus every nanometre on a glass plate. Now that we can actually address flat surfaces, we are getting closer to surface science, but we are not there yet. But I believe that we will be able to close the gap. To follow the comment of Professor Bao, today we can do NMR of paramagnetic species, we have a research program. We can characterize fully surface chemistry of paramagnetic species, so that is not a problem. This is possible today.

Henk Lekkerkerker: About measuring techniques: what about the exciting chemistry of Xinhe Bao? I feel that Rutger van Santen may want to comment on this.

Rutger van Santen: Concerning these very fascinating results of the single iron that converts methane at high temperature to ethylene and possibly acetylene. The amazing part of that system is of course the high stability. A question that one has is: what happens to the surface structure during the catalytic reaction? Is it still the same as it was before the start of the reaction or is there a reconstruction or something else?

Xinhe Bao: The concentration of the iron is normally less than 0.5% totally. At the beginning, this is iron oxide. During the reaction, one or two reaction treatments to methane at different temperatures, by such treatment, we get structures. The temperature is about 900–950°C. Now we know that this structure is that of iron atoms confined in the lattice. We have used silicon as silicon carbide. There we know that the iron is connected to the silicon. The two carbons are connected with the silicon. This is really a stable structure. We have also tested this material under *operando* conditions using synchrotron radiation for a long time. There the signal is still stable. It means that the silicon-iron bond, or the carbon bond, is very stable during the reaction. For the surface, it is difficult to say because the temperature is really too high. I just want to talk to you about one result: before we used fixed-bed reactor, we used silicon powder, but now we are just coating the iron, or the silicon iron on the walls of the tubes. There it works also very well. It means that for very low amounts of iron or another metal, they create the radical that will be directly in the gas phase, very initially.

Rutger van Santen: But your conditions are such that thermodynamics would like to generate aromatics and at the end carbon deposits, better known as coke. So one can imagine that during the reaction at that temperature some acetylene should be made and some higher aromatics. So what will prevent this reaction?

Xinhe Bao: We discussed it with Enrique Iglesia during the coffee break. In our system, we have only detected three products. One is ethylene, another benzene, and another naphtalene. The distribution of reaction products is dependent on the space velocity. If the contact time is very short, then you can get a large distribution of ethylene. If the contact is very long, then you can get the heavier molecular weight species. If it is too long, then you can even get coke. But the coke is not at the surface of the catalyst, it is on the tubes. Of course we have also small amounts of other products, but the dominant products are ethylene, benzene and naphthalene, and coke. Especially if the space velocity is low, then you have coke formation.

Henk Lekkerkerker: I have the feeling that this discussion is not completely ended but we have a very nice occasion tonight that these two people could perhaps sit together. I would now like to turn to the third topic which is electrocatalysis. It is an old field, perhaps, or a new field... What are the feelings about the potentials of electrocatalysis? If I remember, I am not sure, Professor Ertl, don't you have an electrochemistry background?

Gerhard Ertl: If I remember, water splitting, which has been a topic for decades, obviously we still hold the hope that this could be a key for energy conservation. I understand that oxygen electrons are the main problem. There are worldwide attempts to overcome this. Now I ask the experts in electrochemistry: do you still believe that we will be able to make it or should we give up hope?

Henk Lekkerkerker: I think Marc Koper wants to answer this.

Marc Koper: I can try. I guess we should never give up hope. It is a very difficult problem. I think by combining well-defined experiments and theory, we have been able to nail down where the problem is and that is maybe the first step in trying to find a solution and our current understanding is that this is related to the fact that we have multiple intermediates, and each has a scaling and somehow we need to break that scaling. We need to start to think about different types of active sites in order to make this work. If that will work? I don't know but maybe, yes, and it is indeed a very old problem. I would like to point out that actually water electrolysis was first performed in the Netherlands in 1789 in Leiden. So it is 200 years old, yes.

Henk Lekkerkerker: Thank you Marc Koper. Are there any other views or comments on electrocatalysis?

Graham Hutchings: There is another topic which is attracting a lot of people's interest in the present time and that is the origin of life on this planet and where the chemicals came from. That was mentioned briefly in one of the discussions on the black smokers and in the way in which they make chemicals. We now all know that (I think that's Nora De Leeuw's work in which she has shown that you can make carbon–carbon bonds) you can reduce CO_2 on iron sulphide and you can make carbon–carbon bonds making acetic acid and pyruvate. But what do you think? I mean the real step is to try to activate nitrogen or a nitrogen-containing compound and make carbon–nitrogen bonds and also making carbon–carbon bonds. So do you think there is a possibility that this now will be demonstrated?

Ruyhei Nakamura: One of the key step for the origin of life is the C–N bond formation, and actually we cannot understand if the C–N is based on the ammonia and how some CO_2 compounds react with each other. But the problem is that recently people start to say that there is no ammonia in ancient ores; it is too oxidative, so no ammonia... Previously people thought that lightnings make ammonia from nitrogen, but recent geological records say that probably there is no ammonia by lightning and other products are NO, or N_2O, or nitrate. Probably they are the source of nitrogen and they go into the ocean, and then comes the electrochemistry with nitrate reduction or nitrite reduction. Recently we worked on this kind or reactions. What we found is that iron sulphide is not active for nitrate or nitrite reduction in similar conditions or neutral pH conditions. In acidic condition it is active but in neutral conditions there is no activities. But biology uses trace elements for nitrate reduction. And what we found is that trace amounts of molybdenum can reduce nitrate to NO, to N_2O and even to ammonia. And if there is some CO_2, maybe, CO_2 and nitrate can reduce each other and perhaps intermediates meet each other to form C–N bonds. We don't have direct evidences, but this is a possibility to make C–N bonds, not from nitrogen but from nitrate or from nitrite.

Henk Lekkerkerker: I overlooked you, how could I?

Gerhard Ertl: I think your idea for this novel fuel cell is really exciting based on this idea of the deep seas processes, these dark fumors. There was a man in Germany, in the Stetter group, he proposed the iron, the sulphur chemistry taking place there might be source of life. Is this generally accepted nowadays?

Ruyhei Nakamura: There are intense discussions on the origin of life and on the CO_2 reaction. Black smokers is one of them. But since recently, people start to deny black smokers because the temperature is too high to keep the molecules intact. That is one of the reasons why people start to ignore the black smokers or the hydrothermal vents. There is a pioneering work of a German scientist on the origin of CO. He demonstrated the production of hydrocarbons, but not from CO_2. Because the reduction of CO_2 with hydrogen clearly is thermodynamically tough, even under the hydrothermal vent conditions. So people start to think about this. They need one more energy supply like a pH gradient. In black smokers there is no pH gradient, so people started to think: how about in hydrothermal vents? And there, there is almost no iron sulphide. And then, using a pH gradient, and hydrogen energy, you can convert CO_2 to some hydrocarbons. But today, as I mentioned, even black smoker chimneys can generate high energy electrons and the reaction may happen at the cold ocean interface. So I think we can consider black smokers as one of the ideas.

Henk Lekkerkerker: Are there further burning questions about electrocatalysis? From the exchange, I think that's a field that's alive and Marc Koper is more expert than I am, so...

Marc Koper: I just want to comment on the question raised by Graham Hutchings on how to make carbon–carbon, carbon–nitrogen and nitrogen–nitrogen bonds. One thing that we have learned from the CO_2 reduction and also from the nitrate reduction is that the way you make these kinds of bonds is very different from the way you do high temperature heterogeneous catalysis. One of the things that we find is when you make a carbon–carbon bond for instance through the reductive dimerization of CO, and we see something similar with NO. We make a nitrogen–nitrogen bond through reductive dimerization of NO. And maybe we should try something similar with mixtures of CO and NO to see if that could be a way to make carbon–nitrogen bonds.

Graham Hutchings: I think that it would be a very exciting experiment to try. Because it was surprising to people that using iron sulphide as a catalyst with CO_2 under reductive conditions, you could make carbon–carbon bonds. That surprised a lot of people at the time. It is a huge opportunity here to see how big molecules could be formed in these environments. Because there are not just black smokers, there are other hydrothermal vents which involve basic conditions. You have got different environments you can think some molecules have arisen from.

Ben Feringa: What about HCN? Because I always heard that HCN played an important role, hydrogen cyanide at the origin of life. So what about HCN? Is that out of the picture now?

Ruyhei Nakamura: I know HCN is a candidate but maybe it is not the candidate in the ancient ocean. There is only one reason for the people can distinguish what chemicals are really involved in the origin of life. Most important criteria are the geochemical sitting in the ancient ocean. If HCN really exists in the ancient ocean, yes of course there is the power to synthesize but if not, we have to think about other pathways. So CO_2 and nitrate exist. We know that. But HCN, at least I don't know.

Ben Feringa: I don't know about this fence, but certainly in interstellar space, in scenarios about the origin of life, they always take HCN as one of the key molecules. But I don't know about these fence that you talked about.

Ruyhei Nakamura: OK. Of course, from the space it is possible.

Henk Lekkerkerker: I think the discussion is getting now in the interstellar space, which is certainly extremely interesting, but I thought that interstellar space was fairly cold and at low pressure.

Ben Feringa: I just wanted to learn if HCN was also playing a role in your systems, but apparently not.

Ruyhei Nakamura: For chemistry, we need engines, we need energy supply. So a deep sea hydrothermal vent is an ideal system to supply energy to drive chemistry.

Kurt Wüthtrich: Do you need energy or do you need catalysts?

Ruyhei Nakamura: Of course, we need both: energy and catalysts. In the current biology, you have enzymes and also energy, like NADPH to activate reactions. Both are needed.

Henk Lekkerkerker: Christophe Copéret wants to ask a question, but in the meanwhile I ask the audience to fire up their brains because after Christophe Copéret there will be time for only two more questions.

Christophe Copéret: We have discussed many extreme conditions but we have forgotten a bit pressure. So I have a question related to my colleagues Nakamura and Koper basically. What about the pressure in changing the paradigm in electrochemistry, because we know that for example for CO_2 reduction by H_2, pressure is a crucial component and you can go from low activity and low selectivity to 99% conversion, and 99% selectivity to methanol. Is it something to be considered in electrochemistry? Is it something to be considered in deep sea water basically, as conditions to make molecules?

Marc Koper: Certainly this is something to consider. For me high pressure like 200 bar in electrochemical terms is not high pressure. This is a point I wanted to make, but what it mainly means is that I change the concentration. And as a result of the concentration change, I change the population of different species at the surface and I will end up with a different reactivity. But what I will also need is sufficient reductive power to switch on certain reaction paths. There I think pressure may be less helpful; there I would actually use potentials. I would actually like to try to use high pressure, high temperature, and extreme potentials to see what we can make.

Henk Lekkerkerker: Professor Nakamura, would you like to comment on this or do you agree? Who wants to ask the last question?

Takashi Tatsumi: I have a question to Professor Bao. You showed us a system that directly converts CO and hydrogen to light alkanes. Usually methanol that uses carbon monoxide and hydrogen is thermodynamically limited. But in your case produced methanol can be consumed by the following MTO (methanol-to-olefins) reactions. That is a small system but I am just wondering if you have a problem of hydrogenation of once obtained alkenes at very high CO conversion. If there is enough carbon monoxide you might be able to suppress the hydrogenation to alkanes. But at higher conversions, might you have high selectivities to light alkenes? I am just wondering what your thoughts are about this.

Henk Lekkerkerker: Xinhe Bao, could you formulate a short answer?

Xinhe Bao: Yes. I just want to say that the conversion of carbon monoxide in our systems is really very high, the highest conversion achieved is now around 65%. The big trick in our system is the distance from the surface to the zeolite. We have also done different experiments by only changing the distance. If the oxide is very close to the zeolite, then we can only get the paraffins. If we can separate the oxide from the zeolite, then we can get the olefins. The reason I think is that in such systems, we have two different radicals. One is a ketone, but we don't know if it is really the ketone or the radical. There is another radical, which is hydrogen. But the hydrogen is unstable as a radical. The transfer distance is very short. If the distance from the oxide to the zeolite is short enough, then the hydrogen radical will do the hydrogenation and you will get the paraffins. This is very interesting and we can have discussions later after this session.

Henk Lekkerkerker to Takashi Tatsumi: Does this answer satisfy you?

Takashi Tatsumi: Yes.

Henk Lekkerkerker: Thank you. There is time for a last question. If I don't see any raised hand, I am going to conclude the session by thanking the speakers, but also thanking the audience, and the "auditeurs," and all who showed an interest.

Because it is getting late in the afternoon, and yesterday evening we had a dinner. So, I wish you a very pleasant evening, and a fresh start tomorrow morning on enzymes. Thank you!

Session 5

Catalysis by Protein Enzymes

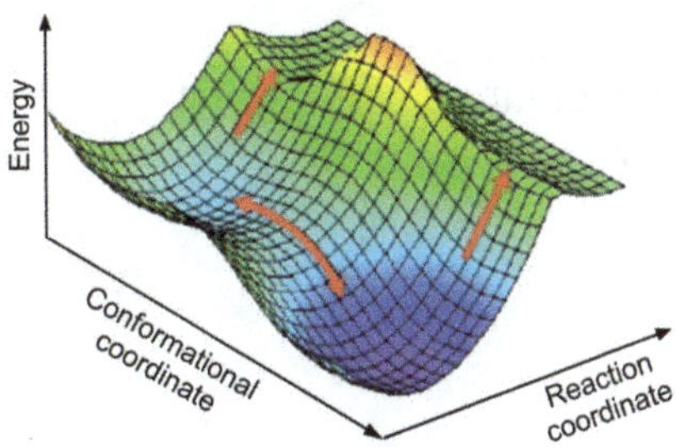

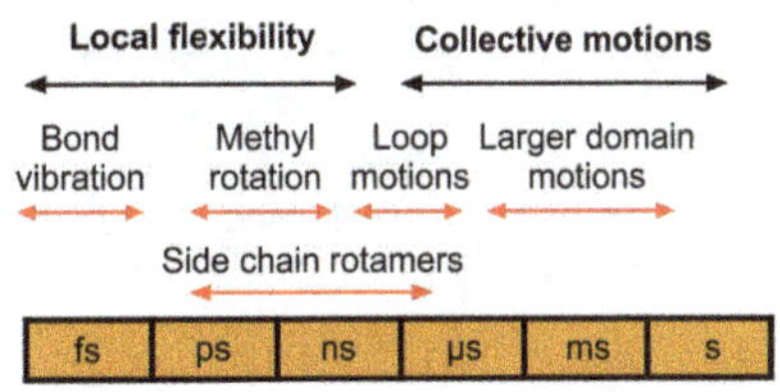

Current view of landscapes and timescales of interest in enzymatic catalysis.
Image by: JoAnne Stubbe, Department of Chemistry, Massachusetts Institute of Technology, USA.

BIOLOGICAL CATALYSIS: UNDERSTANDING RATE ACCELERATIONS IN ENZYMATIC REACTIONS

JOANNE STUBBE

Department of Chemistry, Massachusetts Institute of Technology, Cambridge, MA 02139, USA

Overview of catalysis by protein enzymes

Catalysis is at the heart of the chemical industry and at the center of biology. Chemists have the entire periodic table as their playing ground; in contrast, biologists are limited to C, N, O, P and S, and a few redox active and inert metals (Fe, Cu, Mn, Mg, Mn, Zn). Protein catalysts accelerate the rates of reactions relative to non-enzymatic reactions by factors of 10^6 to 10^{15}, with an upper limit of 10^{23}, under mild temperatures, atmospheric pressure and in aqueous solutions. They also possess striking specificity for the molecules they encounter inside the cell. The amazing rate accelerations have fascinated us all as chemists and inspired many to devote their lives in an effort to understand the basis for this achievement. Every year I teach a course in Enzyme Mechanisms and Catalysis. While exciting progress has been made through development of ingenious technologies (new ways to look, faster and more sensitive ways to look) to identify chemical intermediates and their formation along the reaction pathway, every year I ask myself, has new thinking or new technology lifted the veil to reveal the general factors contributing to these tremendous rate accelerations. Unique to biological catalysis are large macromolecules (predominantly proteins, with some notable exceptions like the ribosome), the solvent water, and evolutionary adaptation. However, even enzymatic reactions as simple as the isomerization catalyzed by ketosteroid isomerase (KSI, Fig. 1A) or peptide bond hydrolysis catalyzed by a protease (trypsin, Fig. 1B), the mechanisms of their 10^{11-12} rate acceleration are still being actively debated: the contribution of electrostatics, low barrier H bonds, general acid/base catalysis, ground state destabilization, over the barrier, through the barrier width, etc. Even the rearrangement of chorismate to prephenate catalyzed by chorismate mutase (CM, Fig. 1C) with an acceleration of a meager 10^6, raises the issue of the importance of a folded protein structure landscape. Why is understanding these enormous rate accelerations such a challenging problem? Why do we care? Can our understanding of protein catalysts lead to an understanding of a few basic principles that will be generally useful? These will be some of the issues addressed by the group convened to discuss biological catalysis.

In this session, we have brought together a group of "enzymologists broadly defined" with diverse backgrounds and interests.

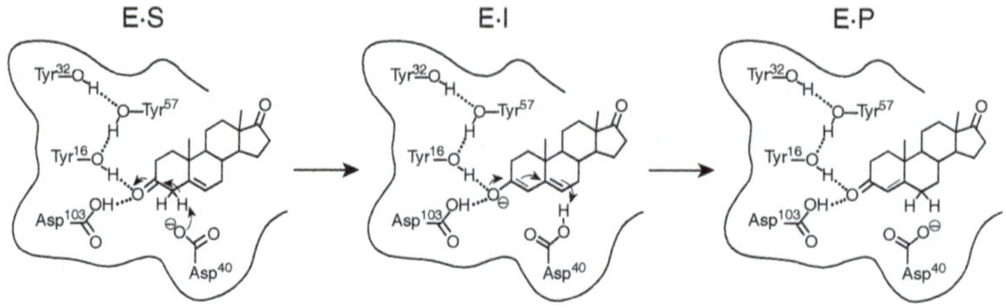

Fig. 1A. The reaction of ketosteroid isomerase [1].

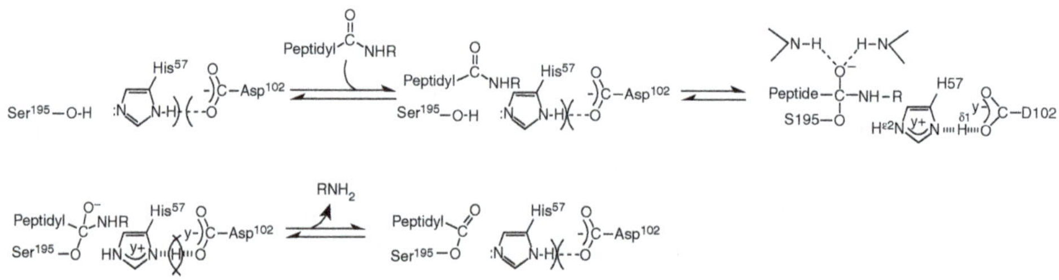

Fig. 1B. The reaction of serine hydrolase [2].

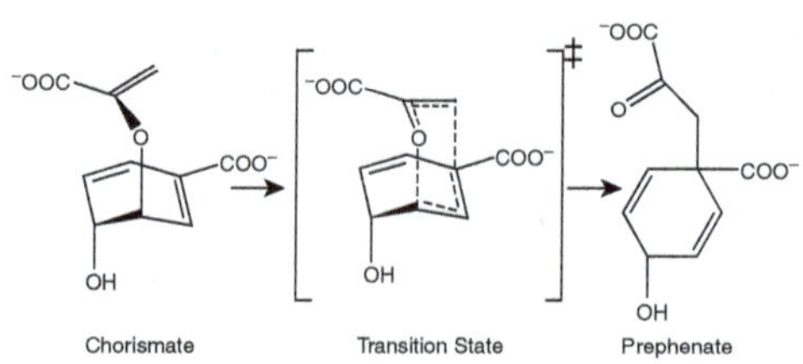

Fig. 1C. The reaction of chorismate mutase [3].

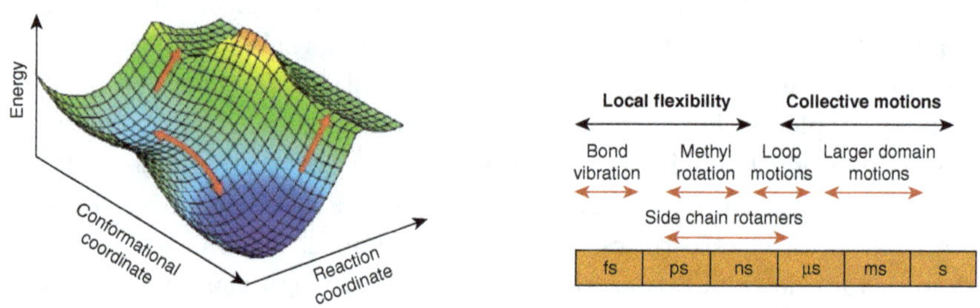

Fig. 1D. Current view of enzymatic landscapes and timescales of interest [4].

Recently the Boxer lab has focused on the importance of electrostatics in rate acceleration using the KSI system (Fig. 1A). Their recent studies using unnatural amino acids site-specifically incorporated into proteins address the importance of LBHB (short strong H bonds) such as those proposed in serine proteases (Fig. 1B) and the short distances between heteroatoms frequently observed in high-resolution protein structures. His lab has measured the Stark Effect on mutant proteins in an effort to understand the basis for KSI rate acceleration [1]. The Herschlag lab also studies KSI and might provide an alternative perspective on the basis for rate acceleration [5].

Electrostatic effects have also been proposed to be important in the CM reaction (Fig. 1C), studied by the Hilvert lab. His group discovered during protein engineering studies, a molten globule CM that is as an effective catalyst as the many natural CMs [3]. The issue of ground state destabilization, mobile protein conformations and how the evolved enzyme maintains its rate acceleration are of general interest to discuss. The ideas about electrostatics and protein dynamics learned from CM and other proteins may contribute to successful catalyst design, one of the exciting frontiers in enzymology. Can we understand enough about protein folding/catalysis and couple it to the powerful evolutionary/engineering methods to make catalysts that are very efficient and can catalyze reactions not found in nature? This information and modern technology could lead to use of robust protein scaffolds to catalyze reactions with, for example, organometallic reagents as shown by Arnold [6], Lewis, and Hartwig among others.

We now know that a reaction coordinate is better described by Fig. 1D which illustrates a conformational axis with many interconverting protein structures. The Dyer lab is interested in protein dynamics and the search for reactive conformations. Small changes in bond order and bond lengths (0.01 Å) are required to understanding bond reactivity and are not accessible through crystal structures. Recently the Dyer lab has carried out studies on lactate dehydrogenase bound to an inhibitor using vibrational spectroscopy to investigate the enzyme's conformational landscape [7]. He will also highlight his lab's methodology development including T-jump and ultrafast flow coupled isotope edited IR to capture and interpret fast dynamics.

Dynamics of proteins, protein motion and conformational gating (local and global), with a focus on its importance in quantum mechanical tunneling in H transfer reactions, has been an important focus of the Klinman lab [4]. A tremendous range of time scales (Fig. 1D) are accessible to monitor changes in protein dynamics with current technologies; however the time scale of microseconds to milliseconds is likely particularly important as it encompasses the time scale associated with many enzymatic reactions. The Klinman lab has studied lipoxygenase as a prototype for C–H activation through H tunneling and the subtle influence of protein motions and active site compaction on catalysis. Recent studies suggest that changes in a single bond distance of 0.1 to 0.2 Å can contribute several orders of magnitude to rates.

The Neese lab has developed computational protocols using structure and molecular dynamics to calculate spectroscopic observables obtained from multiple experimental measurements, providing a critical link between theory and experiment [8]. His lab has focused on identifying catalytically active structures in homogeneous and heterogeneous catalysis and his methods in the case of enzymes, continue to play a very important role in simulation of spectroscopic observables of the reactive intermediates and their identification.

A new direction has taken place in enzyme catalysis in recent years in the form of experiments at the single molecule level. In the case of F1 ATPase, Marcus discusses the recent stalling experiments and controlled rotation experiments in the literature and how "Marcus Theory," designed originally for electron transfer reactions has been adapted to a variety of other processes. He will describe his recent work with Volkán-Kacsó of one such adaptation to interpret and predict ATP binding and release rates at different angles of rotation, their exponential dependence on rotation angle, and potential implications for kinetics-structure relationships [9]. He also will discuss the complementary nature of ensemble and single molecule experiments.

Solvation also plays a critical role in catalysis and nature's solvent is water. Despite the ubiquity of water, its role in enzyme catalysis is still poorly understood. The availability of many high-resolution structures of enzyme homologs from many different sources, often reveals water or clusters of waters in identical locations, suggesting their importance. Havenith–Newen uses THz spectroscopic methods to look at water and water clusters dynamics and the relationship to function in proteins including the temperature and pressure effects on the activity of Matrix Metallo-Protease [10].

Open questions and challenges

What will be the take home messages from our discussions during this session on Biological Catalysis? Can we generate some fundamental rules to guide the future experiments of the community? Can we articulate the next level of experimental tools required to understand rate accelerations? How can we get young people excited about mechanism and catalysis?

References

[1] Y. Wu, S. G. Boxer, *J. Am. Chem. Soc.* **138**, 11890 (2016).
[2] J. Lin, C. S. Cassidy, P. A. Frey, *Biochemistry* **37**, 11940 (1998).
[3] M. Roca, B. Messer, D. Hilvert, A. Warshel, *Proc. Natl. Acad. Sci. USA* **105**, 13877 (2008).
[4] S. Hammes-Schiffer, J. Klinman, *Acc. Chem. Res.* **48**, 899 (2015).
[5] V. Lamba, F. Yabukarski, M. Pinney, D. Herschlag, *J. Am. Chem. Soc.* **138**, 9902 (2016).
[6] F. H. Arnold, *Q. Rev. Biophys.* **48**, 404 (2015).
[7] R. Callender, R. B. Dyer, *Acc. Chem. Res.* **48**, 407 (2015).
[8] D. Maganas, A. Trunschke, R. Schlögl, F. Neese, *Faraday Discuss.* **188**, 181 (2016).
[9] S. Volkán-Kacsó, R. A. Marcus, *Proc. Natl. Acad. Sci. USA* **112**, 14231 (2015).
[10] E. Decaneto, S. Suladze, C. Rosin, M. Havenith, W. Lubitz *et al.*, *Biophys. J.* **109**, 2371 (2015).

TOP-DOWN AND BOTTOM-UP APPROACHES FOR ELUCIDATING THE ORIGINS OF ENZYME EFFICIENCY

DONALD HILVERT

Laboratory of Organic Chemistry, ETH Zürich, CH-8093 Zürich, Switzerland

Chemists and biologists have long dreamed of creating catalysts whose rates and selectivities match those of natural enzymes. Given the complexity of even the simplest enzymatic systems, along with our incomplete understanding of their mechanisms, such an aim has seemed virtually unrealizable. Nevertheless, recent advances in the field of protein design that build on powerful computational and evolutionary methods give cause for optimism.

Research contributions to protein design

Efforts to design new enzymes generally start with Linus Pauling's proposal that proteins accelerate chemical reactions by stabilizing the rate-limiting transition state relative to substrate(s) [1]. To that end, a protein binding pocket in an appropriate scaffold needs to be equipped with sets of functional groups preorganized to achieve shape and chemical complementarity to this transient, high-energy species. Many mechanistic strategies can be exploited for this purpose. Electrostatic, steric, proximity and solvent effects, for example, have all been harnessed for catalysis. Because these factors are not generally independent or energetically additive [2], effective implementation can be difficult. A further obstacle is posed by protein motions, which may be necessary for substrate binding, product release or to shield reactants from bulk solvent. Generating active site environments conducive to reaction may also require consideration of protein conformational sampling and dynamics [3].

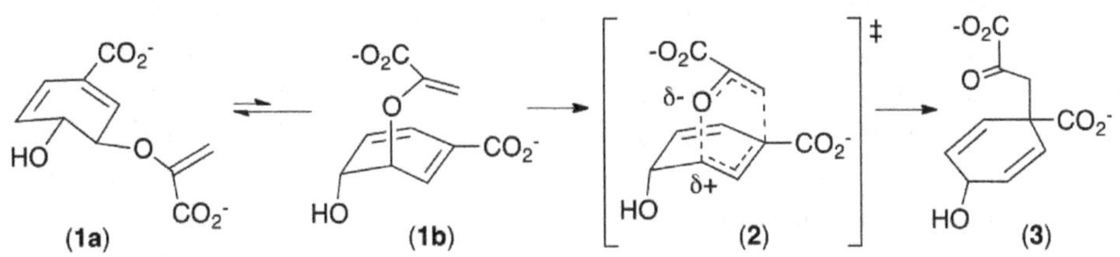

Fig. 1. The Claisen rearrangement of chorismate (1) to prephenate (3) is promoted by the enzyme chorismate mutase.

The enzyme chorismate mutase (CM) is a relatively simple biocatalyst that promotes the Claisen rearrangement of chorismate (Fig. 1.1) to prephenate (Fig. 1.3), a key step in the biosynthesis of the aromatic amino acids tyrosine and phenylalanine. Both the enzymatic and nonenzymatic reactions proceed via a concerted but asynchronous chairlike transition state (Fig. 1.2) [4], but the enzymatic transformation is more than a millionfold faster [5]. Structural studies show that natural CM active sites are buried and contain extensive arrays of hydrogen-bonding and charged residues for substrate recognition and transition state stabilization [6–8]. Because their three-dimensional folds are quite dissimilar, the common mechanistic strategies employed imply strong evolutionary convergence.

Although the chorismate rearrangement has a comparatively modest activation barrier, its catalysis has generated considerable debate. At one extreme, and counter to the standard Pauling model, transition state stabilization has been claimed to be unimportant for this system [9]. In this view, catalytic efficiency derives largely from the ability of the enzyme to populate a reactive substrate conformer that constrains the reactive centers to contact distances. The opposing position holds that electrostatic complementarity with the transition state is crucial [10]. Mutagenesis [11], modeling [12, 13] and structural [6–8] studies have provided strong evidence that a cationic residue — either an arginine or a lysine — is needed to stabilize developing negative charge on the substrate ether oxygen in the pericyclic transition state. Experimentally replacing the key arginine in *Bacillus subtilis* CM with citrulline, a neutral arginine isostere, reduced both k_{cat} and $k_{\mathrm{cat}}/K_{\mathrm{m}}$ by more than four orders of magnitude [14]. Though it binds chorismate in a "near attack conformation" [9], the citrulline variant is ultimately a poor catalyst. The positively charged arginine contributes only 0.6 kcal/mol to substrate binding, but up to 5.9 kcal/mol to transition state stabilization. Thus CM appears to present an archetypical instance of the Pauling paradigm rather than an exceptional departure from it.

So long as the constellation of active site residues is preserved, CM redesign is not difficult. In one case, genetic selection was successfully exploited to isolate a fully functional CM constructed entirely from a restricted set of nine amino acids [15]. The simplified enzyme reproduced key features of its natural counterpart, including rescue of auxotrophic cells lacking CM, but displayed lower stability and greater flexibility. Genetic selection also enabled conversion of a natural, domain-swapped homodimeric CM into a highly active monomer [16]. Although the topologically redesigned enzyme exhibited catalytic activity similar to the native dimer, it behaved quite unexpectedly like a molten globule, a dynamic ensemble of poorly packed and rapidly interconverting conformers [17, 18]. Such a transformation demonstrates that intrinsic disorder is compatible with efficient catalysis. Indeed, computational studies indicate that the molten globule variant accesses a broader range of catalytically competent conformations than the conventionally folded dimer without significant preorganization penalty [19]. Although conformational disorder decreases ligand affinity somewhat, molecular recognition is facilitated, judging by significantly faster substrate binding and product release [20]. That catalysis does not require a stable and persistent fold sheds new light on the role of dynamics in enzyme catalysis and offers interesting opportunities for design.

Redesigning existing CMs is much easier than creating artificial mutases de novo. One approach to new enzymes exploits transition state analogs as antigens to induce an immune response. Murine antibodies with weak CM activity were elicited, for example, with a conformationally constrained CM inhibitor [21]. Antibody 1F7 catalyzed the conversion chorismate to prephenate with a 200-fold rate acceleration over background and high enantioselectivity [22]. Crystallographic analysis revealed that it utilizes many fewer hydrogen bonds and electrostatic interactions for ligand recognition than natural CMs and highlighted the absence of a cationic residue for stabilization of the dipolar transition state as the likely reason for its relative inefficiency [23]. Although the 1F7 antibody was metabolically competent in living yeast cells [24], our efforts to optimize its activity by directed evolution proved unsuccessful.

Computational design, in its turn, offers a potentially more powerful strategy for creating new catalysts [25, 26]. Introduced little more than a decade ago, *in silico* methods for enzyme design are not limited to the antibody scaffold and can be used to construct active sites around more realistic transition state models. Despite notable progress on several reactions, efforts to install the many polar and charged residues needed for the CM reaction in protein folds that do not normally promote this reaction nevertheless have failed to yield detectable activities. Given the modest energetic demands of the chorismate rearrangement and a comparatively good understanding of what is required for efficient catalysis, this result is disappointing. In the future, greater algorithmic robustness and reliability will evidently be needed to successfully recapitulate the CM active site.

Transformations successfully catalyzed by computationally designed proteins include hydrolysis of activated esters [27, 28], an abiological Diels–Alder cycloaddition [29], proton transfer from carbon [30], and reversible aldol reactions [31, 32]. Starting activities tend to be modest ($k_{\text{cat}}/k_{\text{uncat}} = 10^2$ to 10^3), reflecting the relative simplicity of designs that rely on proximity effects and one or two catalytic residues placed in a hydrophobic binding pocket. In contrast to catalytic antibodies, however, these constructs can be optimized by directed evolution. In favorable cases, 10^2 to 10^5 fold increases in specific activity have been attained. The $\sim 10^9$-fold rate accelerations achieved in this way for benzisoxazole deprotonation [33] and a multistep aldol reaction [34] rival the efficiencies of analogous natural enzymes.

Detailed biochemical and structural characterization of the starting computational designs and their evolutionary descendants have helped to identify limitations of current design models. Insufficiently precise molecular ligand recognition and reliance on oversimplified active sites are two obvious deficiencies. During experimental optimization, the designed binding site is usually remodeled to some extent [35]. Improving shape complementarity discourages unproductive binding modes and, at the same time, optimizes interactions between the substrate and key catalytic residues. Partial [33] or even complete [34, 36] reconfiguration of the catalytic apparatus may sometimes be necessary. During laboratory engineering of a computationally designed aldolase, for instance, a reactive lysine was replaced by an unplanned Lys–Tyr–Asn–Tyr tetrad [36]. This hydrogen-bonded network of polar residues activates the substrate by Schiff base formation, and is responsible for mechanistically important proton transfers and stabilization of sequential transition states along the multistep reaction pathway. Such a sophisticated catalytic center is reminiscent of the extended networks of functional residues seen in true enzymes.

Outlook to future developments of research on protein design

As these few examples attest, combining computation with evolution is a powerful strategy for investigating enzyme chemistry and discovering new catalysts. Moving forward, improved computational algorithms, more powerful computer hardware and high-throughput screening techniques can be expected to make catalysis of demanding chemical transformations with complex mechanisms and high energy barriers possible. If successful, a robust and practical source of tailored catalysts for medicine and industry will be at hand.

Acknowledgments

The author is indebted to the ETH Zurich, the Swiss National Science Foundation and DARPA for generous support.

References

[1] L. Pauling, *Chem. Eng. News* **24**, 1375 (1946).
[2] D. Herschlag, A. Natarajan, *Biochemistry* **52**, 2050 (2013).
[3] Z. D. Nagel, J. P. Klinman, *Nat. Chem. Biol.* **5**, 543 (2009).
[4] D. J. Gustin, P. Mattei, P. Kast, O. Wiest, L. Lee *et al.*, *J. Am. Chem. Soc.* **121**, 1756 (1999).
[5] P. R. Andrews, G. D. Smith, I. G. Young, *Biochemistry* **12**, 3492 (1973).
[6] Y. M. Chook, J. V. Gray, H. Ke, W. N. Lipscomb, *J. Mol. Biol.* **240**, 476 (1994).
[7] A. Y. Lee, P. A. Karplus, B. Ganem, J. Clardy, *J. Am. Chem. Soc.* **117**, 3627 (1995).
[8] N. Sträter, G. Schnappauf, G. Braus, W. N. Lipscomb, *Structure* **5**, 1437 (1997).
[9] X. Zhang, X. Zhang, T. C. Bruice, *Biochemistry* **44**, 10443 (2005).
[10] A. Warshel, P. K. Sharma, M. Kato, Y. Xiang, H. Liu *et al.*, *Chem. Rev.* **106**, 3210 (2006).
[11] P. Kast, M. Asif-Ullah, N. Jiang, D. Hilvert, *Proc. Natl. Acad. Sci. USA* **93**, 5043 (1996).
[12] M. Štrajbl, A. Shurki, M. Kato, A. Warshel, *J. Am. Chem. Soc.* **125**, 10228 (2003).
[13] F. Claeyssens, K. E. Ranaghan, N. Lawan, S. J. Macrae, F. R. Manby *et al.*, *Org. Biomol. Chem.* **9**, 1578 (2011).
[14] A. Kienhofer, P. Kast, D. Hilvert, *J. Am. Chem. Soc.* **125**, 3206 (2003).
[15] K. U. Walter, K. Vamvaca, D. Hilvert, *J. Biol. Chem.* **280**, 37742 (2005).
[16] G. MacBeath, P. Kast, D. Hilvert, *Science* **279**, 1958 (1998).
[17] K. Vamvaca, B. Vögeli, P. Kast, K. Pervushin, D. Hilvert, *Proc. Natl. Acad. Sci. USA* **101**, 12860 (2004).

[18] K. Pervushin, K. Vamvaca, B. Vogeli, D. Hilvert, *Nat. Struct. Mol. Biol.* **14**, 1202 (2007).
[19] M. Roca, B. Messer, D. Hilvert, A. Warshel, *Proc. Natl. Acad. Sci. USA* **105**, 13877 (2008).
[20] K. Vamvaca, I. Jelesarov, D. Hilvert, *J. Mol. Biol.* **382**, 971 (2008).
[21] D. Hilvert, S. H. Carpenter, K. D. Nared, M. T. M. Auditor, *Proc. Natl. Acad. Sci. USA* **85**, 4953 (1988).
[22] D. Hilvert, K. D. Nared, *J. Am. Chem. Soc.* **110**, 5593 (1988).
[23] M. R. Haynes, E. A. Stura, D. Hilvert, I. A. Wilson, *Science* **263**, 646 (1994).
[24] Y. Tang, J. B. Hicks, D. Hilvert, *Proc. Natl. Acad. Sci. USA* **88**, 8784 (1991).
[25] G. Kiss, N. Çelebi-Ölçüm, R. Moretti, D. Baker, K. N. Houk, *Angew. Chem. Int. Ed.* **52**, 5700 (2013).
[26] D. Hilvert, *Annu. Rev. Biochem.* **82**, 447 (2013).
[27] D. N. Bolon, S. L. Mayo, *Proc. Natl. Acad. Sci. USA* **98**, 14274 (2001).
[28] F. Richter, R. Blomberg, S. D. Khare, G. Kiss, A. P. Kuzin *et al.*, *J. Am. Chem. Soc.* **134**, 16197 (2012).
[29] J. B. Siegel, A. Zanghellini, H. M. Lovick, G. Kiss, A. R. Lambert *et al.*, *Science* **329**, 309 (2010).
[30] H. K. Privett, G. Kiss, T. M. Lee, R. Blomberg, R. A. Chica *et al.*, *Proc. Natl. Acad. Sci. USA* **109**, 3790 (2012).
[31] L. Jiang, E. A. Althoff, F. R. Clemente, L. Doyle, D. Röthlisberger *et al.*, *Science* **319**, 1387 (2008).
[32] E. A. Althoff, L. Wang, L. Jiang, L. Giger, J. K. Lassila *et al.*, *Protein Sci.* **21**, 717 (2012).
[33] R. Blomberg, H. Kries, D. M. Pinkas, P. R. Mittl, M. G. Grutter *et al.*, *Nature* **503**, 418 (2013).
[34] L. Giger, S. Caner, R. Obexer, P. Kast, D. Baker *et al.*, *Nat. Chem. Biol.* **9**, 494 (2013).
[35] N. Preiswerk, T. Beck, J. D. Schulz, P. Milovnik, C. Mayer *et al.*, *Proc. Natl. Acad. Sci. USA* **111**, 8013 (2014).
[36] R. Obexer, A. Godina, X. Garrabou, P. R. E. Mittl, D. Baker *et al.*, *Nat. Chem.* **9**, 50 (2017).

ELECTRIC FIELDS AND ENZYME CATALYSIS

STEVEN G. BOXER[1], STEPHEN D. FRIED[1,2], SAMUEL H. SCHNEIDER[1] and YUFAN WU[1]

[1] *Department of Chemistry, Stanford University, Stanford, CA 94305, USA*
[2] *Current address: Proteins and Nucleic Acid Chemistry Division, Medical Research Council, Laboratory of Molecular Biology, Cambridge CB2 0QH, UK*

Status of the physical basis for enzymatic catalysis

This field, which is relatively new to my lab, has a long history with numerous textbook chapters and reviews. Despite textbook descriptions of "how enzymes work," the descriptions do not agree and so this continues to be vigorously debated. In particular, the contribution of electrostatic interactions to catalysis, strongly advocated by Warshel [1] and other computational biochemists, has been difficult to evaluate experimentally due to a lack of local probes for measuring intermolecular interactions and electric fields with a consistent physical framework. While it is simple to measure rates and binding constants, both are complicated convolutions of contributions and therefore difficult to interpret within a consistent physical framework, even given X-ray crystal structures and multiple mutants. Much current work in this field is focused on directed evolution for improved or novel function [2], and primarily uses selection and screening strategies. *De novo* design [3], where it has been successful, tends to make extensive use of empirical correlations and heuristic models.

Recent Work from Our Lab on Enzymatic Catalysis

My group has been using Stark spectroscopy to characterize the displacement of charge (change in dipole moment or $\Delta\vec{\mu}$) for electronic and vibrational transitions for many applications [4]. $\Delta\vec{\mu}$ for electronic transitions in dye molecules leads to shifts in the absorption and emission maxima of dyes in solvents of different polarities, typically blue for non-polar solvents and red for polar solvents. A similar solvatochromic effect is observed for vibrational transitions, *e.g.* for the carbonyl group (–C=O). Vibrational Stark spectroscopy (VSS), which probes the effect of a known *applied* electric field (units MV/cm) on a transition, provides a measure for $|\Delta\vec{\mu}|$ (units of cm^{-1}/(MV/cm)) as illustrated in Fig. 1a. Note that $\Delta\vec{\mu}$ is a vector and for relatively localized high frequency vibrations, like –C=O and –C≡N, the direction of this vector is expected and has been shown to lie along the diatomic bond axis (thus its orientation is known by X-ray crystallography when used as a probe in a protein). VSS is, in effect, a calibration step that establishes the sensitivity of the vibrational frequency (measured by IR or Raman spectroscopy) to an electric field. Once calibrated, the probe can be inserted into different environments and observed frequency shifts (Fig. 1b) interpreted in terms of *changes* in the electric field sensed by the probe going from solution into a particular site (*e.g.* in a protein), or the effect of titrating nearby charged groups, making a mutation, folding/unfolding, etc.

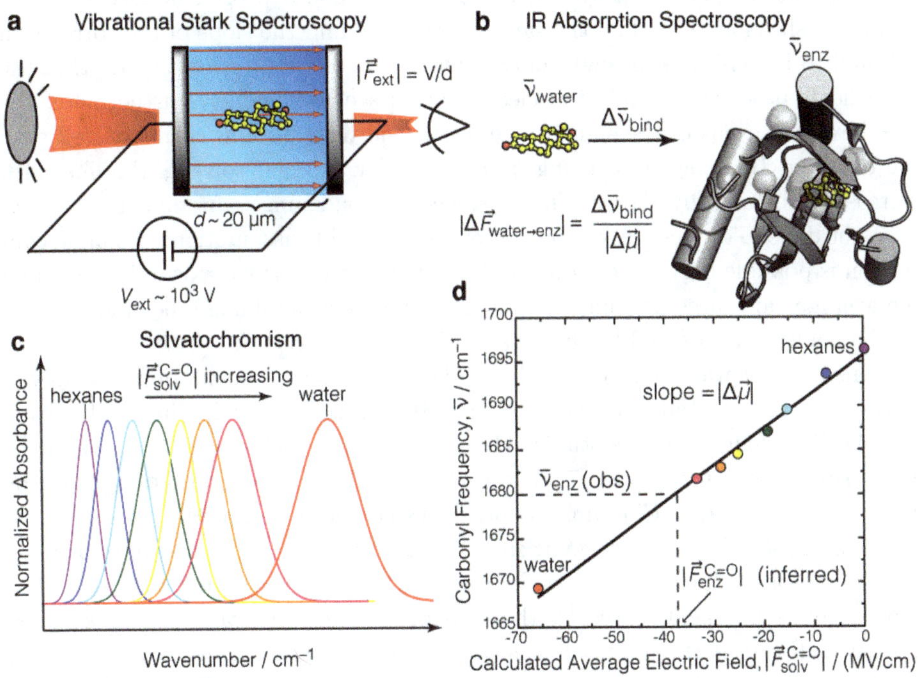

Fig. 1. Vibrational Stark spectroscopy, solvatochromism and frequency/electric field conversion. (a) In VSS a known external electric field is applied to a frozen isotropic sample, and the perturbation caused to the infrared spectrum is analyzed using well-developed methods [4]. (b) The frequency shift, $\Delta\bar{\nu}$, caused by changing the environment surrounding a carbonyl (*e.g.*, by binding from solution, but also by mutation, folding, pH change, etc.) can be translated into *differences* in the electric field projection along the C=O bond, using the Stark tuning rate ($|\Delta\vec{\mu}|$) as the conversion factor. (c) Schematic showing vibrational transitions shift from blue to red going from non-polar to polar solvents (solvatochromism) and accompanying band broadening. (d) The average solvent electric fields can be calculated (horizontal axis) and correlated with the observed solvatochromic frequency shifts (vertical axis, panel c); the slope is the Stark tuning rate ($|\Delta\vec{\mu}|$) [5]. This allows one to map the vibrational frequency in an enzyme active site to an absolute electric field (dashed lines). Figure adapted from Ref. [6].

A key step towards obtaining experimental information on the *absolute* electric fields — the field "felt" by a substrate inside an enzyme active site and potentially lowering the activation barrier for catalysis — was to connect the VSS data to vibrational solvatochromism, illustrated schematically in Fig. 1c. As discussed above, such frequency shifts in solvents of different polarity are well known, but by itself "polarity" is not a useful concept for precisely quantifying intermolecular interactions. This required the use of MD simulations with various force fields, all of which found a linear correlation between the average simulated value of the electric field in the solvent and the observed peak frequency in the IR, as shown for the carbonyl group of acetophenone in Fig. 1d (note that the field fluctuates in a simple solvent and this is reflected in the simulations; plotted in Fig. 1d are the time-averaged values) [5]. Both hydrogen bonded and non-hydrogen bonded carbonyl frequency shifts are found on the same correlation, and the slope of the correlation is found to be $|\Delta\vec{\mu}|$ (see below). This experimental check gives confidence that the force fields used to calculate the electric fields are reasonable. Furthermore, the standard deviations of the simulated electric field distributions correlate well with the observed IR linewidths: going from non-polar to polar solvents the linewidth increases substantially (largely inhomogeneous broadening). Assuming that $|\Delta\vec{\mu}|$ does not depend on field (a key but potentially testable assumption), a simple conversion is possible between the observed frequency (the vertical axis in Fig. 1d) and the average calculated absolute electric field (the horizontal axis), i.e. we can read off the field from observed frequencies (dashed lines in Fig. 1d).

This concept was applied to evaluate the electrostatic contribution to catalysis in the enzyme ketosteroid isomerase (KSI), introduced to our lab in collaboration with D. Herschlag where we studied changes in electric fields (Fig. 1b) around the active site using –SCN probes [7]. Figure 2a shows KSI's mechanism established by work from many labs. Close inspection shows charge is displaced at the carbonyl group of the steroid substrate going from the E•S complex to the first intermediate, and we reasoned that a substrate-like inhibitor's carbonyl group could be used to probe the electric field sensed by the substrate during this rate-limiting step [8]. The inhibitor 19-nortestosterone was bound to the active site of KSI and a very large shift to the red was observed, even compared with water (Fig. 2c). Following the conversion in Fig. 1d, this implies the electric field sensed by 19-NT's carbonyl group is *very* large. Furthermore, the vibrational transition is sharp, suggesting a narrow distribution of electric fields in this equilibrium measurement, i.e. the enzyme creates and holds this very large field projected onto the bond involved in charge displacement during catalysis.

Several mutants of key active site residues were studied for which information had already been obtained on k_{cat} by other labs. As k_{cat} decreased, the observed 19-NT carbonyl group shifts to the blue; this data was recently supplemented by subtler, structure-preserving changes where Cl-tyrosine was introduced at each of the active site tyrosines [9]. As seen in Fig. 2d, a robust linear correlation between the activation free energy (obtained from k_{cat} using transition state theory) and the observed electric field, derived from the measured frequencies using a conversion like Fig. 1d, was observed. The slope of this line gives information on the displacement of charge in the transition state relative to the E•S complex ($|\Delta\vec{\mu}_{rxn}|$, Fig. 2b), and the

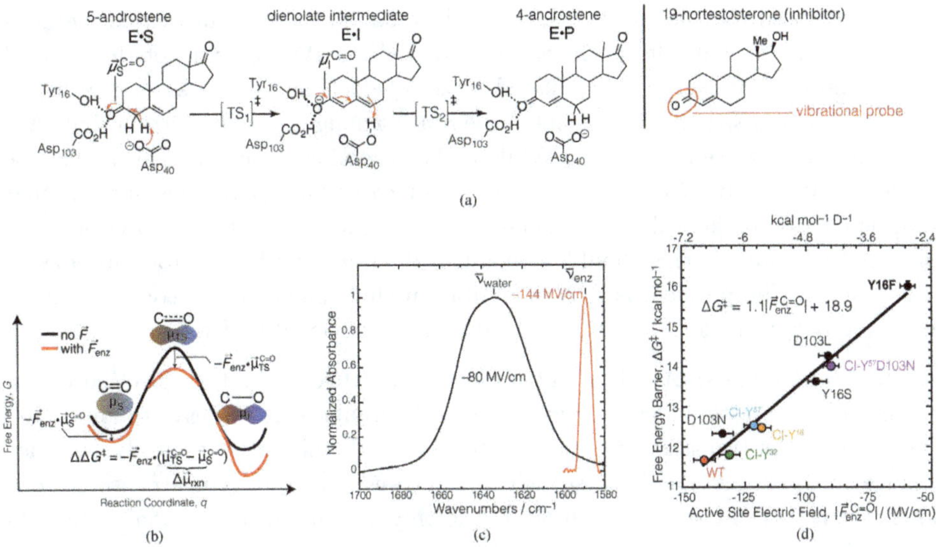

Fig. 2. Mechanism, concept of electrostatic catalysis, and observed electric field catalysis of KSI. (a) 5-Androsten-3,17-dione, the steroid substrate, is converted to the conjugated isomer, 4-androsten-3,17-dione, via the enolization (first step) and reketonization (second step) of the carbonyl. (b) Schematic illustration of the possible effect of the enzyme electric field, $\vec{F}_{enz}$, to differentially stabilize the transition state relative to the bound substrate by interacting favorably with $\Delta\vec{\mu}_{rxn}$. (c) KSI exerts a very large and homogeneous electric field on 19-nortestosterone's carbonyl bond relative to water's reaction field, based on the significant red-shift and band narrowing detected in the IR spectrum [8]. (d) Linear correlation ($R^2 = 0.98$) between KSI's catalytic power (expressed as $\Delta G^\ddagger$ and the electric field its active site projects onto the carbonyl of the inhibitor 19-nortestosterone. Black points represent mutations of the oxyanion hole residues to other canonical amino acids [8]; colored points represent subtle mutations to non-canonical amino acids [9]. Figure adapted from Ref. [6].

intercept, extrapolated to zero electric field (i.e. no net electrostatic stabilization) suggests that more than 70% of the catalytic power of this enzyme is due to the lowering of the transition state free energy by the electric field (the remainder is largely entropic). We believe that this approach is general, and it has now been used to re-interpret data from other groups already in the literature [10].

Outlook to future developments of research on enzymatic catalysis

Technical issues: The frequency-field conversion shown in Fig. 1d should have a slope equal to the independently measured Stark tuning rate (by VSS). However, we have consistently found that they differ by a factor of 2. We believe that the largest source of discrepancy is that the external electric field in the VSS measurement is not the field felt by the probe because of the "local field correction," a long-standing source of uncertainty in a variety of spectroscopies. While a consensus value around 2 appears appropriate, as discussed in depth in [11], some variation has been observed [10]. The discrepancy may also reflect systematic errors in the force fields used to calculate the field in solvents, and more sophisticated methods may be required. This affects the value extracted for $|\Delta\vec{\mu}_{rxn}|$, but not the value of the activation free energy extrapolated to zero field (Fig. 2d). A second technical issue concerns *in situ* measurements of the Stark tuning rate to be certain that $|\Delta\vec{\mu}|$ is unaffected by the environment. While we have good evidence that this is the case in fields spanned by simple solvents [12], and *ab initio* calculations suggest that $|\Delta\vec{\mu}|$ is mostly unaffected for the fields of the magnitude created by environments observed to date [8], this should be confirmed *in situ*, with the probe in enzyme active sites. The problem is S/N, a limitation that might be overcome by use of high intensity tunable IR sources instead of a conventional FTIR.

Conceptual issues: A subtle but important issue concerns the direction of the dipole of a substrate undergoing catalysis in its ground state relative to an altered geometry in the transition state (Fig. 2b). For KSI, this direction likely does not change substantially, but for cases where the transition state geometry is very different from that of the bound inhibitor, one only obtains the projection of the field probed by the inhibitor's geometry [6, 10]. Thus, probe/inhibitor design is critical. Ideally, what is needed is the ability to probe the electric fields as experienced by both ground and transition state analogs. Catalysis by electrostatic preorganization would imply that an enzyme would *maximize* the field projected onto the TS geometry compared to that projected onto the GS geometry. In conjunction with binding measurements and crystal structures, this has the potential to directly report on the functional fields relevant for catalysis from the perspective of the substrate, providing design principles for how to engineer active sites for electric field catalysis. It would be very interesting to use the vibrational Stark effect as a quantitative metric during computational enzyme design or as an observable during screening. It would also be very interesting to investigate the relationship between the evolutionary history of an enzyme and the electric field it creates. Finally, while we have shown that electric fields are correlated with increased rates, a large electric field in and of itself is not diagnostic of a good catalyst. This is clearly exemplified by water, where the average electric field and field distributions are large, though water, in general, is a poor catalyst [1]. The electric fields measured in aqueous solution and in enzyme active sites represent different manifestations of the electric field with respect to catalysis by transition state stabilization, a point that is often misunderstood [6]. One might explain these observations by pointing out that the electric field generated by a preorganized active site specifically stabilizes the transition state, while the field generated by solvent reorganization generally stabilizes the ground state, though this point needs to be investigated further.

Acknowledgments

S. D. F. was supported by the Stanford Bio-X interdisciplinary graduate fellowship. This work was supported in part by the NIH.

References

[1] A. Warshel, P. K. Sharma, M. Kato, Y. Xiang, H. Liu *et al.*, *Chem. Rev.* **106**, 3210 (2006).

[2] H. Renata, Z. J. Wang, F. H. Arnold, *Angew. Chem. Int. Ed.* **54**, 3351 (2015).

[3] H. Kries, R. Blomberg, D. Hilvert, *Curr. Opin. Chem. Biol.* **17**, 221(2013).

[4] S. G. Boxer, *J. Phys. Chem. B* **113**, 2972 (2009).

[5] S. D. Fried, L.-P. Wang, S. G. Boxer, P. Ren, V. S. Pande, *J. Phys. Chem. B* **117**, 16236 (2013).

[6] S. D. Fried, S. G. Boxer, *Ann. Rev. Biochem.* **86,** 317 (2017).

[7] A. T. Fafarman, P. A. Sigala, J. P. Schwans, T. D. Fenn, D. Herschlag *et al.*, *Proc. Natl. Acad. Sci. USA* **109**, 1824 (*summary*), E299 *(full paper)* (2012).

[8] S. D. Fried, S. Bagchi, S. G. Boxer, *Science* **346**, 1510 (2014).

[9] Y. Wu, S. G. Boxer, *J. Am. Chem. Soc.* **138**, 11890 (2016).

[10] S. H. Schneider, S. G. Boxer, *J. Phys. Chem. B* **120**, 9672 (2016).

[11] S. D. Fried, S. G. Boxer, *Acc. Chem. Res.* **48**, 998 (2015).

[12] S. H. Schneider, H. Kratochvil, M. T. Zanni, S. G. Boxer, *J. Phys. Chem. B* **121**, 2331 (2017).

WATER MAPPING IN ENZYMATIC CATALYSIS BY THZ SPECTROSCOPY (THZ CALORIMETRY)

MARTINA HAVENITH

Physical Chemistry II, Ruhr-Universitaet, 44780 Bochum, Germany

My view of the present state of research on enzymatic catalysis

Enzymatic reactions are a core topic in a field ranging from biochemistry to biotechnology as most biomolecular reactions are catalyzed by enzymes. Understanding the structures and functions of proteases, watching proteases in action, and approaching their intricacies in the protease web is critical for basic science as for future drug design. Directional hydrogen bonds, together with polar or charged residues, constitute a framework of non-covalent interactions within the enzyme and in the complex it forms with regulators and substrates. Most strikingly, water, being a generic solvent, acts as a strong competing agent for all these interactions, leading to a delicate balance between functional structures and complete solvation. Characteristic is large, but competing enthalpic (ΔH) versus entropic (ΔS) contributions that mostly compensate and result in subtle energy differences of only a few kJ/mol, which then dictate the biological function [1]. Protein–ligand bounding is found to be favorable when the change in free energy: $\Delta G = \Delta H – T\Delta S$ is negative. A lot of work has focused on affinity measurements and have thus focused on ΔH. However only the subtle compensations of entropy and enthalpy or decides about successful molecular recognition (For a review of the current understanding [2].

For a long time, efficient enzyme catalysis was thought to be attributable mainly to a direct *structural* interaction between the enzyme and the substrate via the "lock and key" or the "induced fit" mechanism. Emerging experimental and theoretical approaches provide a revised, broader view in which *conformational changes* on a wide range of timescales influence or may even dominate enzymatic reaction-rate enhancements [3]. The underlying protein motions cover a broad range of timescales and are span the msec to the psec timescale [4]. More recent experimental results indicate that protein dynamics may be fundamentally linked to biological function [5, 6]. Single molecule fluorescence correlation studies measuring diffusion coefficients detected indications of sudden heat release effects during enzymatic turnover, which traditional bulk calorimetry is unable to probe [7]. As yet, almost no information is available on the molecular details that link psec protein functional motion with enzymatic turnover. However, it is now becoming widely accepted that their functions are mediated by their dynamic character and their interactions with the solvent. *These fast fluctuations on the free energy landscape during molecular recognition cannot yet be accessed.*

Exploring the unique role of water for life has been named as one of the top future challenges in chemistry [8]. Whether fast protein motion and solvent dynamics are correlated at the very heart of enzymatic reactions is still under heated debate. The underlying molecular mechanism of enthalpy-entropy (H/S) compensation in protein–ligand binding remains controversial and there are still no models available for ligand-binding that are both quantitative *and* predictive. Systematic studies under steady state conditions revealed that differences in the structure and thermodynamic properties of the waters surrounding the bound ligands are an important contributor to the observed H/S compensation [9]. New insights were obtained by calorimetric measurements with specific ligand which underlined the decisive role of the water network motions in cavity [10]. The future impact of these studies is based on the belief that *"Changes in protein and solvent dynamics are not mere epiphenomena, but have a vital role in substrate binding and recognition: they are more cause than consequence"* [11].

My recent research contributions to enzymatic catalysis

Our group has pioneered Kinetic Terahertz (THz) absorption spectroscopy as a new tool to probe changes in the low frequency spectrum of the solvated protein, which gives a direct access to the collective, (sub-)psec hydrogen bond dynamics, thereby opening a new window to directly probe the coupled protein/hydration dynamics [12–14]. THz light extends the dielectric regime from nsec motions down to (sub-)psec motions, i.e. the timescale where vibrational, translational and intermolecular, collective motions of hydration water, as well as large amplitude motions of biomolecules, come into play. In frequency domain, this corresponds to low-frequency modes in the frequency range between 1 and 10 THz play. These large amplitude modes, which include *e.g.* skeleton modes, play a crucial role in controlling conformational changes, and appear to be responsible for triggering many biochemical reactions and energy transport. Kinetic THz Absorption Spectroscopy (KITA) allows to probe the coupled protein solvent dynamics during biological function in real time [13].

Correlated structural kinetics and retarded solvent dynamics at the metalloprotease active site

The main focus of enzymology lies on the enzyme rates, substrate structures, and reactivity, whereas the role of solvent dynamics in mediating the biological reaction is often left aside owing to its complex molecular behavior. We used integrated X-ray and THz based time-resolved spectroscopic tools to study the link between water and protein dynamics during proteolysis of collagen-like substrates by a matrix metalloproteinase in real time [5, 13, 14]. More specifically, we investigated the protein/solvation dynamics of human membrane type-1 MMP (MT1-MMP) using kinetic THz absorption spectroscopy. The specific binding of the substrate at the active site is accompanied by structural rearrangements at the active site during the formation of productive enzyme–substrate intermediates. We recorded changes in THz absorption during peptide hydrolysis by a zinc-dependent human metalloprotease. The probed changes of the protein–water coupled motions were found to be tightly correlated with structural rearrangements at the active site during the formation of productive enzyme–substrate intermediates, and interestingly, were different from those in an enzyme–inhibitor complex. Accompanying MD simulations showed a gradient of fast-to-slow coupled protein–water motions in the catalytic domain of a model enzyme (MT1-MMP), with the solvent dynamics at the metalloprotease active site being retarded the most. *Importantly, this evidence suggests the existence of a heterogeneous hydration dynamics paired with a less flexible water network when approaching a functional site: the "hydration funnel."*

The hydration water in close proximity between the active site and between the substrate and the enzyme is found to be retarded in the hydration bond dynamics (with typical hydrogen bond times varying between 1 psec for bulk water and 7 psec for water at the active site), see Fig. 1. Our results indicated that water retardation occurs before formation of the functional Michaelis complex. *We propose that the protein–water coupled motions gradient assists enzyme–substrate interactions via water retardation mechanisms remotely mediated by chemical fluctuations of enzyme surface and active site.*

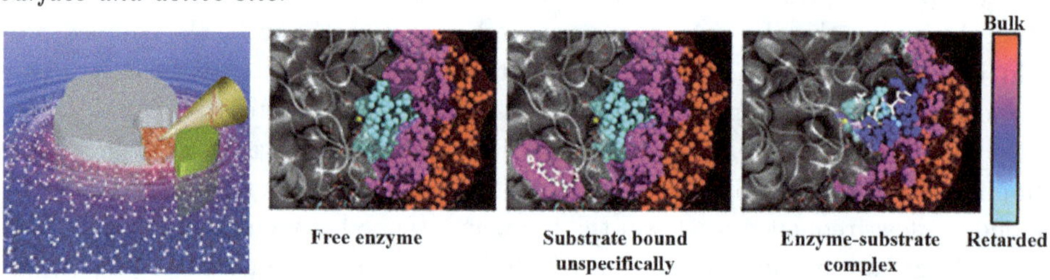

Fig. 1. A gradient of water motions (from retarded water molecules (cyan) to bulk (red)) is detected in the unbound MT1-MMP when the substrates are at a distance d $\sim$ 7 Å from the zinc ion. When the substrates approach closer to the zinc ion (d $\sim$ 4.9 Å), an additional decrease in the water HB lifetime at the active site for both substrates is observed. The gradient of water motions depends on the substrate: τ_{HB} is steeper when the MT1-MMP is bound to the triple-helical substrate (right) than when it is bound to the single-strand substrate (middle). Taken from Ref. [5].

Fluctuations of charges upon binding are found to have a direct influence on the specific water–protein coupling as the cause of the observed dynamical hydration heterogeneity. The retardation of the hydration bond dynamics is connected with a decrease in the entropic cost for the desolvation of the water at the active site upon substrate–enzyme binding. Since molecular binding depends on a subtle balance enthalpy and entropy ($\Delta G = \Delta H - T\Delta S$). A less flexible water network (lower entropy) at the active site compared to the hydration water at the unspecific binding sites on the enzyme lowers the overall entropic cost for binding, thereby assisting molecular recognition processes. Further insights on the underlying mechanism of binding will be provided by quantitative studies of the entropy and enthalpy of binding of water in the binding pocket.

Future research direction: THz-calorimetry

THz-calorimetry is the science of measuring the low frequency spectrum of solvated solutes for the purpose of deriving the calorimetric quantities such as the change in heat capacity, C_p, entropy S and enthalpy H. Our new approach is based on *spectroscopic* instead of traditional *calorimetric* measurements and can thus in principle be extended to time resolved measurements "THz-Calorimetry" makes it possible to reveal locally resolved entropy changes of the protein and the solvent. This concept can be transferred to water mapping during molecular recognition processes, i.e. under non-equilibrium conditions. THz calorimetry will allow to derive the calorimetric changes associated with biological processes, such as enzymatic reactions, in real time. Thus, these concepts go beyond traditional calorimetry which yield enthalpy and entropy values as ensemble averages. THz calorimetry, as an inherent spectroscopic technique allows to detect entropic and enthalpic solvation changes with envisioned time resolutions of up to nanoseconds. We hope that with our novel proposed THz calorimeter will screen the origin and key factors driving fast dynamics on the free energy landscape during biological processes, and thus provide a novel insight of molecular recognition processes. We aim to detect signatures of sudden heat release from catalytic sites [7] and fluctuations in the subtle entropy/enthalpy compensation by THz calorimetry.

The methodology of THz calorimetry is based on the idea that under ambient, physiologically-relevant conditions 90% of the modes which contribute to the total entropy of the protein solvent mixture are in fact captured by the low frequency modes of the protein and solvent (including protein domain modes, and collective water network modes), i.e entropy can be deduced from the vibrational density of states (VDOS) in the frequency range between 0 and 10 THz (300 cm^{-1}).

Whereas THz spectroscopy is unable to directly record the vibrational density of state (VDOS) in the low frequency range, instead probes the VDOS multiplied with the transition dipole moment. THz spectroscopy records spectral fingerprints of changes in the hydration water in the frequency range of the intermolecular water water band (at around 200 cm^{-1}). In a first proof of principle experiment, we have recorded the low frequency spectrum of solvated alcohols. We were able to connect changes of the intermolecular modes between water water pairs in the hydration shell as probed by THz spectroscopy with local changes in heat capacity, enthalpy and entropy. As a result, we found that local solvation changes in heat capacity, enthalpy, and entropy can be probed by exact spectroscopic measurements of the solute and temperature dependent spectrum between in the frequency range between 50 and 300 cm^{-1} (2–10 THz) [15].

We found a remarkable agreement between calorimetric data ΔCp, ΔS, ΔH and ΔG as based on traditional heat transfer measurements and those based on THz spectroscopy.

THz calorimetry transcends the theoretical approach of water mapping and allows a detailed insight into the molecular nature of solvation and the local entropic and enthalpic contributions of the solvent.

We propose that THz-calorimetry will enable a characterization of local calorimetric quantities such ΔH, ΔS, and ΔG of individual enzymes and their mode of binding to their substrates and endogenous inhibitors with great precision in real time during enzymatic catalysis. Such precision is required to map potential allosteric small molecule drugs targeting the non-homologous enzyme surface away from the structurally homologous active site. On a long-term perspective, we hope to open the door to a controlled fine-tuning of the contribution of the solvent free energy to molecular recognition processes, which can be used for optimization of ligand and drug design.

Acknowledgments

We acknowledge financial support from the Cluster of Excellence RESOLV (EXC 1069) funded by the Deutsche Forschungsgemeinschaft. MH acknowledges an ERC Advanced Grant 695437 THz-calorimetry.

References

[1] M. J. Todd, J. Gomez, *Anal. Biochem.* **296**, 179 (2001).
[2] R. Baron, J. A. Mc Cammon, *Annu. Rev. Phys. Chem.* **64**, 151 (2013).
[3] S. J. Benkovic, S. Hammes-Schiffer, *Science* **301**, 1196 (2003).
[4] K. A. Henzler-Wildman, M. Lei, V. Thai, S. J. Kerns, M. Karplus *et al.*, *Nature* **450**, 913 (2007).
[5] M. Grossman, B. Born, M. Heyden, D. Tworowski, G. B. Fields *et al.*, *Nat. Struct. Mol. Biol.* **18**, 1102 (2011).
[6] T. I. Igumenova, K. K. Frederick, A. J. Wand, *Chem. Rev.* **106**, 1672 (2006).
[7] C. Riedel, R. Gabizon, C. A. M. Wilson, K. Hamadani, K. Tsekouras *et al.*, *Nature* **517**, 227 (2015).
[8] G. M. Whitesides, *Angew. Chem. Int. Ed.* **54**, 3196 (2015).
[9] B. Breiten, M. R. Lockett, W. Sherman, S. Fujita, M. Al-Sayah *et al.*, *J. Am. Chem. Soc.* **135**, 15579 (2013).
[10] P. W. Snyder, J. Mecinovic, D. T. Moustakas, S. W. Thomas III, M. Harder *et al.*, *Proc. Natl. Acad. Sci. USA* **108**, 17889 (2011).
[11] P. Ball, *Nature* **478**, 467 (2011).
[12] S. Ebbinghaus, S. J. Kim, M. Heyden, X. Yu, U. Heugen *et al.*, *Proc. Natl. Acad. Sci. USA* **104**, 20749 (2007).
[13] S. J. Kim, B. Born, M. Havenith, M. Gruebele, *Angew. Chem. Int. Ed.* **47**, 6486 (2008).
[14] V. Conti Nibali, M. Havenith, *J. Am. Chem. Soc.* **136**, 12800 (2014).
[15] F. Böhm, G. Schwaab, M. Havenith, *Angew. Chemie Int. Ed.* **56**, 9981 (2017).

THEORY OF SINGLE MOLECULE EXPERIMENTS OF F_1-ATPASE: PREDICTIONS, TESTS AND COMPARISON WITH EXPERIMENTS

SÁNDOR VOLKÁN-KACSÓ[1,2] and RUDOLPH A. MARCUS[1,2]

[1] *Chemistry Department, California Institute of Technology, Pasadena, CA 91125, USA*
[2] *Noyes Laboratory of Chemical Physics, Pasadena, CA 91125, USA*

There is now a substantial body of single molecule experiments on biomolecular motors, supplementing the extensive structural and ensemble data on these systems. Our current focus is on the motor, F_1-ATPase, a component of the enzyme ATP Synthase that uses proton gradients to synthesize ATP. Our interest is in seeing what information single molecule studies may provide on the relation between structural changes and the rates of various processes involved in the ATP binding, release, hydrolysis and synthesis. Our approach is to utilize a class of kinetic-thermodynamic relations used for chemical reactions and originally coming from the field of electron and other transfer reactions. To adapt those relations to biomolecular motors we incorporated into the kinetic-thermodynamic free energy expressions an elastic interaction between the rotor and stator subunits of the motor, and coupled the rotor to the chemical and physical processes. The theory is aimed at adding insights to the different of rates in terms of protein structure. On the more technical side another aim is to relate one type of single molecule experiment to another and to ensemble experiments. Theory-based predictions are made and compared with experiment. Some of our recent PNAS studies are summarized, together with more recent results.

Our view of the present state of research on single-molecule systems

Ensemble (bulk system) measurements have, until the past couple of decades, been the only way of studying reaction rate mechanisms in chemical and biological systems. Such studies have led to a deep understanding of many problems including reaction mechanisms in these systems. What new information, then, can single molecule experiments provide? In ensemble measurements one obtains averages over events from many different molecular trajectories. Single molecule methods can provide direct experimental information at the molecular level on the individual events. They tell us how fast a molecule undergoes a particular physical or chemical process under different particular conditions, such as at each angle of rotation of the bio molecular motor. We know that the subunits of the protein ATPase have "pockets" that open and close during the course of a binding or release of an

ATP molecule and during its reaction. Single molecule methods provide details on these individual processes, details that are "fodder" for computer simulations that can permit a deeper insight into the mechanisms of these and of similar systems. At certain angles of rotation of the motor the channels to the entry into the binding pockets in the motor must close and one might hope to obtain information, by inference from the data, describing such openings and closings, not accessible by ensemble methods. In our experience, however, single molecule methods by themselves do not provide all the answers. Indeed, in our own analysis we have needed to draw upon ensemble information, combining it with single molecule information, in an effort to obtain a further understanding.

There are now extensive experimental studies of single molecule systems, ranging from the intermittent fluorescence of illuminated semiconductor nanoparticles to the quite different observations of biomolecular motors. The latter includes the subject of our present focus, F_1-ATPase [1–9]. One question that arises is how to treat this body of kinetic-thermodynamic data theoretically in a way that has the possibility of providing added insight into the phenomena, and relating different kinds of single molecule studies to each other and to ensemble data.

To this end we describe a chemical-mechanical theory [10–12] that we have formulated to treat several types of single molecule experiments on a biomolecular motor [8, 9]. The theory is adapted from a theory of electron and other transfers in solution [13, 14] by including the elastic effects in the motor in the various processes. We focus on the four types of single molecule experiments that have been performed on the F_1 component depicted in Fig. 1(a) and 1(c) of the enzyme, F_1F_o-ATPase. This theory incorporates elastic effects arising from rotor-stator interactions that are coupled to the sequence of different physical and chemical processes [15] involved in the operation of the motor.

These single molecule experiments are: (1) *Stepping experiments,* [1–3] in which it was shown that under the influence of ATP the motor rotates from one "dwell" angle to another in successive 80° and 40° steps; (2) *Stalling experiments,* [6–8] in which the rotor is held fixed by magnetic tweezers at some specified stalling angle θ for different amounts of time and then allowed to relax, moving either backward to the initial dwell angle or forward to the next dwell angle; (3) *Controlled rotation experiments,* [5] in which magnetic tweezers are used to rotate the rotor at a fixed angular velocity and the probability of a site occupancy is observed for small (10°) successive angular displacements using a fluorescent probe attached to the nucleotide, ATP or ADP, to detect whether it is bound or not to a subunit; and (4) *controlled rotation experiments on fluorescent nucleotides,* [9] in which the rotor is rotated at some fixed angular velocity by magnetic tweezers and the site occupancy detected by the fluorescent probe-labeled nucleotide. In each of the experiments (2)–(4) rates of the various processes can be measured and the results analyzed in terms of theory.

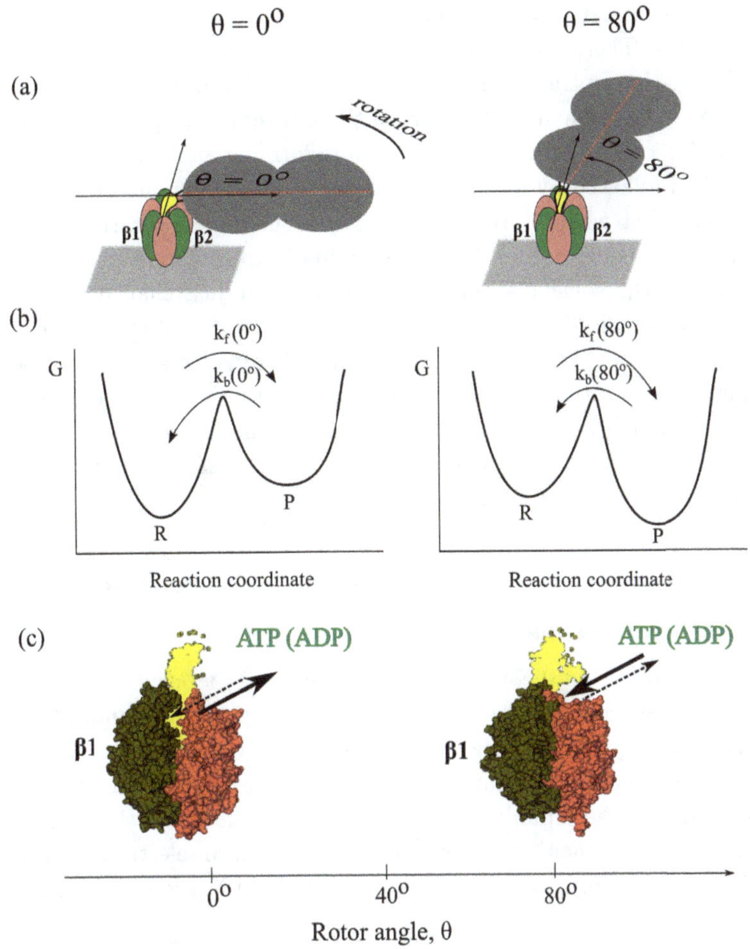

Fig. 1. (a) F_1-ATPase in single molecule imaging and controlled rotation experiments at two rotor angles: 0° and 80°. A double-bead is attached to the γ rotor shaft (in yellow) and rotated against the stator ring (active subunits $\beta 1$–$\beta 3$). (b) Free energy profile for nucleotide binding (k_f) and release (k_b) rate constants at the two angles. (c) Open-to-close changes in the nucleotide binding $\beta 1$ subunit as a function of rotor angle.

The analysis includes the comparison of the predicted and observed effect of the free energy driving force on the rate of the process involved (*e.g.* the nucleotide binding, seen in Fig. 1(b)), and predicting the results of one type of experiment, with no adjustable parameters, from those of another in the angular regions where the two sets of results overlap. The analysis provides additional insight into various processes, indicating from a "turnover" behavior an internal diffusion control within the F_1-ATPase, when a channel for the entrance or exit of a nucleotide from a subunit has presumably narrowed, as seen in Fig. 1(c).

We consider first the reaction-elastic equations used in this series of single molecule studies. There are numerous technical details given there in this description. However, the central theme present at all of the discussion is the "give and take" nature of the mechanisms. In any of these reactions or processes come for example, breaking one set of hydrogen bonds and forming another, the first sets don't break and then followed by formation of the second set, but rather one anticipates a cooperative effect whereby these events are occurring cooperatively rather than sequentially. This effect occurs, for example, in electron transfer reactions between two ions, where the solvent around one ion does not just change its distribution from its initial set of configurations of oriented solvent molecules to its final set appropriate to its new ionic charges but rather there is some change around both ions simultaneously rather than consecutively. Another example, discussed below, is in a bond breaking-bond forming reaction where the breakage and formation occur simultaneously rather than consecutively, with a considerable reduction in the energy barrier, as described below.

Our recent research contributions to single-molecule systems

Free energy relations and rates

The starting point of the theory [10–12] is two-fold: (1) an equation describing the dependence of the rate constant of any particular step on the standard free energy reaction of that step, and (2) the introduction into a kinetic-thermodynamic free energy equation a mechanical elastic coupling term arising from the interaction with each other and with the subunits and the spindle of the motor of the F_1-ATPase and between the subunits themselves. We recall first from an electron transfer theory for reactions in solution [13, 14] the free energy barrier for the reaction, ΔG^*, is given by

$$\Delta G^* = \frac{\lambda}{4}(1 + \Delta G^o/\lambda)^2 \tag{1}$$

where, for simplicity of presentation in this article, we have omitted the "work terms" that occur in transitioning from a pair of reactants (r) or products (p) in solution to form a collision complex. In Eq. (1), ΔG^0 denotes the standard free energy reaction of the ET reaction and λ denotes the "reorganization energy" for the reaction. The rate constant k is given by

$$k = A \exp\left(-\Delta G^*/kT\right) \tag{2}$$

where A depends on the nature of the process, *e.g.*, whether the reaction is first or second order.

In Eq. (1) the free energy barrier in the series of reactions having the same λ but different ΔG^o at first decreases when ΔG^o becomes increasingly negative, then vanishes when $\Delta G^o/\lambda = -1$ and then increases again when $\Delta G^o/\lambda$ is made increasingly negative. This predicted "inverted effect" was experimentally confirmed. This effect is a now well-known consequence of the topology of the two "nearly intersecting" free energy parabolas for the reactants and the products when the magnitude of $\Delta G^0/\lambda$ becomes large. Indeed, there is this expected topological difference from the difference between weak-overlap reactions and strong-overlap reactions.

Strong overlap reactions

The equations of the "weak-overlap" theory of electron transfer reactions were adapted to treat "strong overlap" reactions, such as the transfer of an atom B in the atom transfer reaction $A + BC \rightarrow AB + C$. In this case the topology is such that there is in effect only one relevant potential energy surface (the lower adiabatic surface) instead of two diabatic surfaces. The simplest adaptation is to use Eq. (1) but, because of the different topology of the surfaces for weak and strong interaction, to set $\Delta G^* = 0$ when $-\Delta G^0 \geq \lambda$. That is, whereas the ET reaction rate decreases at high negative ΔG^o the "strong overlap" reaction rate reaches some limiting maximum value, a result clear from the topology of the adiabatic surface, instead of experiencing an inverted effect. When, instead, ΔG^o is very positive we use for this limiting case $\Delta G^* = \Delta G^o$.

Using a bond energy-bond order model [16] for the above atom transfer reaction, in which the sum of the bond orders of the AB and BC bonds is assumed equal to unity throughout the reaction step, one can obtain a simple analytic form containing a log cosh function of $\Delta G^0/\lambda$ instead of the quadratic function in Eq. (1). Again, omitting work terms for brevity of presentation we have [16]

$$\Delta G^* = \frac{\lambda}{4} + \frac{\Delta G^0}{2} = \frac{1}{2}\frac{\Delta G^0}{y}\ln\cosh y \tag{3}$$

where $y = (2\Delta G^0/\lambda)\ln 2$.

The two functions in Eqs. (1) and (3) both have a Brønsted slope, $\partial\Delta G^*/\partial\Delta G^0$ of 0.5 when $\Delta G^o = 0$. They show a rather similar exponential dependence of the rate constant k on ΔG^o, when the absolute value of the latter is small relative to λ. They differ when the magnitude of $\Delta G^0/\lambda$ in absolute value becomes large (e.g., approximately equal to or larger than unity), as expected from the topological difference between the free energy profiles of the weak and strong overlap reactions. They both show a cooperative effect during the reaction. For example, Eq. (1) exhibits a compensation effect that occurs in all of the processes we shall consider. As an example, for an isotopic exchange electron transfer the free energy barrier to reaction given by Eq. (1) is $\lambda/4$. Had there been no reorganization prior to the electron transfer the barrier to reaction would have been λ, the vertical distance of the two parabolas in the usual two-parabola plot for fast reactions. (Such a vertical ET could occur only by the absorption of light, whereas we are treating a thermally-induced electron transfer.) So for isotopic exchange electron transfers ($\Delta G^0 = 0$) the reorganization prior to the actual electron transfer, required by the application of the Franck Condon principle, provides a reaction barrier that is one quarter the value it would have had, had there been no reorganization prior to the transfer. In the case of Eq. (3) for an atom or group transfer a similar cooperative effect is seen. Here, the activation energy when C is transferred to A is roughly 10% of the BA bond energy. It is easier to break a chemical bond if at the same time a new chemical bond is also forming, in fact, in this case the process is 10 times easier in terms of energetics. This cooperation is tacitly assumed when applying Eqs. (1) and (3) to these kinds of processes, including the physical processes of binding a nucleotide and release, which involve the breaking and forming of bonds such as hydrogen bonds.

Application to processes in single molecule studies of the F_1-ATPase

Turning now to the application of these concepts to the different kinds of single molecule experiments on F_1-ATPase we focus initially on the binding and release of ATP, for which there are the most extensive data obtained by different methods [2, 5, 17, 18]. In the ATP binding after an ATP first forms a collision complex with the F_1-ATPase, is followed by an ATP binding as the next step (an 80° step) in the overall hydrolysis and then in the same subunit the latter is followed by an ATP hydrolysis step (a 40° step) and by a successive release of ADP (in an 80° step) and then of Pi (inorganic phosphate) in a 40° step. In the meanwhile, related out-of-step processes are occurring in the other two β subunits, as depicted in Table 1 of Ref. [10].

When an ATP enters a channel in an empty β subunit and finally locks into place in the subunit some hydrogen bonds are broken while new hydrogen bonds are formed. In that case there is some analogy to the situation that led to Eqs. (1) and (3) for the activation free energy ΔG^*. Accordingly, we have used this idea in applying Eqs. (1) and (2) in some recent PNAS articles [10–12] on the binding and release of ATP in single molecule stalling and controlled rotation experiments. In this adaption of Eqs. (1) and (2) it is necessary to include the elastic properties of the ATPase motor, noting that the dwell angles 0°, 80°, 120°, 200°, 240°, 320° and 360° are angles of rotation of the rotor with respect to the stator that are angles of local stability.

We introduced into the free energy curves of the reactants and of the products an elastic coupling between the rotor and the stator [3, 19] involved in each of these steps. In the two-state model originally used to derive Eq. (1) we added to each parabolic free energy curve an elastic component. The elastic component of the free energy equations for the reactants and products of an individual step in a subunit are given in Eq. (4), so yielding Eq. (5) for the free-energy barrier as a function of θ. Equation (5) is a function of the rotor angle θ for that step since the standard free energy of reaction is also given as a function of θ in Eq. (6),

$$w^r = \frac{k}{2}(\theta - \theta_i)^2 \quad \text{and} \quad w^p = \frac{k}{2}(\theta - \theta_f)^2 \tag{4}$$

$$\Delta G * (\theta) = W^r + \left[\lambda + \Delta G^0(\theta)\right]^2 / 4\lambda \tag{5}$$

where

$$\begin{aligned}\Delta G^0(\theta) &= \Delta G_0^0 + w^p(\theta) - w^r(\theta) \\ &= \Delta G_0^0 - k(\theta_f - \theta_i)[\theta - (\theta_f + \theta_i)/2]\end{aligned} \tag{6}$$

and where W^r is the work term for attaching the ATP from solution to the exterior of the F_1-ATPase, ΔG_0^0 is the free energy change of a step in the ATPase, namely, for that step the free energy of the final dwell state minus the free energy of the initial dwell state.

Application of theory to the single molecule experiments

Stalling and controlled rotation experiments provide the θ-dependent rate constants and equilibrium constants for the steps in the overall process. The $\Delta G^0(\theta)$ is related to the θ-dependent equilibrium constant $K(\theta)$ for that step by $\Delta G^0(\theta) = -kT \ln K(\theta)$. The equation for the $\ln K(\theta)$ is seen in Eq. (6) to be linear in θ, in agreement with the experimental data in Fig. 1. With experimental data from ensemble [17, 18] and free rotation [2, 3, 5] experiments Eqs. (5)–(6) were used to predict [10] for the stalling experiments the Brønsted slope $\partial \Delta G^* / \partial \Delta G^0$ of 0.47, which compares with the value of 0.48 in the stalling experiment for the rate of ATP binding and release over the θ range studied. For the binding rate (k_f) the slope of the $\ln k_f(\theta)$ versus θ was also predicted for these nucleotides from the Brønsted slope and the $\ln K(\theta)$ versus θ data, and found to be in reasonable agreement ($\sim$ within 10%) with experiment [10]. For the spring constant k of the rotor that appears in Eq. (4) the value of $k = 16$ pN nm rad^{-2} obtained from the stalling experiments and used in the prediction of the absolute value of the binding and release rate constants in Figs. 2 and 3 below for the fluorescent nucleotides in controlled rotation experiments. With no adjustable parameters, a good agreement was found between these experiments and our calculations, in which we also used the binding rate constant from the free rotation experiment for the fluorescent ATP species [5].

The controlled rotation experiments are complementary to the other forms of single molecule experiments in several respects. There is a region of angles θ where the controlled rotation and stalling experiments overlap, so that with the aid of ensemble data it is possible to predict results of the controlled rotation experiments with no adjustable parameters. The comparison is given in Fig. 2, covering the θ-range, as noted earlier, $\sim -40°$ to $\sim +40°$. In the controlled rotation experiments were bound fluorescent nucleotides that become non-fluorescent upon release they are used to observe individual binding and release events, but there will be certain missed events when the rate of release is very fast. It was possible to take into account these missed events using the theory embodied in the previous equations, and the results, after correcting for these events, are given in Fig. 2. The observations

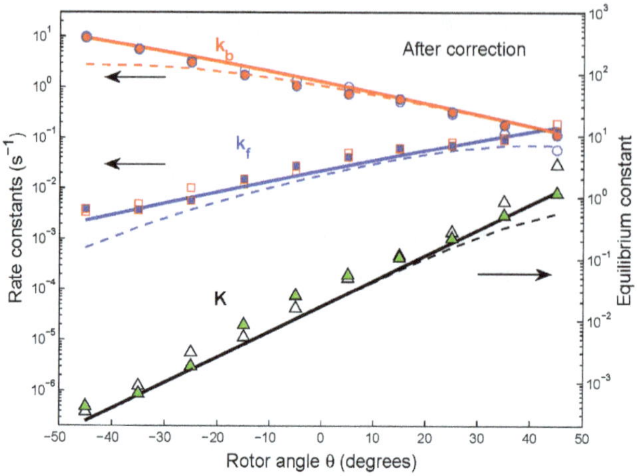

Fig. 2. Corrected binding and release rate and equilibrium rate constants versus θ angle for Cy3-ATP in the presence (solid squares, circles and triangles) and absence of Pi (open symbols) in solution adapted from Ref. [16]. The experimental data of Adachi *et al.* [9] corrected for missed events (and an error due to replacing the time spent in the empty state by total time of a trajectory) are compared with their theoretical counterparts (solid lines). Dashed lines show the data without corrections.

of the rates without this correction for the missed events is given by the dashed lines in that figure.

The controlled rotation experiments provide rate constant data over much of the 360° range of rotor angles, and so the treatment has been extended beyond the overlapping range between stalling and controlled rotation experiments [12]. An observation seen in Fig. 3 is a symmetry about $\theta \sim -40°$. To the right there is the increased rate of ATP binding with increasing θ. To the left, with the motor rotating in reverse, there is again an ATP binding whose angular rate dependence is approximately the symmetric image of the behavior to the right. Similarly, if one looks at the ATP release rates in the plot they again show a mirror-like symmetry about this value of θ. Nevertheless, there is a marked difference when P_i is added to the solution, and the results are not shown in the plot but shown in Ref. [10]. For the data to the right of $\theta \sim -40°$ the rate of ATP binding is unaffected by the presence of P_i, as indeed should be, if the P_i in solution is not to interfere with the motor. In contrast, to the left of $\theta \sim -40°$, when the motor running in the forward direction, ADP is the first to be released after a hydrolysis followed by the release of the P_i. Thus, there is some tendency for the P_i to be bound to the subunit in this region of θ. Accordingly, when the motor is running in reverse in this region any tendency to bind P_i interferes with the binding of ATP. This basis of an asymmetry in the effect of an added P_i offers an explanation of the different effect of the addition of P_i on the ATP binding rate and it is of interest to explore its relation to structural differences.

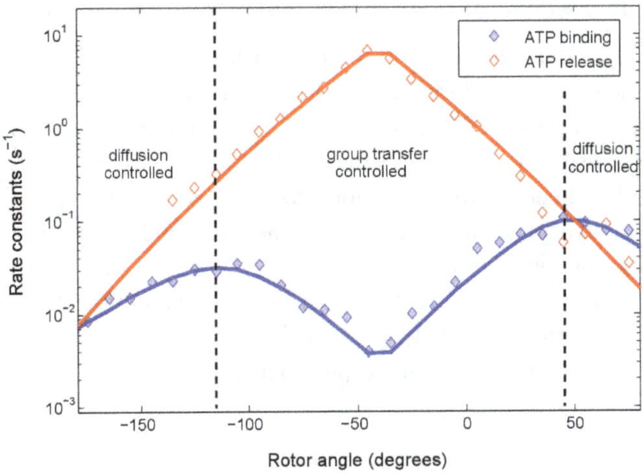

Fig. 3. Angular dependence of rate constants monitored in controlled rotation experiments in F_1-ATPase. Red open and blue closed symbols are data points for binding and release processes for ADP (circles) and ATP (squares) monitored in experiments with corresponding theoretical curves for ADP binding in matching colors.

Another feature seen in this figure is the turnaround in binding rates at $\theta \sim +40°$ on the right side and at $\theta \sim -130°$ on the left hand side of the plot. In these two regions there is a tendency for the β subunit that is binding the ATP via a channel for the latter to become narrow, as seen in X-ray crystallography structures [20, 21], and eventually to close at some angle θ. In such angular regions the rate of binding should decrease with increasing θ on the right side of the plot and decrease with decreasing θ on the left side of the plot due to the rate now being limited by diffusion [22, 23] in the channel entrance to the β subunit in these regions of the plot ($\theta > 50°$ on the right-hand side of the plot and $\theta < -140°$ on the left-hand side of the plot). Indeed, in these two regions there's a tendency for the binding and released rates to be about the same. If there is a diffusion control through the canals into the subunit, and if there is not a large bias in the diffusion, then indeed one expects these two rates to be comparable, depending on the amount of bias. In a steady-state approximation the reciprocal of the rate constant in this region $1/k$ equals the sum of the reciprocals of the diffusion rate constant and the activation rate constant and so k becomes equal to the diffusion rate constant when the latter is sufficiently small.

Another feature seen in Fig. 2 is that the maximum rate of ATP release occurs at $\theta \sim -40°$ while the minimum rate of ATP release occurs at $\theta \sim -220°$, roughly 180° apart. If the opening and closing of this β subunit is due to the asymmetric [20] spindle (the γ subunit) pressing the most and pressing the least, respectively, against the subunits. In that case, one expects a 180° difference in angles for the two extrema.

Outlook to future developments of research on the present theory

For some steps the data are less complete than for others. For example, the ATP binding and release data are extensive. In the stalling experiments, the step involving the hydrolysis of ATP has only limited data, namely data at relatively high percent conversions, and so it would be helpful to obtain data at shorter reaction times, if possible. Then a Brønsted slope for that step can be informative about the nature of the hydrolysis mechanism more accurately determined. At present its tentative value appears to be unity.

The focus in our present work has been on a rotary motor [10–12]. More generally, on looking to the larger field of linear biomolecular motors, we are exploring the possible extension of the present type of theory to those motors.

Acknowledgments

The authors would like to acknowledge support from the Office of the Naval Research, the Army Research Office, and the James W. Glanville Foundation.

References

[1] H. Noji, R. Yasuda, M. Yoshida, K. Kinosita, Jr., *Nature* **386**, 299 (1997).
[2] R. Yasuda, H. Noji, M. Yoshida, K. Kinosita, Jr., H. Itoh, *Nature* **410**, 898 (2001).
[3] H. Sielaff, H. Rennekamp, S. Engelbrecht, W. Junge, *Biophys. J.* **95**, 4979 (2008).
[4] D. Spetzler, R. Ishmukhametov, T. Hornung, L. J. Day, J. Martin *et al.*, *Biochemistry* **48**, 7979 (2009).
[5] K. Adachi, K. Oiwa, T. Nishizaka, S. Furuike, H. Noji *et al.*, *Cell* **130**, 309 (2007).
[6] R. Watanabe, R. Iino, H. Noji, *Nat. Chem. Biol.* **6**, 814 (2010).
[7] R. Watanabe, H. Noji, *Nature Comm.* **5**, 3486 (2014).
[8] R. Watanabe, D. Okuno, S. Sakakihara, K. Shimabukuro, R. Iino *et al.*, *Nat. Chem. Biol.* **8**, 86 (2012).
[9] K. Adachi, K. Oiwa, M. Yoshida, T. Nishizaka, K. Kinosita, Jr., *Nature Comm.* **3**, 1022 (2012).
[10] S. Volkán-Kacsó, R. A. Marcus, *Proc. Natl. Acad. Sci. USA* **112**, 14230 (2015).
[11] S. Volkán-Kacsó, R. A. Marcus, *Proc. Natl. Acad. Sci. USA* **113**, 12029 (2016).
[12] S. Volkán-Kacsó, R. A. Marcus, *Proc. Natl. Acad. Sci. USA* **114**, 7272 (2017).
[13] R. A. Marcus, N. Sutin, *Biochim. Biophys. Acta* **811**, 265 (1985), and references cited therein.
[14] R. A. Marcus, *Les Prix Nobel*, T. Frangsmyr (ed.), 63 (Almqvist & Wiksell, 1993).
[15] P. D. Boyer, *Biochim. Biophys. Acta* **1140**, 215 (1993).
[16] R. A. Marcus, *J. Phys. Chem.* **72**, 891 (1968).
[17] A. E. Senior, *Cell* **130**, 220 (2007).
[18] J. Weber, A. E. Senior, *Biochim. Biophys. Acta* **1319**, 19 (1997).
[19] O. Panke, D. A. Cherepanov, K. Gumbiowski, S. Engelbrecht, W. Junge, *Biophys. J.* **81**, 120 (2001).
[20] K. Braig, R. I. Menz, M. G. Montgomery, A. G. Leslie, J. E. Walker, *Structure* **8**, 567 (2000).
[21] J. E. Walker, *Biochem. Soc. Trans.* **41**, 1 (2013).
[22] R. A. Marcus, *Discuss. Farad. Soc.* **29**, 129 (1960).
[23] R. M. Noyes, *Progr. React. Kinet.* **1**, 129 (1961).

INTERPLAY OF HIGH-LEVEL SPECTROSCOPY AND QUANTUM CHEMISTRY: A POWERFUL TOOL TO UNRAVEL ENZYMATIC REACTION MECHANISMS

FRANK NEESE

Department of Molecular Theory and Spectroscopy, Max Planck Institute for Chemical Energy Conversion, 45470 Mülheim an der Ruhr, Germany

My view of the present state of research on enzymatic reaction mechanisms

The importance of transition metals for all areas of catalysis (homogeneous, heterogeneous and biological) hardly needs to be emphasized. In many respects, biology provides benchmark catalysts in the sense of unprecedented efficiency and selectivity while operating under the mildest chemical conditions without producing harmful waste. Hence, biological catalysts are widely aspired to in the design of new catalytic systems. In fact, for almost all particularly difficult chemical transformations, nature has evolved metalloproteins. For example, such reactions include the reactions of central importance for energy research: (1) activation of water (2) activation of dioxygen (3) activation of CO_2, (4) activation of dinitrogen (5) activation of methane (6) formation of dihydrogen. For many of these reactions, industrial scale catalysts that challenge their biological counterparts in terms of efficiency and selectivity are missing. Hence, it appears to be critically important to learn lessons from nature in the design of future catalysts.

The desire for the rational design of catalysts is deeply routed in the reductionist approach to science, where one tries to understand a complicated system in terms of the interactions of its constituent parts at a deeper and deeper level. By understanding the most important factors that govern the behavior of the target system, one eventually hopes to be able to construct the desired systems in a rational, targeted approach. Obviously, this is not the only possible route to catalyst design as witnessed by the successes of trial-an-error, high-throughput screening or guided evolution approaches. For the sake of this short essay, however, the reductionist point of view is taken.

Biochemistry has been highly successful in the reductionist approach by being able to understand the behavior of cells in terms of the interaction of the properties of the constituting proteins, nucleic acids, membranes, vitamins, hormones to name only a few. Enzymologists strive to understand the behavior of enzymes in terms of individual functional groups and substrate or product molecules. In chemistry, the ultimate level of understanding involves unraveling the underlying mechanisms that govern the making and breaking of bonds in terms of the behavior of the electrons and nuclei that participate in a given chemical reaction.

This latter understanding, however, requires the laws of quantum mechanics, which describes the behavior of all matter in an, as far as we know today, exact manner. In fact, Dirac's famous quote from 1929 states that the laws of a large part of physics and *all* of chemistry are known and that the only principle problems lies in the fact that the application of these laws leads to equations that are too complex to be solved [1]. This statement contains the implicit promise that, *provided* we would be able to solve the quantum mechanical equations that Dirac refers to exactly, we would be able to predict chemistry exactly and, by inference, the same would apply to biochemistry. This is obviously a very bold promise that formulated a distant goal to aspire to for generations of scientists to come.

In fact, 90 years after Dirac's promise, remarkable progress has been made in applying quantum mechanics to chemistry and, in many instances, also to biochemistry. It is nowadays commonplace in all areas of chemistry to supplement experimental studies with quantum chemical calculations or to find purely computational chemistry studies in high-profile chemical journals.

However, despite all this progress, one should not forget that we are still far from being able to compute all aspects of enzymatic catalysis exactly. In fact, enzymatic catalysis is governed by the free-energy profile of the reaction. Without complicating matters too much, the free energy is comprised of the inner energy, zero-point energy corrections, enthalpic contributions and entropic contributions. If one focuses on a core of an enzyme active site, one also has to deal with solvation and longer range electrostatic effects as well as weak intermolecular interactions. Not all these contributions can be computed to "chemical accuracy." The latter is defined by the commonly accepted goals of computing energies to within 1 kcal/mol. If we would be able to reach this goal reliably, the predictions for reaction rates and equilibrium constants would be within one log unit of experiment and redox-potential to within 56 mV. While this is moderate accuracy for chemical applications, where measurements are often much more precise, it is an heroic goal to reach for quantum chemistry since it requires an accuracy in solving the molecular Schrödinger equation to an accuracy of about 0.0001%. While one can reach this goal for small molecules in the gas phase, it is not currently attainable for enzymatic reaction profiles.

This discussion serves to motivate the main theme of this essay, which is that one is well advised to enter a self-critical dialogue between theory and experiment rather than to do either theory or experimental in isolation and unconditionally rely on the results of such studies. In fact, theory can be wrong for many reasons. In addition to possibly insufficient intrinsic accuracy of the quantum chemical approximations, this may involve wrong structures or overlooked reaction pathways. Experiment viewed in isolation may face the challenge that the results are too complex to be interpretable or that oversimplifying assumptions based on "fingerprinting" are relied on. The latter may have been derived from studies on much simpler systems and may not hold for complex metalloenzyme active sites.

Hence, there is ample opportunity for true synergy between theory and experiment. The common meeting point should, however, be *observable* properties of the system. The three big areas in this respect are kinetic data, thermodynamic data and spectroscopic properties. Of these, kinetic data are complex, since a detailed comparison to theory requires the rates of the elementary reaction steps to be known, rather than just the rate at which product emerges or substrate is consumed. Thus it requires the assumption of a reaction network, which is often not unique and involves very extended studies to be established. However, where available, this kinetic data is invaluable. Thermodynamic quantities are scarcely available beyond the overall energy change of a given chemical reaction. However, since the catalyst itself is not part of this energy difference this provides very limited information upon which theory can be gauged. Again, where available, this thermodynamic information is highly useful.

However, it is the wealth of information provided by all forms of spectroscopy that, in our opinion, is the most useful meeting ground for theory and experiment in chemistry. While both, thermodynamics and kinetics are based on the small differences between very large total energies, spectroscopic observable are sensitive to intimate details of the geometric and electronic structure of the system under investigation. Hence, they provide highly specific local probes. Examples include electron paramagnetic resonance (EPR) g-values, EPR hyperfine couplings, nuclear magnetic resonance (NMR) chemical shifts, vibrational frequencies measure by infra-red (IR) or resonance Raman (rR) spectroscopy, UV/visible absorption and emission band shapes and energies, X-ray absorption pre-edge energies and intensities, X-ray emission features at various edge, to name only a few. Spectroscopic experiments can be performed in a time resolved fashion to detect transient intermediates, on liquid or frozen samples or even *in situ.* Thus, they provide information about geometric and electronic structures that are impossible to obtain by X-ray crystallography.

It is this wealth of experimental data that is so amply amenable to quantum chemical studies. In many cases, calculations can help to assign spectroscopic features and thus enable to interpretation of the experiments in terms of geometric structures. The comparison of experimental and calculated spectroscopic features provides a detailed feedback concerning whether the computed geometric and electronic structures are realistic. Finally, once it has been established firmly that the computations are correctly describing the experimental situation, they can be used to derive a wealth of chemical information and to draw far-reaching cross-correlations between different experimental facts. In fact, the electronic structure of molecules is not only reflected in the spectroscopic properties, it also dictates reactivity. Hence, by studying spectroscopic observables one can obtain the type of deep insight into the principles governing the reactions that one is studying. It is emphasized that the analysis derived in this way is deeply rooted in experimental reality and hence is to be preferred over conclusions that are solely derived from computations without any connection to experiment.

There are a few important requisites for the success of such studies: (1) It is absolutely critical to understand the accuracy limits of the quantum chemical methods one applies for the properties one is studying. Valid conclusions can only be drawn, if established error bars exist for the given combination of method and property. (2) Importantly, a correctly computed spectrum does not *prove* the correctness of a given structure. First, one needs to establish how sensitive the property reacts to the different structural alternatives that one wishes to discriminate. Secondly, the desire to deduce structure from property is, mathematically speaking, an *inverse* problem, which is underdetermined. With the possible exception of NMR (or modern high-resolution EPR variants), the amount of spectroscopic information is never enough to fully reconstruct a complete three-dimensional structure. Hence, by proving agreement between theory and experiment, one only proves that the suggested structure is *consistent* with the experimental observation, not its correctness. A corollary to this statement is that one is well advised to seek as much and as diverse spectroscopic information as possible in order to arrive the most substantiated conclusions possible.

My recent research contributions to enzymatic reaction mechanisms

The previous paragraphs laid out a research strategy that our group has been following for several decades. The work can be broadly categorized in three overlapping areas: (1) development of new theoretical methods, (2) applications to chemical problems (3) spectroscopic experiments. Significant progress has been made in theoretical method development: (a) methods that allow for the calculation of a broad range of spectroscopic properties and nearly all properties in common use in (bio)inorganic chemistry, (b) the development of more efficient methods for studies on larger and larger systems, including entire macromolecules and (c) the development of ever more accurate approximations that will allow more precise predictions and, hence, also more precise conclusions to be drawn. The combination of these three goals is an open-ended research program at the heart of theoretical chemistry. Our results are available free of charge to the scientific community via the ORCA program [2].

The importance of choosing the right property that is sensitive to details of the structure one is interested in, is nicely demonstrated by the example of the nitrogenase active site (the FeMoCo) [3]. Early on, Einsle *et al.* [4] have identified a light atom in the center of the active site, the identity of which could not be established by X-ray crystallography. In the following decade, many theoretical and experimental studies were published that were largely inconclusive and all possible variants (carbon, nitrogen or oxygen) were proposed by various researchers. Conclusive evidence for the detection of the first carbide center in biochemistry was obtained by X-ray emission spectroscopy, pioneered in (bio)inorganic chemistry by the DeBeer group [5], in combination with quantum chemical calculations [3]. By constructing models of the FeMoCo with either carbon, nitrogen or oxygen in the center and using a previously calibrated density functional theory (DFT) based theoretical protocol, the comparison of experimental and computed valence-to-core X-ray emission spectra showed, that only carbon is consistent with the experimental observations. This conclusion could have hardly been obtained from total energy calculations alone and was subsequently substantiated by further studies that revealed the biochemical pathways responsible for carbon insertion. The original paper was published back-to-back with the study of Einsle and co-workers, who came to the same conclusion on the basis of sophisticated crystallographic experiments.

A second important study demonstrated how the analysis of spectroscopic data in terms of structure can contribute to the understanding of an important reaction mechanism, namely the water oxidation reaction by the oxygen-evolving complex in photosystem II [6]. Experimentally, it has been known that in the so-called S2 state, an established reaction intermediate, EPR signals can arise around g = 2 (the famous "multiline signal") or g = 4. The carefully calibrated DFT studies came to the conclusion that this must arise from two isomers of the system with total spin equal to 1/2 and 5/2 respectively. These isomers arise from the switching of a single oxo-bridge between to manganese centers. The calculations showed that the two isomers are connected by a low-energy transition state and are thus interconvertible. Interestingly, the calculations predicted that an isolated oxo-bridge Mn_3Ca cube should be intrinsically coupled ferromagnetically to a S = 9/2 state. This was impressively demonstrated in subsequent model studies by the Cristou and Agapie laboratories. Furthermore, it became evident that the two isomers are highly relevant for the reaction mechanism of the OEC since the system initially arrives in the S2 state in the S = 1/2 state but proceed, under water binding, to the S3 state in the high-spin configuration. Hence, there must be "two-state-reactivity" in the OEC [7].

We note in passing, that a detailed recent study showed that, at the present level of theoretical sophistication, EXAFS spectra can not be *predicted* from first-principles with sufficient accuracy in order to allow proposed models of the OEC to be differentiated reliably. This conclusion holds more generally and is not confined to the OEC. Note that this does *not* question the reliability of EXAFS derived *fits* for metal–ligand bond distances. It does, however, underline the importance of using carefully calibrated theoretical protocols [8].

The sensitivity to structural details is nicely demonstrated by a study on the ribonucleotide reductase system [9, 10]. One of the defining features of this important enzyme is a very long range electron transfer through the protein, that is accomplished through several intermediates, one of which is a pair of Tyrosine residues, Tyr_{730} and Tyr_{731}. The Stubbe lab has succeeded to chemically modify this pair of tyrosines by selectively placing NH_2 groups on the aromatic ring. This substitution changes the redox-potential of the site such that it can be trapped in the tyrosyl radical state. This state has been investigated in deep detail by Bennati and co-workers using high-resolution ENDOR spectroscopy. The wealth of hyperfine coupling information could be successfully interpreted on the basis of DFT calculations of the g-tensor and all hyperfine couplings in the system. The results demonstrated that there must be a hydrogen bond between the oxygen of the Tyr_{730} and a water molecule. The further analysis of the quantum chemical calculations revealed that this arrangement potentially has important implications for the reaction mechanism of the enzyme. Specifically, the slight alterations in the electron transfer energetics can alter the electron transfer rate into the active site by orders of magnitude. Hence, the hydrogen bond could de facto act as an on-off switch in this system.

As a final example, that illustrates how electronic structure and reactivity can be linked to spectroscopy, a recent study on a high-valent Fe(IV)-oxo system can be quoted [11]. These systems are prominently involved in hydrogen atom abstraction reactions. A detailed analysis has shown that these systems can react via four different reaction channels depending on the geometry of the attack (pi versus sigma) and two-different multiplicities (triplet or quintet) and that there can be triplet-quintet crossings along the reaction profile. On the basis of MCD spectroscopy coupled to high-level calculations, it was shown that two different ligand systems, an aliphatic amine ligand framework (by Que and co-workers) and a carbene ligand (by Meyer and co-workers) induce different orbital orderings in the respective Fe(IV)-oxo complexes (both of which feature a triplet ground state). The analysis implies that in the carbene system the iron based $d_{x}^{2}-_{y}^{2}$ orbital is pushed up very high in energy due to the exceptional sigma-donor strength of carbene versus amine. The consequence of this is, that the carbene does not posses a low-lying quintet state, while the amine system does. Subsequently, the analysis showed that the amine system must react via a two-state mechanism involving the quintet state while the carbene system must react via single state reactivity on the triplet surface. Thus, the experimentally calibrated electronic structure study showed that, despite the fact that the two systems are geometrically very similar and have nearly identical $Fe = O$ bonds in their electronic ground state, their reactivity is fundamentally different.

Outlook to future developments of research on enzymatic reaction mechanisms

Starting from the methodological side: despite all progress in theoretical methodology, there is ample room for improvement. Specifically, high-accuracy electronic structure methods must become widely available, must be made ever more precise and ever more efficient. They must run on modern highly parallel computer platforms. All of these tasks are in reach now and will probably become reality for mainstream use within the next few years. However, computing an accurate electronic energy is only one part of the problem. It becomes increasingly evident that the overall accuracy of quantum chemical predictions will soon be limited by the errors made in the other parts of the calculation. Most prominently, solvation energies and entropic contributions are two quantities that one cannot calculate with sufficient accuracy to allow overall, systematic chemical accuracy in the predictions. Both problems, in principle, can be addressed by molecular dynamic approaches. However, their combination with high-accuracy quantum chemistry and sufficiently converged sampling times is not within reach presently. The treatment of macromolecular systems will require extended development work as well in order to become efficient and user-friendly. However, perhaps even more daunting than the complexities of the electronic structure problem, are the problems of conformational complexity in biological systems (this is intimately related to the folding problem). These systems feature high-dimensional, highly complex potential energy surfaces with countless shallow local minima and consequently, very complex dynamics. How to accurately, reliably and efficiently maneuver theoretically through these potential energy surfaces is a problem of the "grand-challenge" class, perhaps for a long time to come.

However, as was emphasized throughout this article, the ultimate aim is not to try solving all chemical problems by theory alone, but rather to engage a productive and inspiring dialogue between theory and experiment. This requires significant dedication on both sides. The theoretician needs to be interested in the experimental observables that are measured and the experimentalist should be willing to take inspirations from theory and incorporate them into the experimental design. In order to succeed, one must be able to speak a common language. This language is necessarily a chemical language and not the language of quantum mechanics. Hence, while some achievements have been made [12, 13], new ways to connect the results of quantum chemical calculations to the language of chemistry must emerge and taken up by the computational chemistry community.

Rather detailed insight has been obtained in many important reaction mechanisms in biochemistry and also in molecular chemistry. However, how to turn these insights into widely applicable catalytic system, ready for large-scale industrial use is still an unsolved problem. In particular, one can dream to combine the selectivity achievable in finely tuned molecular systems with the efficiency attainable in heterogeneous systems and to understand these systems at the same level of detail that can be reached in molecular systems. It is probably fair to say that at this point in time, this is a great dream to be pursued in future research.

Acknowledgments

The generous financial support of our research by the Max Planck society is gratefully acknowledged.

References

[1] P. A. M. Dirac, *Proc. R. Soc. A.* **123**, 714 (1929).
[2] F. Neese, *WIREs Comput. Mol. Sci.* **2**, 73 (2012).
[3] K. M. Lancaster, M. Roemelt, P. Ettenhuber, Y. Hu, M. W. Ribbe *et al.*, *Science* **334**, 974 (2011).
[4] O. Einsle, F. A. Tezcan, S. L. A. Andrade, B. Schmid, M. Yoshida *et al.*, *Science* **297**, 1696 (2002).
[5] S. DeBeer George, T. Petrenko, F. Neese, *J. Phys. Chem. A* **112**, 12936 (2008).
[6] D. A. Pantazis, W. Ames, N. Cox, W. Lubitz, F. Neese, *Angew. Chem. Int. Ed.* **51**, 9935 (2012).
[7] N. Cox, M. Retegan, F. Neese, D. A. Pantazis, A. Boussac *et al.*, *Science* **345**, 804 (2014).
[8] M. A. Beckwith, W. Ames, F. D. Vila, V. Krewald, D. A. Pantazis *et al.*, *J. Am. Chem. Soc.* **137**, 12815 (2015).
[9] T. U. Nick, W. Lee, S. Kossmann, F. Neese, J. Stubbe *et al.*, *J. Am. Chem. Soc.* **137**, 289 (2015).
[10] T. Argirević, C. Riplinger, J. Stubbe, F. Neese, M. Bennati, *J. Am. Chem. Soc.* **134**, 17661 (2012).
[11] S. Ye, C. Kupper, S. Meyer, E. Andris, R. Navratil *et al.*, *J. Am. Chem. Soc.* **138**, 14312 (2016).
[12] W. B. Schneider, G. Bistoni, M. Sparta, M. Saitow, C. Riplinger *et al.*, *J. Chem. Theory Comput.* **12**, 4778 (2016).
[13] M. Atanasov, D. Ganyushin, K. Sivalingam, F. Neese, *Structue and Bonding* vol. 143, *Moleular Electronic Structures of Transition Metal Complexes II*, D. M. P. Mingos, P. Day, J. P. Dahl (eds.), 149 (Springer-Verlag, 2012).

PROTEIN DYNAMICS IN ENZYMATIC CATALYSIS

R. BRIAN DYER[1], MICHAEL J. REDDISH[1] and ROBERT CALLENDER[2]

[1]*Department of Chemistry, Emory University, Atlanta, GA 30322, USA*
[2]*Department of Biochemistry, Albert Einstein College of Medicine, Bronx, NY 10461, USA*

Biological catalysis: understanding the role of protein motions in enzymatic catalysis

Enzymes are dynamic molecules whose structures are in constant flux, in thermal equilibrium with the surrounding solvent. It is now well established that such protein motions are essential to enzyme function. For example, loop motions sometimes control the approach to reactive conformations and different loop conformations may impose a distribution of activation barriers in the Michaelis state. Thus, enzymatic reactions are better described by a multi-dimensional reaction landscape rather than by a classic single-barrier reaction coordinate [1, 2]. But how does the protein architecture (its atomic structure and dynamics) shape the energy landscape of catalysis? The multitude of available protein conformations exist on the landscape as minima separated by low barriers representing small-scale atomic level motions, by high barriers representing large-scale protein domain rearrangements, or anything in-between [3, 4]. Therefore, motions on many timescales are important, from slow conformational changes that control the search for reactive conformations to fast motions coupled to crossing the transition state. Many enzymatic reaction pathways connecting the various minima might also be available on the landscape. This idea has been explored theoretically in various enzymes [5–7], implied experimentally by single-molecule studies on enzymes revealing various reaction rate populations [8], and studied experimentally with vibrational spectroscopy [9].

My recent research contributions to enzymatic catalysis

We have examined these questions in the model enzyme lactate dehydrogenase (LDH), an oxidoreductase enzyme that catalyzes the reduction of pyruvate to lactate using a NADH cofactor. There are multiple variants of LDH within the same organism, such as heart, muscle, and sperm, with different reactivity. We have studied porcine heart LDH, which exists as a homotetramer [10], as a model for enzyme dynamics and catalysis because it has been previously well-studied and contains a significant structural change produced by the binding of substrate or substrate analog [11–15]. This structural change is characterized by the closing of a small loop section of the enzyme over the active site. The loop closure brings a catalytically relevant arginine (Arg-109) into position to stabilize the reduction site carbonyl on the substrate pyruvate. The loop closure event has previously been studied theoretically and experimentally to show that the loop can close into a variety of states [16–21]. Here we focus on how the Michaelis sub-state distribution and catalytic efficiency of LDH are controlled by the energy landscape of the catalytically important loop motions.

Temperature jump studies of enzyme dynamics

We have pioneered the development of temperature-jump (T-jump) and ultrafast microfluidic mixing methods coupled with structure-specific spectroscopic probes for the study of protein dynamics. We have applied these new approaches to study the dynamics of the Michaelis complex of LDH [21]. Figure 1 shows isotope-edited T-jump IR data for the live LDH:NADH:pyruvate system, poised in equilibrium with the lactate product side of the reaction. In this case, the T-jump shifts the equilibrium towards the lactate product side, and the observed relaxation kinetics report on the conformational reorganization as well as the chemistry step.

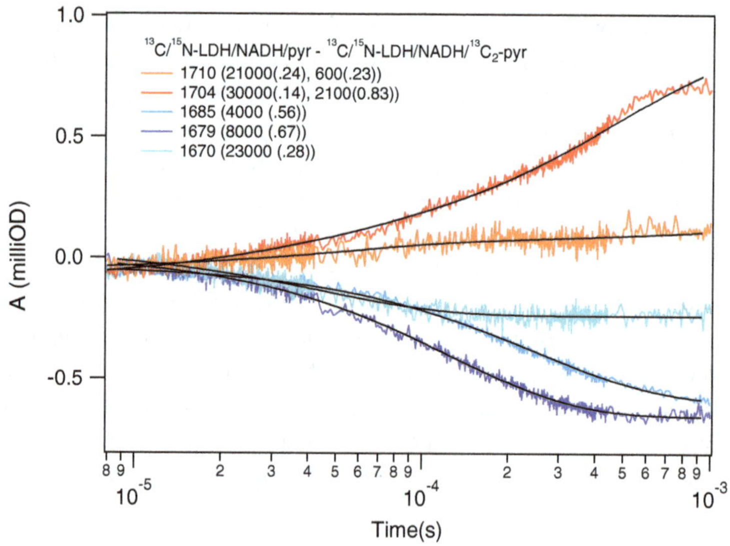

Fig. 1. Time-resolved isotope-edited infrared transients of LDH/NADH/pyruvate in response to a laser induced temperature jump, probed at the indicated frequencies (cm^{-1}). The black lines are exponential fits to the data and the $1/\tau$ values from the fits are indicated, along with the relative amplitudes.

The infrared experiment views the reaction from the perspective of the pyruvate $C_2 = O$ stretch, which provides a quantitative measure of the polarization of this bond as the system moves from the initial encounter complex, to the Michaelis ensemble and ultimately to product. The isotope-edited FTIR spectrum shows a $C_2 = O$ stretch at 1704 cm^{-1} for the encounter complex, only slightly perturbed from its solution value of 1710 cm^{-1}. The protein collapses from there to a distribution of more strongly bound sub-states that we describe as basins on the Michaelis complex energy landscape, having varying degrees of electrostatic and H-bonding interactions of the keto group with key residues of the active site. The broad multi-component band near 1679 cm^{-1} is due to this distribution of Michaelis conformations, with the low frequency side representing the more polarized conformations. We find that the collapse to the Michaelis ensemble occurs on the sub-millisecond timescale, as shown in the T-jump IR transients in Fig. 1. Furthermore, the Michaelis sub-states do not appear to interconvert directly but rather only indirectly, by returning to the encounter complex. Finally, the chemical step occurs with varying rates from the different Michaelis sub-states; the fastest relaxation rate is observed for the most polarized sub-state (lowest frequency $C_2 = O$ stretch at 1670 cm^{-1}). Thus, the T-jump IR method provides direct insight on the dynamics of the Michaelis ensemble, its formation and the sub-state dependent rate of the chemical step.

We have used T-jump to perturb the equilibrium of the "live" system as a means to probe the conformational dynamics, but these experiments are limited by the small temperature dependence of the conformational equilibria, resulting in small signals. An alternative approach is to use ultrafast mixing to initiate the reaction starting far from equilibrium, to generate much larger changes. Since we expect conformational search dynamics important to catalysis to occur on the microsecond to millisecond time scale, we have developed new ultrafast mixing methods to study these microsecond events [22, 23]. This approach is complementary to the T-jump method; the latter has a faster time resolution (as fast as 10^{-11} s) and can reveal hidden dynamics in cases where slow processes are followed by fast ones. But the population change that can be achieved by a T-jump is often small, so that the system is always near equilibrium, whereas the flow reactor produces a perturbation that is two orders of magnitude larger. Our specific approach is to employ microfluidic continuous flow fast mixing for reaction initiation and either fluorescence or IR spectroscopy to probe the reaction progress.

A fluorescence image of the flow system is shown in Fig. 2. We have shown that the ultrafast mixer can achieve diffusional mixing in the lamellar regime with mixing times as short as 50 μs. We developed methodology for calibrating the time resolution using fluorescent beads, and the reactor mixing time using KI or pH jumps to quench indicator dye fluorescence. Furthermore, we have extended the probe fluorescence excitation wavelengths deeper in the UV than what is generally available to commercial confocal fluorescence imaging systems. We can now directly excite tryptophan (Trp) and observe its fluorescence ($\lambda_{\text{pump}} = 290$ nm, $\lambda_{\text{probe}} = 350$ nm) using a modified commercially available confocal microscope. The flow system and the modified confocal imaging system are designed to be simple and low cost, using off the shelf parts (*e.g.* standard HPLC capillary tubing and fittings from Upchurch) so that the technology can be easily transferred to the broader biophysics community interested in fast protein dynamics. We have applied the ultrafast mixing technique to study reaction kinetics of LDH. Figure 2 shows ultrafast mixing results for the reaction of the LDH:NADH binary complex with pyruvate substrate or with oxamate inhibitor. The progress of each reaction is monitored using fluorescence resonant energy transfer (FRET) from Trp to NADH cofactor. The major kinetics phase observed in both cases (a large decrease in FRET efficiency) is related to pre-chemistry conformational changes as the system searches for the Michaelis complex. The apparent similarity between the transients indicates that observed search process occurs in both cases, not just the chemically competent pyruvate case.

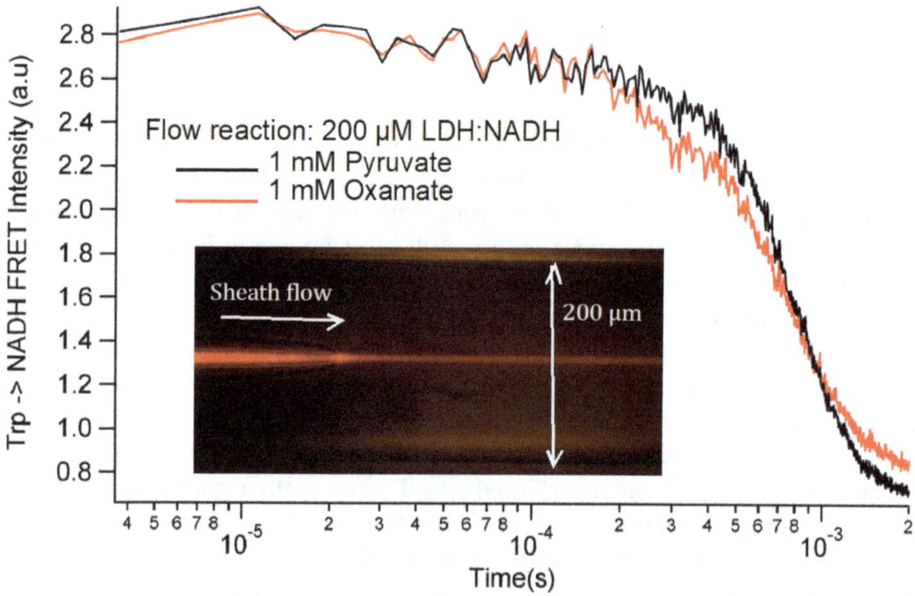

Fig. 2. Continuous flow fast mixer reaction of substrates with lactate dehydrogenase: NADH binary complex, followed by monitoring FRET from Trp to NADH.

Outlook for resolution of the role of protein dynamics in catalysis

While protein dynamics are clearly important on the timescale of the approach to reactive conformations (μs to ms), the question of whether they matter on the timescale of crossing the transition state is still controversial and an active area of research. In LDH, crossing the transition state occurs in about 1 ps, therefore the relevant motions must be low frequency collective motions of proteins that occur on this timescale. Such motions have been observed directly in various proteins using THz spectroscopy and fs stimulated Raman spectroscopy. They have also been observed theoretically, and in the case of LDH, one such motion has been observed using transition path sampling methods. This collective motion of LDH is predicted to act as promoting vibration, assisting the hydride transfer step. This idea has been difficult to verify experimentally, and its overall importance to the catalytic power of enzymes is not known. One approach to address this question has been to begin to decouple such motions from the chemistry by making the heavy version of the enzyme (uniform stable isotope labeling with ^{13}C, ^{15}N and ^{2}H). This method does not perturb the electrostatics, but it does slow the protein motions. Preliminary results for LDH indicate a significant slowing of the chemistry step in the heavy enzyme. Another approach to this issue is to use ultrafast (fs) T-jump to observe fast energy flow through the enzyme structure as a means to observe coupling of collective protein motions to the enzyme active site. These methods have begun to show some promise for sorting out the contribution of collective protein motions to chemistry step.

Acknowledgments

This work was supported by a grant from NIGMS, GM068036.

References

[1] W. Min, X. S. Xie, B. Bagchi, *J. Phys. Chem. B* **112**, 454 (2008).
[2] J. P. Klinman, A. Kohen, *Annu. Rev. Biochem.* **82**, 471 (2013).
[3] V. L. Schramm, *Acc. Chem. Res.* **48**, 1032 (2015).
[4] J. P. Klinman, *Chem. Phys. Lett.* **471**, 179 (2009).
[5] K. Arora, C. L. Brooks, *Dynamics in Enzyme Catalysis*, J. Klinman, S. Hammes-Schiffer (eds.), 165 (Springer, 2013).
[6] S. J. Benkovic, S. Hammes-Schiffer, *Science* **301**, 1196 (2003).
[7] I. Zoi, M. W. Motley, D. Antoniou, V. L. Schramm, S. D. Schwartz, *J. Phys. Chem. B* **119**, 3662 (2015).
[8] B. P. English, W. Min, A. M. van Oijen, K. T. Lee, G. Luo *et al.*, *Nat. Chem. Biol.* **2**, 87 (2006).
[9] R. Callender, R. B. Dyer, *Acc. Chem. Res.* **48**, 407 (2015).
[10] J. J. Holbrook, A. Liljas, S. J. Steindel, M. G. Rossmann, *The Enzymes*, Vol. 11, P. D. Boyer (ed.), 191 (Academic Press, 1975).
[11] S. J. F. Vincent, C. Zwahlen, C. B. Post, J. W. Burgner, G. Bodenhausen, *Proc. Natl. Acad. Sci. USA* **94**, 4383 (1997).
[12] J. E. Basner, S. D. Schwartz, *J. Phys. Chem. B* **108**, 444 (2004).
[13] J. J. Holbrook, H. Gutfreund, *FEBS Lett.* **31**, 157 (1973).
[14] J. Pineda, R. Callender, S. D. Schwartz, *Biophys. J.* **93**, 1474 (2007).
[15] S. McClendon, N. Zhadin, R. Callender, *Biophys. J.* **89**, 2024 (2005).
[16] H. Deng, D. V. Vu, K. Cinch, R. Desamero, R. B. Dyer *et al.*, *J. Phys. Chem. B* **115**, 7670 (2011).
[17] S. McClendon, D. M. Vu, K. Clinch, R. Callender, R. B. Dyer, *Biophys. J.* **89**, L7 (2005).
[18] H.-L. Peng, H. Deng, R. B. Dyer, R. Callender, *Biochemistry* **53**, 1849 (2014).
[19] N. Zhadin, R. Callender, *Biochemistry* **50**, 1582 (2011).
[20] N. Zhadin, M. Gulotta, R. Callender, *Biophys. J.* **95**, 1974 (2008).

[21] M. J. Reddish, H.-L. Peng, H. Deng, K. S. Panwar, R. Callender *et al.*, *J. Phys. Chem. B* **118**, 10854 (2014).
[22] K. S. Burke, D. Parul, M. J. Reddish, R. B. Dyer, *Lab Chip* **13**, 2912 (2013).
[23] D. P. Kise, D. Magana, M. J. Reddish, R. B. Dyer, *Lab Chip* **14**, 584 (2014).

HOW CLOSE ARE WE TO EXPLAINING ENZYME CATALYSIS?

JUDITH P. KLINMAN[1,2,3], SHENSHEN HU[1,3] and ADAM OFFENBACHER[1,3]
[1]*Department of Chemistry*, [2]*Department of Molecular and Cell Biology*, [3]*QB3, University of California, Berkeley, CA 94720, USA*

My view of the present state of research on catalysis by protein enzymes

During the last several decades, the field of enzymology has undergone a profound period of "deconstructionism" — from a half century of consensus regarding the origins of catalysis to a long stretch of dissension and lack of agreement among many researchers. The early agreement phase was in the context of a widely accepted hypothesis of "transition state stabilization," while the disagreement phase has occurred in the context of defining the *precise physical parameters* that actually give rise to catalysis. Once achieved, this understanding is expected to lead to significantly improved success in the *de novo* design of catalysts capable of achieving the spectacular rate accelerations of enzyme, upward of 10^{25} fold in relation to their solution counter-parts [1].

Perceptions of the physical phenomena underlying catalysis vary greatly from one investigator to another, and include proposals that focus exclusively on electrostatic stabilization [2], to others that focus on the compounding roles of *e.g.*, protein conformational landscapes [3], the existence of protein networks for conveying thermal activation from solvent to the enzyme active site [4], quantum mechanical (QM) barrier penetration (especially in the case of C–H activation) [5] and the emerging evidence that proteins can bring about small changes in active site, inter-nuclear distances that may contribute to large rate enhancements [6, 7].

In the spirit of an attempted reconciliation, a brief flow chart will be presented at the end of this chapter, as a possible frame of reference for future *de novo* design and as a context from which to integrate the various experimental findings regarding Nature's most powerful catalysts.

My recent research contributions to catalysis by protein enzymes

Enormous kinetic isotope effects at RT

The dominant importance of C–H activation reactions in the chemical industry is paralleled in the field of enzymology by virtue of the number of enzymes (estimated as *ca.* one third) that catalyze either a proton, hydride or hydrogen atom abstraction. Independent of the nature of the transferred hydrogen, there is a large and growing body of evidence implicating QM barrier penetration (H-tunneling) in course of these processes. A particularly powerful experimental system for exploring the detailed features of H-tunneling is the soybean lipoxygenase (SLO), a member of a diverse and biologically important class of enzymes that convert fatty acids to specific oxygenated products. In the case of SLO, an active site ferric-hydroxide carries out a regio- and stereo-selective hydrogen atom abstraction to generate a substrate derived pentadienyl-free radical and ferrous-water [8].

Beginning with the first reports in the 1990's of grossly inflated kinetic isotope effects (KIEs) in the SLO reaction [9], there has been a strong push toward formulating a theoretical basis for the understanding of this behavior [10]. In particular, the goal has been to find a theoretical treatment that makes testable predictions that can be further examined experimentally, followed by further refinement of the theory as new data appear. In this context a modified form of the Marcus equation for electron tunneling, that accommodates the greater dependence of H-tunneling on donor acceptor distance (DAD), has emerged as a highly viable approach. Many reviews have been published describing this type of analysis [5] that focuses on three key physical processes, each of which is expressed within a separate exponential form that represents:

- A hydrogenic wave function overlap term that is independent of temperature but dependent on the mass of the transferred particle and is the origin of the KIE.
- A "Marcus" term that expresses the dependence of the tunneling rate on the reaction driving force, ΔG^o, and the environmental reorganization term λ. This is largely independent of the mass of the transferred hydrogen but is temperature dependent.
- A correlation between the tunneling efficiency and the DAD, where the "effective distance" reflects a balance between repulsive interactions that oppose a reduction in the DAD and the increased tunneling rate as the DAD gets smaller.

The latter term is relatively unimportant in most native enzymes, attributed to a separate conformational sampling coordinate that transiently creates a family of protein substrates with reduced ground state geometries. The DAD sampling term is most active in mutants that perturb the ability of a protein to replicate the tunneling optimized, active site configurations. The absence of significant DAD sampling in native enzymes produces the commonly observed temperature independent KIEs, $\Delta E_a \sim 0$, one of the signature features of enzymatic C–H activation. For mutated enzymes, this term becomes both temperature and mass dependent, the latter a consequence of the shorter wave length for D versus H, $\Delta E_a > 1$.

A number of recent findings have provided a critical confirmation of the above physical picture. First, in a quest to determine how far the geometry of an enzyme active site can be disrupted before tunneling will be precluded, we generated a series of double mutants of SLO in which pairs of bulky hydrophobic side chains were simultaneously replaced by alanine. The most interesting DM to emerge was the one in which *both* of the side chains abutting the reactive carbon C-11 of linoleic acid (LA) had been reduced in size {L546A and L754A}; characterization of this variant (DM-SLO) by X-ray crystallography indicates a greatly enlarged cavity in the vicinity of the reactive carbon of substrate and no measureable changes elsewhere. Remarkably, full kinetic characterization of DM-SLO shows an elevation of the already inflated deuterium isotope effects of 80 for the WT enzyme to a value between 500 and 700! This enormous KIE has been seen under a variety of conditions initially in the steady state at 30°C and under single turnover conditions at 35°C and, more recently, at six temperatures between 5°C and 30°C [6, 11]. Most significantly, analysis of the KIE for DM-SLO at either a single temperature or across the full temperature range provides a ready explanation for the observed behavior: This involves a combination of an elongation of the H-transfer DAD, together with a reduced ability of the enzyme to sample the "native" tunneling distance via an unexpected rigidification of the DM-SLO.

Note that the DM-SLO differs from all other characterized packing defect variants of SLO, which across the board lead to more flexible active sites and increased DAD sampling. Regardless of the exact structural origin of the reduced flexibility of the DM-SLO (currently under investigation), DM-SLO is concluded to provide an unexpected and especially robust corroboration that:

(i) Non-adiabatic analytical expressions describe the H-tunneling process well;
(ii) Short, tunneling compatible DAD distances, achieved either by conformational sampling (WT) or via the onset of additional DAD sampling (observed for all mutants with the exception of DM-SLO) are an integral part of this behavior;
(iii) Small increases in DAD away from the tunneling-ready state can produce significantly decreased rates (*e.g.*, estimated as ca. 10^2 for an 0.15 Å increase in the DAD for DM-SLO).

Defining tiers of donor acceptor distances and their relationship to protein motions

The critical role of donor acceptor distance during C–H activation can be conceptualized within three basic tiers that progressively bring the DAD from its initial ground state to the tunneling-ready state. The first tier represents the dominant ground state distances for bound substrates, detectable in X-ray structures as well as other biophysical probes and via MD simulations. These almost always reveal inter-nuclear distances that are within the van der Waals limit. The second tier is the time-dependent sampling among a large family of protein substates (the protein's conformational landscape), to achieve a family of configurations that are optimal for catalysis. This is expected to take place over a hierarchy of time scales and may involve fairly large and distal parts of the protein; it is also one of the most challenging to detect via experiments that allow a correlation with the rate of chemical catalysis at the active site. The third tier, by contrast, can be readily inferred from the temperature dependence of the KIE (discussed immediately above); this represents the final tuning of the reaction coordinate via a local DAD sampling on a nsec to psec time scale. Although there are numerous X-ray structures for SLO, none has yet been reported for an enzyme substrate complex, raising the question of whether this model enzyme system may somehow be capable of producing DAD distances shorter than van der Waals in its dominant ground state. This point has now been laid to rest via a series of elegant ENDOR measurements (with Brian Hoffman) that examine the coupling between site-specifically labeled ^{13}C-LA and an active site paramagnetic metal ion in SLO. The studies show that while the active site of SLO is deeply buried, the reactive carbon of substrate is located within van der Waals distance of the activating metal center. We can, thus, rule out unusual bond "shortening" effects in the dominant ground state of SLO [12].

Defining key regions of protein that control conformational sampling

Given the challenge of detecting the conformational sampling process as it relates to chemical catalysis at an enzyme active site, we have entered into an exploration of the use of hydrogen exchange by mass spectrometry (HDXMS) as a tool toward this end. In addition to high pressure [13], experiments under development involve an examination of rate constants for HDX under variable conditions that can be related to a similar interrogation of catalytic rate constants. These HDXMS experiments are generally conducted under the "EX2" condition, where the rate of exchange of solvent-derived deuterium into the peptide backbone of a protein is dependent on an equilibration of *local segments of protein* between open (exchangeable) and closed (non-exchangeable) forms. We have been particularly interested in the possible link of motions within solvent accessible surfaces to catalysis and, further, the degree to which these regions will be connected to the active site via discrete protein networks.

Despite its large size (94 kDa), a full HDXMS characterization of the catalytic domain of SLO has recently been completed. *Of great interest, a single surface loop,* $\geq$ *15 Å from the active site*, shows trends in its behavior that correlate directly with those describing catalysis of C–H activation. Further, structural and site-specific mutagenesis studies support a well-defined network that extends from the identified loop to hydrophobic residues in contact with bound substrate [14]. These exciting results will be highlighted in my verbal presentation.

Generalities and concluding comments

One question that frequently arises is whether the observations discussed above will be limited to C–H activation reactions with their inherent QM behavior. This appears not to be the case, from the burgeoning literature on the role of protein motions in enzyme catalysis [15] as well as recent work from our own group that is extending many of the above concepts to the methyl transferases [7]. In Scheme 1, we present a working scheme that both helps to formalize our current thinking about enzyme design and provides a context for combining many of the extant ideas regarding the origin of protein catalysis. A number of features that appear unique to protein catalysts include (i) the role of a *very large number of functional groups* that are precluded by collective entropic barriers from occurring in comparable solution reactions, (ii) the contribution of protein scaffolds to *the (transient) generation of reactant configurations with short inter-nuclear distances*, and (iii) the likelihood of *a funneling of thermal activation from solvent surfaces to the reacting entities at the active site.*

→ **Getting started:**

I. Find a monomeric and stably folded protein into which a ligand substrate can be modeled.

II. Based on the chemical reaction desired, introduce a large number of functional groups.

→ **But, more needed:**

III. Perform biophysical probes, e.g., HDX as a function of temperature, to identify regions of enhanced thermal activation within the chosen protein.

IV. Find one such region that resides near the solvent interface and create a network of interactions that connect solvent to the active site.

V. Introduce bulky hydrophobic groups behind the reacting bonds of substrate, to maintain close packing/compaction and to promote communication with the designed dynamical network.

→ **Ready to execute:**

VI. Push the button (in spectrophotometer and/or computer) and see what happens.

Scheme 1. Proposal for the design of highly active enzymes that goes beyond traditional approaches (I and II).

Outlook to future developments of research on catalysis by protein enzymes

The prevalent use of theoretical methods that are uncoupled from experimental results and that fail to make testable predictions has impeded progress, giving the false impression of increased understanding. In some instances, experimental approaches have been ridiculed because they do not agree with theory. Future progress will be critically dependent on a synergistic approach that enables theoreticians and experimentalists to work together productively.

An important direction is the expansion of our ability to detect remote regions within proteins that are intimately linked to the generation of precise electrostatics and inter-nuclear distances within enzyme active sites. This approach needs to distinguish the extent to which identified regimes contribute in static versus dynamic ways. In the latter context, tool development (both experimental and computational) is needed that will enable the integration of the spatial resolution of catalytically relevant motions to the time scales of such motions (*e.g.* via time dependent X-ray methodologies).

It will be important to see whether the strategies outlined in Scheme 1 can serve as a guide to design systems that begin to approximate the upper limit of enzyme catalysis.

Acknowledgments

This work is supported by the NIGMS of the NIH (R35GM118117 to JPK and F32GM 113432 to AO).

References

[1] A. Radzicka, R. Wolfenden, *Science* **267**, 90 (1995).
[2] A. Warshel, P. K. Sharma, M. Kato, Y. Xiang, H. Liu *et al.*, *Chem. Rev.* **106**, 3210 (2006).
[3] S. J. Benkovic, G. G. Hammes, S. Hammes-Schiffer, *Biochemistry* **47**, 3317 (2008).
[4] S. W. Lockless, R. Ranganathan, *Science* **286**, 295 (1999).
[5] J. P. Klinman, A. Kohen, *Annu. Rev. Biochem.* **82**, 471 (2013).
[6] S. Hu, S. C. Sharma, A. D. Scouras, A. V. Soudackov, C. A. M. Carr *et al.*, *J. Am. Chem. Soc.* **136**, 8157 (2014).
[7] J. Zhang, H. J. Kulik, T. J. Martinez, J. P. Klinman, *Proc. Natl. Acad. Sci. USA* **112**, 7954 (2015).
[8] M. H. Glickman, J. P. Klinman, *Biochemistry* **35**, 12882 (1996).
[9] T. Jonsson, M. H. Glickman, S. Sun, J. P. Klinman, *J. Am. Chem. Soc.* **118**, 10319 (1996).
[10] M. J. Knapp, K. Rickert, J. P. Klinman, *J. Am. Chem. Soc.* **124**, 3865 (2002).
[11] S. Hu, A. V. Soudackov, S. Hammes-Schiffer, J. P. Klinman, *ACS Catal.* **7**, 3569 (2017).
[12] M. Horitani, A. R. Offenbacher, C. A. M. Carr, T. Yu, V. Hoeke *et al.*, *J. Am. Chem. Soc.* **139**, 1984 (2017).
[13] S. Hu, J. Cattin-Ortolá, J. W. Munos, J. P. Klinman, *Angew. Chem. Int. Ed.* **55**, 9507 (2016).
[14] A. R. Offenbacher, S. Hu, E. M. Poss, C. A. M. Carr, A. D. Scouras *et al.*, *ACS Cent. Sci.* **3**, 570 (2017).
[15] S. Hammes-Schiffer, J. P. Klinman, *Acc. Chem. Res.* **48**, 899 (2015).

SESSION 5: CATALYSIS BY PROTEIN ENZYMES

CHAIR: JOANNE STUBBE
AUDITORS: G. BRUYLANTS[1], W. VERSEES[2]
[1] *Université libre de Bruxelles, Engineering of Molecular NanoSystems, CP 165/64, 1050 Brussels, Belgium*
[2] *Structural Biology Brussels, Vrije Universiteit Brussel, and VIB-VUB Center for Structural Biology, Pleinlaan 2, 1050 Brussels, Belgium*

Discussion among the panel members

JoAnne Stubbe: I would like to start the discussion by asking anyone what he or she thinks about the question I posed at first at the beginning of this: do you think we will ever achieve the goal to understand these different kinds of contributions to these tremendous rate accelerations that we see?

Judith Klinman: I have had theoreticians calling me and say: what should we do to make an enzyme? And obviously, the first thing that you need is a binding pocket. The next thing you need is lots of functional groups. We have seen that over and over again. But that is just the beginning and that this is where we have been stuck. The role of protein dynamics, the need to do a sampling of many conformers on both a more global and local time scale and positioning is essential. And I think that is part of the explanation why proteins are so large. But there is also the issue that if we want to incorporate that into design, how do we use that principle? The finding of pathways for thermo-transduction from a protein-solvent interface to the active site might help. If I had a protein scaffold and I wanted to know how to use its motional effect, its global motional effect, I would probably look for the most flexible region and then try to incorporate a network going from the protein-water interface to the active site. We need to get the energy focused into the active site. It's going to be difficult. I also think distance is very important. We have seen with these tunneling systems that reduction of distance from the optimum, that is expanding the donor-acceptor distance by 0.1 to 0.2 Å, can reduce the rate about a hundred fold. That is just one distance within an enzyme active site. So, I do think in addition to hydrogen bonds we need to focus on hydrophobic side chains and close packing. You may want to say something about that Don?

Donald Hilvert: I just wanted to make the comment that not all enzymes catalyze such difficult reactions. The enzymes for which we see these massive rate accelerations of 10^{27} are those that have extraordinarily slow background reactions. For an aldol reaction, for protein hydrolysis, you don't need a 10^{27} fold acceleration to get a decent activity, you need 10^9, that is enough. That is what you see for that kind of enzyme. So I agree that the challenges are enormous, even for some of those systems, because you do have to control distance and probably, control the conformational landscape. But I think that the challenge of design is perhaps not as dramatic except for these very, very slow reactions.

Rudolph Marcus: As far as the question that Don raised: one sees in these single molecule studies that, even within the same enzyme, and depending on the angle, there are changes of two orders of magnitude in rate. So, probably the answer depends on the particular class of enzymes, the answers are quite different. But in that particular case, and perhaps in cases for other enzymes when studies are performed under very controlled conditions then, with enough information about structure, and supplemented by calculations because structures aren't available for all these intermediate configurations, one might get additional answers to the question.

Frank Neese: In terms of understanding, the analysis of many enzyme reactions has proceeded to a very advanced stage. However how to incorporate that into design is a different story because once that understanding is advancing, we realize that subtle things in the second coordinate sphere are decisive for turning something into an amazing catalyst. We have not been able to incorporate these things into the design of low molecular weight catalysts, and probably never will. This is really the biggest open question: how to convert insight into design, if we just don't want to do large scale screening, which is another valid approach, of course.

Brian Dyer: This is more of a question for Don. Clearly, the combinations of *ab initio* design and directed evolution is a powerful approach. How close do you have to get in your *ab initio* design and then what do you learn about what you did wrong from the directed evolution?

Donald Hilvert: The better your initial design is, the easier it is to evolve. But typically today the models that have been implemented are very simple: one or two catalytic groups that are often not very accurately placed in a scaffold, and so the evolution is an absolutely essential component. What we see as we evolve the catalyst is that typically the active site becomes much more complex, much more highly ordered. I think that is really going to be important: you need a very precise molecular recognition of the transition state and that is still very difficult to do with current computational methods. As computational power increases, as the force fields improve, the designs will get better. Current design force fields, for example, are very primitive: they typically have a Lennard-Jones potential, an oriented H-bonding term and an implicit solvent model. This is not very accurate, so you don't really expect to get very much out of it. We need better methods still.

Marina Havenith: I just want to add this: this is like a difficile entropy-enthalpy compensation, and since the solvent does contribute a lot and we see that the enzymes, which were active, had a bigger impact on the solvent than those which are not. So the question is, which is speculation: do these enzymes also have optimized these inhomogeneous hydration maps around them to specifically support this transition? And if this is so, because it is different — it is quite inhomogeneous — we see retardation towards the active site termed hydration funnel. If this is the case, then I think also for "designer enzymes", this kind of water mapping and optimization of the solvent at the catalytic site have to be included in drug design.

Frank Neese: For cytochrome P450, these changes of water molecules that are on the proton delivery pathway have been firmly established. So there is quite some understanding of what the ordered water molecules do in active sites.

Steven Boxer: I wanted to turn the question around a little bit and go back to Don. This is something we talked about a little bit privately but I would be interested to put these thoughts here in a more public setting. If you look at the PDB — protein data bank — there are now thousands of structures with unknown function, a fair fraction of those I presume are enzymes, the question is: why is it so fundamentally difficult to figure out, given the structure, what kind of a catalyst it is, unless you know some that are very similar. Why is it so hard? Isn't there an algorithm that allows you to do the inverse design-problem? Why isn't there?

Donald Hilvert: I think this would be fantastic but you are trying to fill a space with a set of molecules. There are an infinite number of possibilities and searching that space is not so trivial.

Steven Boxer: Just a follow-up, don't the people who do drug design, for example, do exactly that?

Donald Hilvert: They are trying, and they have some limited successes, but this is very difficult. I mean it is usually accompanied by some kind of bio-informatic data, the location of the gene in the cluster. You take all the information that you can find, and try to make plausible guesses about what might fit into the binding pocket. But it is difficult.

JoAnne Stubbe: I would like to make a comment. People have done this: a group at UCSF has taken something with no known function and then they looked at all the metabolites inside the cell, because some of those might be related to the substrate. It took them five years to figure out that they had an adenosine DNAse, I think it was. They went through the structures of all these metabolites and fitted them into the active site to try to figure out what the function is. But I think it is a major unsolved problem: how do you define the substrate for the huge number of enzymes where we don't know what the function is?

Judith Klinman: Even when you do that you don't know if you have the right substrate. You may have hierarchies of activities. I would like to ask a question. I would like to see if we could agree that the role of precision is essential. That this is what happens when you take your designed enzyme and allow it to evolve. That raises the question to what extent is that static or dynamic? Because this is a very important issue in enzyme catalysis: to what extent those protein motions contribute to protein catalysis? I have some thoughts myself but perhaps you want to say something first, Don. No? So we have been collaborating with Todd Martinez at Stanford, who is doing QM (Quantum Mechanics) calculation on proteins trying to see how far out he needs to go in order to replicate a structure as seen by X-ray crystallography. He is using GPU (Graphic Processing Unit) methodology and what he is finding is that he has to go out to 300 to 600 atoms in order to converge to a structure that matches what is seen experimentally. What this means is that we have to go far out of the active site and we need to use lots of residues in order to mimic just energetically and distance-wise what is seen, and the distance that is seen in this case in a crystal structure is much shorter than the van der Waals distance. Once again, this is a very short distance and it requires a lot of the protein to achieve. What we haven't been able to figure out is to what extent is that static or dynamic. Because you can't use the GPU methodology to look at a large number of atoms and, at the same time, do molecular dynamics. You need improvements in your computing technology. I know that precision is important, I know that short distances can be achieved in an enzyme active site, and sorting out under what circumstances, is it happening dynamically or statically, is very important and is relevant to the design objective.

JoAnne Stubbe: Ok, let's have a coffee break now and then come back for the general discussion.

General Discussion

JoAnne Stubbe: Do we have any questions from the floor?

Henk Lekkerkerker: First of all, as a physical chemist, I was thrilled to see the wonderful physical chemistry that is put into this enzyme field. But I have a question for Steven Boxer: it is wonderful to see that you can relate the electric field strength which you measure through infra-red shifts of the carbonyl group so well with transition state energies. Is this now highly specific for enzymes that involve this CO-stretching or is it of broader relevance?

Steven Boxer: This is a great question. We were talking about it a little bit a minute ago. Obviously carbonyl chemistry is widespread in biology and in chemical chemistry, so that it makes it a very useful probe. The criteria that we have are whether you can see it and you saw from Brian Dyer a very nice way to see it via difference spectroscopies and by isotopic substitutions, which we also do. We isotopically substituted that inhibitor, but the most important thing is that the mode needs to be a relatively high frequency mode, so the mode itself is relatively simple. So, we originally worked with nitriles, nitriles have very localized modes. Carbonyl groups are, this is based on calculations, relatively localized. But as soon as you go to lower frequencies, where let's say one wants model tetrahedral transition states or something like that, where you would have a single bond, the modes become very floppy and very difficult to observe. That is really a great limitation of the approach we are taking and we are trying to solve it.

Henk Lekkerkerker: If you allow me a short related question: in the field apparently you use MV/cm, I like mV/nm. So, let's say 1 MV/cm is 100 mV/nm and multiple thereof that you see. These are pretty high fields. Is there any specific reason in the local environment that these are so strong?

Steven Boxer: The dominant contributors are those two hydrogen bonds that Don Hilvert was referring to. In fact, what I did not have the chance to say, but it is really interesting and something you can do in simulations, is: if you look at the IR spectra of water for example, you see these very broad absorption bands. You get the exact same thing when you do a simulation, in fact you recapitulate the linewidth quite accurately: it is inhomogeneously broadened by the electric field distribution of the solvent. But in the simulations you can actually look exactly at what it is that produces the extremes. And what is at the extremes is, in fact, two hydrogen bonds to the carbonyl group and short hydrogen bonds, very much like what you see at the active site of the enzyme. That is a dominant but not exclusive contributor, there are more global contributions as well, which is much more difficult to parse.

Avelino Corma: When I was listening to you I tried to catch, to extrapolate to what we were doing the other days on homogeneous catalysis and heterogeneous catalysis. Excuse me if I am too schematic but this is the way I try to rationalize, to extrapolate. I thought that first of all, you have the hard core of your enzyme where your molecule is going to interact first. When you have that already you get some activity that can be two orders of magnitude, for instance, three orders of magnitude more than without catalyst. But now you have the second type of interactions, which are the ones that really make the difference. I understand that you could rationalize how your reactant is when it absorbs with the core and with what groups is it going to interact with: the amino acids that you have and the proteins that you have. So, at the end, what we do is to take care mostly of the first part. We make centers that can be quite active, but we do not have much power to do the second part that you are doing. So from the point of view of organocatalysis, it is possible to do more than enzymes do because enzymes, if I understand well, they do the things they have available in nature. We talk about proteins because they make proteins but in the lab we can make other things to make those interactions. Could we make that step: going for instance into transition metal complexes, into organocatalysts, were we have the core, that is what you are doing now with the first interaction. And see how we can modify these now to get the second type of interactions, and therefore to decrease further the activation energy. And we should be able to do this. Makes this sense to you?

JoAnne Stubbe: I want to stop you. I just wanted to say one thing, I was going to make this point the other day too. If you take an enzyme that sticks a phosphate on into glucose using ATP, it has a binding site for glucose, but if you replace it with water — which hydroxyls have the same pKa's — the water gets phosphorylated but the rate of the reaction gets down to by a factor of 10^{10}. And again, it is this non-productive binding. So you have a cavity and it can't figure out the right way to organize everything to make the reaction work. So that's why the size of these cavities and what you can do, I think, could have potentially a big effect, although the rigidity on what you are making can make it hard to get the flexibility in.

Donald Hilvert: People ask the question why are enzymes so big. And I think one of the reasons is that they can completely envelop the substrate and control the reaction environment in which the transition state is formed beautifully. Much better than we can do with organocatalysts, much better than we can do, perhaps, with single metal sites with a ligand. I think that is one of the advantages. It's potentially also a disadvantage because it means that enzymes tend to be much more selective for a particular transformation than their corresponding chemical counterparts, small molecule counterparts.

Frank Neese: I take your point that you can, of course in principle, synthesize molecules that are very different from proteins, more complicated, covered with other elements or other functional groups and so on to get even better catalytic activity. However, in the field I am working in, bio-inorganic chemistry, it has been tried for the last 40 years and there is not a single larger scale catalyst that came out of these efforts. So, what I'm trying to allude to is that the analysis of the enzyme reactions invariably points to very subtle interactions in the vicinity of the active site that you just struggle in synthetic chemistry to implement into a low molecular weight catalyst. I would love to be able to see that somebody takes that concept and create that amazing marvelous catalyst, maybe Ben Feringa can, but so far it hasn't happened.

Donald Hilvert: But people are trying this with design. They're trying to incorporate metals into proteins, they're trying to put in unnatural cofactors to extend the chemistry. I think this is the next stage for design.

Judith Klinman: There are so many new nanomaterials being synthesized, and what nanomaterials can do is to confine things to very close distances. They are often quite rigid. It is very interesting because there is a need to get close distances in enzymes and how you use a lot of the protein to get this, these close approaches and precision. It's kind of a challenge: can anyone who is designing nanomaterials think of a way to have both the confinement and the potential for flexibility. I mean, it is a different approach from what most people are doing.

Ben Feringa: Let me first comment to Frank Neese's remark about the second coordination sphere. We should not forget that there are approaches, which are pretty successful. Of course it's early days but then you look at some of this supramolecular catalysts where they can control hydrogen bonding interaction, secondary interactions, second metal sites, etc., where the selectivity for instance in hydroformylation completely switches from branched to linear and there are some beautiful examples appearing in the literature now. The same with some cavities that are made artificially. Of course the activities are not anywhere in the range of what enzymes do but we get to that stage and you will see remarkable examples in the years to come. We learn from these enzymes. But I have seen some beautiful examples in supramolecular catalysis. So, I'm fairly optimistic.

Stephen Buchwald: Can I say something about Avelino's comment about like organic catalysis? One of the things I think that differentiate trying to do it synthetically compared to enzymatically, is that you have evolution with a particular substrate. Whereas when we're trying to develop a technique you're looking for breath, as opposed to the fact that then usually we don't have one reaction with one substrate that we're trying to do and working on that over and over and over and over again. And if you were doing that, you would have a better chance of getting these high activities, and you still might be able to do it, but you have a better chance. Whereas if you're trying to fit in a lot of different substrates, you need to have, again, flexibility, size and activity at the same time that makes it a more difficult problem.

Kurt Wüthrich: I feel that we are starting to understand increases in rates of 10^3–10^5, maybe, but we are lacking the 10^{10} factor that we see in enzymes. I think that it is quite irrelevant to go after small changes in distances between atoms, which are measured in crystal structures at liquid nitrogen temperature. In 1974 we have shown that aromatic rings flip over inside proteins. This has not been very popular because these are events which are very difficult to simulate and only in 2010 has Shaw's group (David E. Shaw, Columbia University) in New York been able to simulate what we had measured in 1974, namely that these ring-flips happen and that protein molecules are constantly in fluctuations which have an amplitude of about 1.5 Å. That's probably also why we have these big scaffolds of protein. But this is not material that we are normally used to. These are big scaffolds of constantly fluctuating material. It has been shown experimentally in hundreds of proteins since 1974 that these ring flips are everywhere. This year we found a protein with 15 tyrosines among 55 residues, all but two of these tyrosines flip over at rates of 10^6 per second or higher. I am convinced, of course also in my own interest, that this is key to understand the 10^{10} acceleration of rates in proteins. We should not look for small changes in bond lengths. We should not look for small changes in interatomic distances measured at liquid nitrogen temperature. We should see how these things behave in relation to a "carrier frequency" with an amplitude of about 1.5 Å that goes across the proteins. We have also shown that there are no interior hydration waters in proteins that have a lifetime longer than 10^{-5} per second. I would like to remind you of an experiment done by Bryan Matthews with T4 lysozyme. He replaced an interior phenylalanine with an alanine, which left a cavity in the interior of the protein. Then he saturated the solvent water with benzene, and NMR showed that the benzene was just traveling in and out of the protein, again at the rate of 10^5 per second or higher, depending on the temperature, going in and out of this cavity. I think that it is the dynamic nature of the protein scaffold which has to be taken into account. I think that the secret of the 10^{10}, that we are missing to explain, lies in these basic fluctuations of the protein scaffold.

JoAnne Stubbe: So can I ask a question: what protein did you look at? Did you look at different proteins. Did you look at more than one protein for this tyrosine flipping? Because I thought there are data in literature were the tyrosines do not flip like that.

Kurt Wüthrich: Well, the one in 1974 was BPTI, basic pancreatic trypsine inhibitor. There are eight aromatics and frequencies of the ring flips are between 1 per second and 10^8 per second. There is very recent work on an enzymatic reaction linked to an interface with a number of aromatics. It could be shown that the activation volume for the ring flips of the aromatics in the interface between the two proteins fitted with an explanation of a big change in rates of the resulting reaction.

Judith Klinman: Which temperature range, did you say?

Kurt Wüthrich: 4 to 45°C, except BPTI, which is stable up to 90°C.

Steven Boxer: So I may turn that question around. You don't think that, in Don Hilvert's protein, if he put a tyrosine and looked at it, he wouldn't see those same things? I bet he would. And yet some of them are very good catalysts but the ones he started with weren't.

Kurt Wüthrich: Of course he has his tyrosine flipping. However, he relies on crystal structures and the crystallographers can't see ring flips. The populations of other states than the two indistinguishable lowest-energy orientations of these symmetric rings are so small that they escape detection in crystal structures.

Steven Boxer: So you would consider that to be a design criterion then?

Kurt Wüthrich: Absolutely. I am absolutely convinced that the key to the 10^{10} that we are missing is in these general dynamic properties of the protein scaffold. I see it as being the equivalent of carrier frequencies in communication technology.

Judith Klinmann: The key here is to link it to the catalysis. That's what we're all struggling with. I mean you have seen dynamics all over the place. We see different time scales, different regions of proteins but it's coming up with experiments that link the dynamics directly to bond cleavage steps. And that is, I think, where the field needs to focus.

JoAnne Stubbe: So I work on systems with tyrosines that do electron transfer over 35 Å and we can see spectroscopically using highfield BPR that the tyrosines stack. Some of the tyrosines stack, which enhances the rate by huge amounts and we can reproduce that using chemical models. So, I think this orientation of the tyrosines probably does play a role but it is specific for a defined catalytic transformation.

Kurt Wüthrich: I'm afraid, we are now getting the message wrong. It has nothing to do with specific functions of tyrosines. We see the same ring flipping for phenylalanines. We don't need the OH function. The symmetric aromatic rings of Phe and Tyr are just reporter groups, which show to us that the protein scaffold fluctuates with high amplitude. The amplitude needed to have a ring flip is about a displacement of 1.5 Å of the surrounding atoms. And it typically involves 300 to 500 atoms to let a ring flip.

Gabor Somorjai: Thank you, there is quite a revolution going on in our three fields of catalysis because, mostly, the instrumentation allows us to look at the molecular level and time resolution, etc. Now I'm looking at it from a point of view of maybe integrating the three fields on a molecular level. I know the least about enzyme catalysis and that is not an advantage. But the other two fields, I'm quite familiar with. So, what you have been telling has been a revelation to me. Especially JoAnne's talk that was an overview of this and I think that we are much closer to integrate on a molecular level our three fields in a following way: you have a limitation in enzymes, it is the conditions of how the enzymes work. They're very restrictive. And part of the structure of the enzyme was put together to allow it to work at room temperature or 40 degrees, in water and under very limited conditions of phase. So if you move away from that part of the enzyme structure that makes it possible to work under very restrictive conditions and focus on the rest of molecular science, the amplification of the turnover is absolutely crucial. Why? How do you explain this? Now, it appears that I heard many ingredients to that: charge transfers, which are similar to acid-base and covalent catalysis in heterogeneous catalysis, the water effects, which are crucial. So, what I miss is that you focus too much on how the enzymes work as they were given to us instead of taking model systems, not to poison the catalyst, but just remove some of the acceleration and find out the ingredients that give you that fantastic acceleration. So, instead of focusing on how the enzymes work, which is complicated, are there ways of modifying, decelerating the enzyme to see what decelerates the enzyme the most and thereby getting the ingredients that make up the enzyme. The model-system approach would be great. And I suggest that as a physical chemist: increase the temperature by two or three degrees and see what happens. Slow down the turnover somehow and then we can compare the structure that you have to put in place because of the limited conditions of use of the enzyme from the major physical chemistry uses that decelerate it. Now, if I'm wrong there are many reasons for it, I mean lack of knowledge. Thank you very much.

Judith Klinman: Gabor, we do it all the time, that is what site-specific mutagenesis does. There are very few site-specific mutants that accelerate the rate, they almost always decrease the rate. And a lot of what we are trying to understand is why they decrease the rate: e.g. when we lose an acid catalyst what its contribution is, or when we are creating packing defects like I was talking about today. So, I think that, at the moment, this is a standard tool that we apply all the time.

Erick Carreira: It sounds like the small molecule community should talk to you guys more often and you should talk to us because we deal with 10^{10}-fold accelerations all the time, right? You can mix two olefins and wait till the cows come home and they'll never metathesize on their own, or an olefin and an amine and they'll never add to each other. And so we're dealing with reduced systems, in a commentary to what you just said, where things are very simplified, but we get these phenomenal rate accelerations.

Steven Boxer: Let's return at the beginning. I mentioned that we are now looking in fact at some homogeneous catalysts using exactly the same approach. The irony in some ways is that you would think that they are simpler, but they're actually more complicated because the pieces aren't held together in the same way. There are all kinds of complications in the solvent that are largely excluded by the active site of an enzyme. A chemist can make modifications beautifully, but we can also make modifications pretty successfully by site-directed mutagenesis, and now there is this tool to put in non-canonical amino acids. So I think that you're right and several people came up to me during the break and were interested to know if this could be used in zeolites. I think the answer is "yes" and I bet that the data is already out there.

JoAnne Stubbe: So, I just wanted to say that this is why I was asking the question of what the turnover numbers are. To try to get a feeling for how good your catalysts are. Are enzymes really different? Without knowing that, it's hard to compare the two.

Dan Herschlag: I have a question for Brian Dyer, and Don Hilvert may want to jump in on this as well. You showed beautifully these different rates between forms of the enzyme and conformers of the enzyme that weren't interconverting rapidly. What are the physical models for what's different? So, since it's hard to design something *de novo* that is performing better, are there any ideas about how to make changes to have higher populations of the more active form and ideas about how you might go about doing that?

Brian Dyer: That's a very interesting question. So the most reactive conformations are the ones in which the substrate is very tightly bound and in fact it is so tightly bound that once the product is formed, the product releases quite slowly. And so in these enzymes the transition state is very far along the reaction coordinate and the product release is actually the rate limiting step. And it turns out that there is a balance. Most of the flux goes to less reactive, less tightly bound states that actually react more slowly but then release the product more quickly. The main structural feature that seems to determine the tightness of binding and the reactivity is the loop. So, the loop is completely closed in the states where the enzyme is most reactive and the loop is not quite completely closed in the less active state. So you could imagine playing with the loop and modify it a little bit to try to tweak the tightness of binding. But this comes back to Professor Wüthrich's point about the flexibility: the motion of the protein is so important in the conformational selection process. Crossing the transition state takes a picosecond, why than does it take the enzyme a millisecond to turn over? This is because there is this long conformational search to find those reactive conformations, and in fact when you make the protein more and more rigid, it gets stuck in the wrong conformations.

Donald Hilvert: In the case of the molten globular chorismate mutase, we think it's just wobbling around. Basically what you see in the crystal structure, just a much more dynamic species. Calculations suggest though that you can have ground states bound in different landscapes and that the reorganization energy to get through the transition state is basically the same from all of them.

Rudolph Marcus: I wonder if I may make a comment on these tremendous differences in rates. In the field of electron transfer reactions, a much simpler field, in reactions one can see differences of 10^{10} or 10^{11} difference in rates. These differences are understood. So, having huge differences in reaction rates is not necessarily a surprise.

Philip Bevilacqua: So along those lines, in the various talks (in JoAnne's talk as well) there is quite some discussion about working towards enzyme design and *de novo* design and so on. In the papers that I have read on this, e.g. concerning David Baker's work, the folding of these proteins is often not cooperative in contrast to the situation in natural proteins. To me that reflects very strong secondary structures. So, to what extent could that be hindering the enzymatic activity of some of these designed proteins? So, that they are overdesigned and they are too stable.

Donald Hilvert: For the design of enzymes, that is not an issue, because there you take an existing scaffold and you graft an active site, an idealized active site, into the binding pocket. The computationally *de novo* designed proteins are indeed much more rigid, and we're just now exploring whether we can functionalize them and what kinds of activities we can get out. It will be interesting to see whether they are over-designed and too rigid for efficient catalysis.

Ben Feringa: Let me briefly come back to something I was fascinated about and this is the role of the water. Many of you alluded to the network of motions from the solvent to the active site. So why do many enzymes work so fantastically in organic solvents? Recently, I read an article "How wet can one get." So how many water molecules do you need to have an active enzyme? Why do they work in organic solvents at all? What can we learn from that? Or is that not very relevant?

Martina Havenith: I would say that you would need one or two coverages of it with water. But on the other hand, all the other hydrogen bonds will have these picosecond motions, a hydrogen bond breaking. So, if you have another solvent, you also have room temperature motions which could act differently. I haven't looked into this, but obviously in many synthetically synthesized enzymes you see that even ΔH can increase, but $T \cdot \Delta S$ can increase as well. So I would say it doesn't exclude that they also work in other solvents and because you would have to look in detail into the motions there.

Judith Klinman: I think that in the early work on the role of organic solvents in enzyme catalysis the concentration of the organic solvent was really important, and the retention of a water hydration shell around the protein was concluded to be important. But that does raise the question of how far out into the bulk water you need to go in order to conserve the properties. In general, the proteins are less active in organic solvents but they are active. They are just less active.

Donald Hilvert: It also depends on what the organic solvent is. Enzymes are typically most active in solvents that are immiscible with water. So it's really a heterogeneous process. If you have a competitive solvent like DMF or DMSO then you can denature the protein which can be a problem. Some enzymes like subtilisin can be very stable.

Ben Feringa: Did anybody systematically study what happens, in a particular solvent for instance, if you systematically go down in the amount of water and just leave a few water molecules maybe in the active site?

Donald Hilvert: Klebanov did that experiment in the 1980s.

Ben Feringa: This systematic study by going down in a particular solvent to remove the last water molecule for instance?

Donald Hilvert: Klebanov did that study in the 1980s and he found about 0.2% of water was what you needed for some lipases. If you went below, then it was a problem.

JoAnne Stubbe: Yes, I think that there are many crystal structures that Petsko and Ringe did studies on like this. You can put your proteins in 10–15% DMSO and most proteins are perfectly happy. You can even put them in acetone. But again it's concentration dependent before they crash out of solution. But they turn over pretty well.

Graham Hutchings: I've been trying to get in for some time to ask this point. It's not a problem, but the discussion has gone away from this. But I'd like to get back to this baseline between what chemo-catalysis can do if we think about homogeneous and heterogeneous catalysis grouped together and enzymes. So I would be interested to see where we have got direct comparisons. Pyruvate to lactate was one of the examples today, where we can do that with a chemo-catalyst. We got rate data and we can do this at different temperatures. So it would be interesting to compare directly such reactions, because then we can get maybe an idea of what effect this confinement or the binding aspect of it really has on the enzyme compared to the chemo-catalyst. Because that's the bit we don't have. But to give you an example where we've tried to emulate methane monooxygenase, which is not a very fast enzyme, I suspect. If we use a zeolite for confinement, which is one of the points that's been made, and we adventitiously make the dioxygen diiron oxo-bridged species in there then we can make something which has much the same rate as the enzyme for methane to methanol, when you add the hydrogen peroxide. So this is a system in which the two are quite close together possibly. And so we may not be miles apart once we try and get this confinement.

Judith Klinman: What is the temperature at which you get this?

Graham Hutchings: The temperature for that is 35–50 degrees.

Judith Klinman: Thank you.

Graham Hutchings: We compared it under the same conditions as methane monooxygenase.

Christophe Copéret: I would like to build up on these questions. So, now we're trying to bridge the gap between homogeneous, heterogeneous and bio-catalysis. But we are talking about adapting that. We know how to do homogeneous catalysis, transformations similar to enzymatic catalysis. My question is: what about doing hydro-amination metathesis in a totally artificial way from enzymes? Are people working in that field where we can design a new reaction with enzymes rather than doing the same reaction over again? What is the state of the art in the field of trying to make new reactions?

JoAnne Stubbe: Frances Arnold sort of started a field where you can use heme-dependent systems to do carbene reactions. A lot of people that are coming from a physical organic background are now trying to put in organo-metallic cofactors into some site, like biotin synthase or something like that, and are getting turnovers. It is at the very beginning of the whole of this. We're not good at it yet. But I think you're going to see an explosion in this area.

Stephen Buchwald: Well, I would argue that those are not necessarily enzymatic reactions. I mean, you're choosing a different ligand. Your ligand is now the enzyme with something else in it. Is that really an enzymatic reaction? Or is that just semantics in asking the question? I don't know.

JoAnne Stubbe: Well you would have to look at the rate of turnover, you would have to look at the specificities. I mean that is what they are after.

Judith Klinman: John Hartwig published a paper recently in Science where he uses P450 Heme and iridium to do a reaction. And he says he's getting close to enzyme rates. But if you look at the rate acceleration, the protein is only twenty-fold faster than the reaction in solution, and what he's really doing is using the protein to get regio- and stereo-specificity. So I think in terms of this "*de novo* design," that you're referring to: yes in terms of unusual reactions and unusual metals and then achieving these stereo-chemistries, but the rate accelerations aren't necessarily much faster.

Bert Weckhuysen: I'm coming from heterogeneous catalysis. I must say I admire the way you can study the active site. I would like to come back to a discussion, which we have had earlier on the active site. I have two questions. The first is: how unique is the active site? If you take one enzyme and you measure it 50 times, 50 crystals: do you then always have the same active site? In other words: is protein folding always leading to the same active site?

JoAnne Stubbe: You selectively crystallize something out of solutions. So if you look at the structures a lot of times you see the same structure but there could be other structures that didn't crystallize. Because you have this issue when cloning and over-producing enzymes inside the cell which is a highly crowded environment where you have lots of equipment that fold proteins correctly. And so I think we don't really know. But if you really overproduce an enzyme then it crashes out and you get no folding.

Bert Weckhuysen: So, if you have single crystals and you would have ten labs measuring them and you would have different crystals of the same enzyme you would all measure the same structure. Ok, that's an important thing from a synthetic point of view.

Steven Boxer: Can I just say one little detail? There are many proteins that have been crystallized in many different space groups and the structures are the same. Just one more point: there are parts of the protein that are relatively disordered and those parts might be different depending on what the space group is. Those are usually near the surface.

Bert Weckhuysen: We had a discussion that an enzyme is a huge molecule. If you count the active sites per volume then you are not that high compared to, for example, heterogeneous catalysis. And then comes the question on how much you can cut out of that protein and make sure that you still have your active site and still have the functionality to do the job. Maybe with a little bit lower turnover, I accept that. And that brings me to something that has been done years ago, actually at Leuven university, Professor Jacobs and I also have been active in this. We have been active in what we call zeozymes, i.e. zeolitic enzymes. Then we used the cavities and we wanted to see, can we for example have amino acids around a transition metal ion and how much you can still have. We were for example able to make an entatic state of an enzyme. We were never ever able to catalyze the reactions however. At least not for this example. So my question is: how much can you cut? And can you do that to reduce your protein and still have an active site which can have a certain activity?

Rudolph Marcus: I wonder if I can make a comment on this heterogeneity question. To the extent one can do single molecule experiments, one can look at the catalyst many times and see how much variation there is. One might anticipate that, if it is a tightly knit catalyst, such as this ATPase, they'll look very much alike. But if it's a catalyst where there's extensive flexibility, one may well see a distribution of rates. So it might be possible, at least in some cases, to obtain a direct answer to this question.

JoAnne Stubbe: Does this answer the question about how much you can take away? People spent a lot of time studying that. There was a famous paper on serine proteases, published in PNAS 20 years ago or something, where they said they could make a peptide that was analogous to the catalytic triad and they said they were getting activity. It was immediately disproved by the community. A hundred people ran out and repeated it but nobody could repeat the results. So I think a lot of people have been trying to figure this out. I mean: the question is how important the network is.

Brian Dyer: If I could comment. If you take an enzyme like hydrogenase, you can make a film of hydrogenase on an electrode and you can measure the exchange in current density and you can get a rate of hydrogen production per enzyme. Then you can make the same measurement with the platinum electrode and it turns out that you can get about the same current exchange, current density, with the protein film as you can with a platinum surface. Now, we don't know how big the platinum active site is. It is probably at least two platinum, maybe more. So even though the hydrogenase is very large, it can function at essentially the same efficiency as a platinum electrode. If you take the active site out of the enzyme, it doesn't work at all. People have then tried to take the active site and start to build a peptide scaffold that mimics some of the functions of the protein. The protein has to funnel the electrons and protons to the active site. It has very sophisticated proton and electron transport chains that bring those to the active site. They're very difficult to mimic, and so far very little progress has been made on doing that. But people are trying to build up second and third coordination spheres to create proton and electron transfer relays to do that. So I think the jury is still out. But the enzyme works about as well as it can work. And so I don't know why you would want to cut it down. I mean, even if you take this disadvantage of the size, it is doing as well as the best catalyst, platinum, for that reaction.

Donald Hilvert: I think this is also an issue that one can explore by computational design, because in principle you can imprint any active site on any scaffold. What is the best scaffold for an aldol reaction? What is the best scaffold for an S_N2 reaction? We don't know but we can compare and contrast and see whether they're all equally viable and evolvable.

Karen Goldberg: You're talking about cutting away and getting down to the basics. A lot of work is being done now with trying to add these secondary coordination spheres. So this is kind of moving from the opposite direction: taking the principles that you have identified as being important in catalyzing a reaction and then adding that to our secondary coordination sphere to move closer towards our goal.

David Lilley: Can I give a couple of thoughts from the RNA-world, which is tomorrow I know. Firstly, we've learned the hard way: cutting down RNA enzymes too far gives you wrong answers, it definitely changes its properties. Secondly, in terms of "do we always get the same structure?": we've crystallized RNAses in multiple radically different space groups, orthorhombic, tetragonal, hexagonal,..., and we always get the same structure. But, a third thing is the hairpin ribosome, there we did single molecule experiments a number of years ago that showed that the dynamics of the ribosome can vary over two logs between two different molecules that we're observing. So, showing a wide range of kinetic properties for individual molecules.

Joachim Sauer: I enjoyed the selection of the talks in this session very much and I learned that what you are after is to understand what you are calling "enzymatic acceleration." The first question would be: would this include the binding constant or are you always discussing this k (rate per second). And what is the reference? I understand that the reference is a system in solution, not in the enzyme. So in that case, some of the effects that have been discussed may be related to the binding constant. For example, yesterday I heard for the first time "conformational gating." This I would say, would probably show up in the binding constant. Then there were different other effects discussed and it is not clear to me whether the different researchers believe that they can come up with just one. From my experience in heterogeneous catalysis on zeolites, I would say that it is not just one, and neither is it a static structure, it is also dynamic with the possibility to relax. Then we have heard that there are electric fields, which may be used as a descriptor for the, I would call it, intrinsic rate. But then I believe I have heard someone saying that electrostatics account for 60% of the acceleration? Did anybody say this or did I get it wrong? How has such a conclusion been reached? I would appreciate some opinion about this.

JoAnne Stubbe: So you asked a hundred questions. We need to start back at the beginning. So does somebody want to answer the question of how we define rate acceleration, this tremendous rate acceleration?

Judith Klinman: Dick Wolfenden measured lots of reactions in water and then he measured them on the enzyme. He was looking at k_{kat}/K_M which is a second order rate constant and he calculated an effective molarity. There was a reason to do that. His goal was to get effective molarity, because he dealt with solution reactions as unimolecular reactions and the enzyme is a bi-molecular reaction. It's just a question of what you want as your frame of reference. But I think the same principles apply, independent of whether you look at first or second order reactions. But the issue of how tight a substrate should bind to the enzyme is really important because there are two aspects to that. First of all, if it binds too tightly you get good capture but then the rate is limited by the release of product, because product and substrate are similar to one another. Clearly, in enzymes we have this balance between binding, catalysis and release and all the rate constants tend to be the same. So Nature seems to have evolved to a point where everything is compromised. You get good binding but not too good binding. The other question is: what is happening in the cell? Are the enzymes in the cell operating under substrate conditions that are below saturation or above saturation? It used to be that people concluded they are under a k_{cat}/K_M or low saturation situation. The more recent literature suggests that a lot of enzymes are saturated in the cell. So that will also influence the evolution of binding constants. But there is this balance that has to be achieved.

JoAnne Stubbe: Steven do you want to come back to the electrostatic, or Dan?

Steven Boxer: I was the one who said this, and that was for a particular case. Our approach to this problem has been to put in place a methodology that can be applied to a diverse collection of cases. And we've already done that, because again it turns out that the literature is already filled with data. Mostly Raman data. Outstanding, beautiful data. It's been out there for a long time and a multitude of explanations have been invoked. In some cases the electrostatic contribution is much smaller and in some cases it's much larger. I think it's a really challenging problem to calculate what the electrostatic contribution is. That is in part due to the force field, and this is something we've talked about a little bit before. Let me give you an example from the case of the enzyme that I've talked about. When I say there is a large field, you might ask "can you compute that field?" So we've done that. We put things in, using the coordinates that come from many crystals structures, and all of them are pretty much the same including the ones that we've obtained. Initially the field is very close to what we measure experimentally, which sounds very nice. But what happens when you do molecular dynamics is that all kinds of other compromises come into place. In the case of that steroid that I was showing, the force field doesn't like those short hydrogen bonds and so they make a compromise. You can just watch it in the simulation: they move in and out, and the field actually drops down substantially. So I think that the force fields for macromolecular systems are still very far from being widely useful. I think we need to have experiments that provide input into the development of force fields, and that's been a longstanding problem.

JoAnne Stubbe: Ok, so Professor Wüthrich would like to add word.

Kurt Wüthrich: Well there are two things; we have reached the hour for lunch. The other thing is that I have completely missed the term "allosteric control" of enzymatic rates in this discussion. Allosteric control is probably more unique to proteins, and possibly protein-nucleic acid combinations, than anything else. It's too late now but we should, if at all possible, take it up in tomorrow's session. Is there anything that would correspond to allosteric control in heterogeneous catalysis? I think in homogeneous catalysis we can forget it. But in heterogeneous catalysis it might be, I don't have the insight. This is a remark to the chair of tomorrow's session, because now we are out of time. JoAnne, do you want to close the session?

JoAnne Stubbe: The session is closed.

Session 6

Catalysis by Ribozymes in Molecular Machines

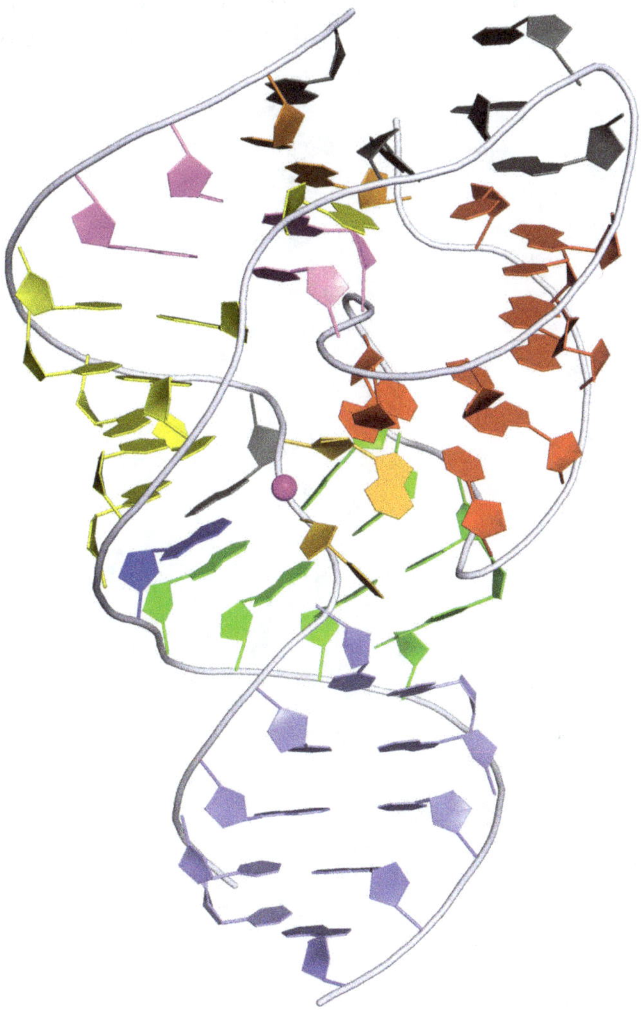

Structure of a twister ribozyme, a recently discovered class of RNA that undergo self-cleavage or ligation at a particular site.
Image by: David M. Lilley, University of Dundee, UK.

MECHANISTIC ORIGINS OF RNA CATALYSIS

DAVID M. J. LILLEY
Cancer Research UK Nucleic Acids Research Group, School of Life Sciences, University of Dundee, Dundee DD1 5EH, UK

RNA catalysis

RNA catalysis is important! Two of the most important reaction of the cell are catalyzed by RNA, i.e. the condensation of amino acids in the peptidyl transferase center of the ribosome [1–4], and splicing of mRNA in eukaryotes [5]. RNase P processes tRNA in all domains of life [6, 7], and some of the nucleolytic ribozymes are widely dispersed in genome sequences [8–14]. Recently the pace of discovery of new ribozymes has picked up significantly [13, 14].

A second reason why ribozymes are more than a curiosity is that they very probably played a key role in the emergence of life on the planet. In the postulated RNA world [15, 16] RNA served two functions; a genetic role to store and replicate information, and a chemical one to accelerate reactions.

When it was first proposed that RNA could function as an enzyme [15] there was no experimental evidence for this, and on the face of it seems rather improbable. Yet we now know that RNA can take this function, and ribozymes can accelerate reactions by a million fold or more. This presents a challenge to understand the chemical origins of the catalysis. Compared to a protein, RNA occupies a small fraction of chemical space. It is limited to four chemically-similar heterocyclic nucleobases, the 2′-hydroxyl group of ribose and a charged phosphodiester with associated hydrated metal ions. Moreover the pK_a values of these potential players are typically far from neutrality in the absence of perturbation. So how does RNA use these rather limited chemical resources to achieve the observed rate enhancement?

The great majority of natural ribozyme catalyze phosphoryl transfer; transesterification or hydrolysis of phosphate esters. For example, the nucleolytic ribozymes generate site-specific cleavage of RNA by nucleophilic attack of O2′ on the adjacent P3′ with departure of O5′ to leave a cyclic 2′,3′ phosphate. In appropriate circumstances the reverse ligation reaction is also catalyzed. In principle this can be catalyzed using several strategies. (1) Achievement of an in-line trajectory so that the O2′ nucleophile, the scissile P and the O5′ leaving group are colinear. (2) Transition state stabilization; this could include structural and electrostatic interactions with the phosphorane. (3) Nucleophile activation. Removal of the proton from the OH by a general base generates a much more potent alkoxide nucleophile. (4) Protonation of the O5′ oxyanion to make a superior leaving group. Ribozyme catalysis is generally multi-factorial.

Protein enzymes have evolved two catalytic strategies, and the same mechanistic division is found in the ribozymes [17]. Many nucleases and polymerases employ divalent metal ions to activate nucleophiles, stabilize transition states and position reacting groups. The self-splicing introns use this strategy [18, 19], and crystallography has revealed metal ions in the active centers of both group I [20] and group II [21] introns. By contrast, nucleases like RNaseA use two histidine side chains in concerted general acid-base catalysis [22, 23]. The same strategy is used by the nucleolytic ribozymes to carry out the same reaction, frequently using the RNA nucleobases like the imidazole side chains of histidine in the enzymes.

The catalytic mechanisms of nucleolytic ribozymes

Guanine nucleobases are employed in the mechanisms of most of the nucleolytic ribozymes [24–29]. However, in contrast the HDV ribozyme uses the nucleobase of a cytosine as a general acid in cleavage [30–32] together with a hydrated metal ion to activate the nucleophile either as a general base [30] or perhaps Lewis acid [33]. Experimental and computational studies suggests that a 2′-hydroxyl group may act as the general acid in the hammerhead ribozyme [25, 34, 35], using a coordinated metal ion to order to lower the pK_a. Finally, in the GlmS ribozyme (also a riboswitch) bound glucosamine-6-phosphate ligand serves as the general acid [36, 37].

The hairpin and VS ribozymes both use a combination of guanine and adenine nucleobases as general base and acid respectively [29, 38–40], and despite having very different overall structure [41, 42] their active sites seem to have converged on essentially the same topology [27]. A significant fraction of the rate enhancement arises from the general acid-base catalysis, and therein lies the main limitation to reaction rates achievable. Since the pK_a values of the nucleobases normally lie below or above neutrality, only a small fraction of the RNA is active at a given time. Using the measured pK_a values, the rate calculated for the ribozyme in the required ionization state is comparable with that of RNaseA [39]. Thus RNA is intrinsically a good catalyst, but limited by nucleobase pK_a.

The relatively recently-discovered twister ribozyme [13] provides an interesting comparison with these ribozymes, and a case study in dissecting the contributions to catalysis. We solved the structure of this ribozyme by X-ray crystallography [43]. The global fold of the ribozyme was based on a double-inverted pseudoknot, and the structure accounted for the role of each conserved nucleotide. The overall structure was in good agreement with those from two other laboratories [44, 45], yet there was a problem in that the deduced position of the O2′ nucleophile was far from in-line. However, we found that a simple, unhindered rotation of U-1 (immediately 5′ to the scissile phosphate) brought the O2′ in-line (Fig. 1); this feasibility was supported by MD calculations [46].

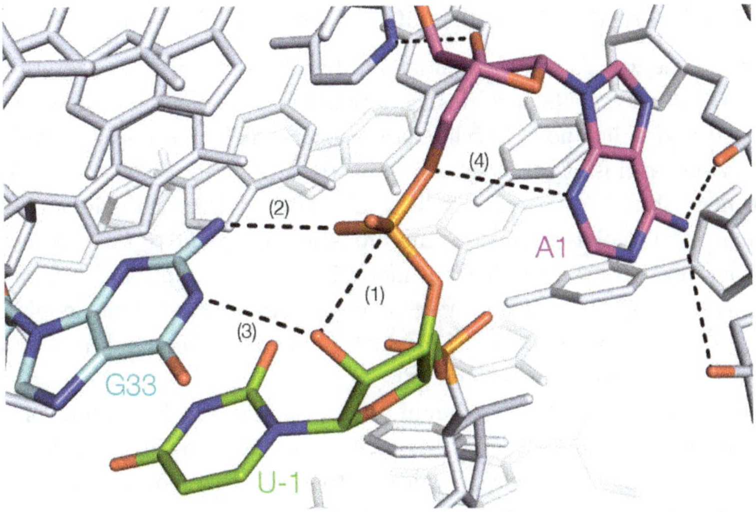

Fig. 1. The active center of the twister ribozyme. The crystal structure [43] was remodeled by a rotation of U-1 to place it under G33 [47].

The remodeled structure shows that the ribozyme can readily achieve an in-line geometry required for S_N2 attack. U-1 lies under G33, which mechanistic study suggested acts as general base [43]. Indeed the N1 is poised to remove the proton from the 2′OH of U-1. In addition, the *pro*R O of the scissile P is hydrogen bonded to G33 N2. We measured a 100-fold effect of replacing this oxygen atom by sulfur, but the stereospecific effect depended on the presence of the exocyclic amine of G33 [47]. We conclude that this interaction stabilizes the transition state.

That leaves the question of what might serve as general acid in the cleavage reaction, taking the role of the adenine in the VS and hairpin ribozymes [40]. The pH dependence of the reaction suggested a group with a $pK_a = 6.9$, so an adenine with an elevated pK_a is a possibility. All conserved adenines except A1 were excluded structurally or by insensitivity of catalysis to substitution. Participation by the nucleobase immediately adjacent to the scissile P is unprecedented, but even more surprisingly the structure indicated that the acidic N3 rather than the more usual N1 was involved in the proton transfer [47]. Atomic substitution of N1 by CH had a small effect, but the corresponding substitution at N3 lowered cleavage activity at pH 7 by 32,000 fold. A1 sits in a pocket in the ribozyme, with its exocyclic N6 H-bonded to two adjacent phosphate groups, accounting for the high pK_a. This in turn compensates for the lower probability of protonating N3.

Thus the conformation of the active center of the twister ribozyme generates the in-line conformation for nucleophilic attack (process 1), stabilizes the transition state (process 2) and juxtaposes the general base and acid (processes 3 and 4). The whole structure of the ribozyme has evolved to create this environment, and to activate A1 for catalysis.

Outlook to future developments of research on RNA catalysis

We are still some way from a full understanding how ribozymes generate their impressive rate enhancements despite such a paucity of chemical resources. Moreover, each new ribozyme has novel mechanistic aspects, and presents new challenges to understand the origins of catalysis. We have recently determined the crystal structure of a new ribozyme we call TS [55]. To date we have no evidence for direct involvement of nucleobases, but our mechanistic data point to the participation of a metal ion, and we observe a divalent metal ion bound as an inner-sphere complex close to active center, directing a water molecule towards the O2′ nucleophile. Much work remains before we have anything like a general understanding of the varied mechanisms of this class of ribozyme.

In parallel with our efforts to gain a mechanistic understanding of these ribozymes, in many cases we also lack knowledge of their biological roles. Some of the nucleolytic ribozymes are very widespread, and their locations often hint at a role in genetic regulation. Indeed, recent discoveries of new ribozymes have resulted from bioinformatic searching focused in the vicinity of known ribozymes and protein-coding sequences in bacterial genomes [13, 48]. Yet assigning a function to these ribozymes has so far resisted all efforts.

Lastly we might ask if is there a greater spread of ribozyme activities as yet undiscovered. With the exception of peptidyl transferase, all the natural ribozymes carry out related phosphoryl transfer reactions, but clearly a viable RNA world would have required a much greater range of reactions to be catalyzed including the formation of carbon-carbon bonds. A greater diversity of RNA catalysts has been created by *in vitro* selection experiments [49–51], offering a kind of proof of principle that such ribozymes can function. The riboswitches [52] suggest that RNA could use bound small molecules as co-enzymes, with GlmS as a clear precedent [36, 37]. The most common class of riboswitch binds TPP [53], and we can envisage this might be used to expand the range of catalytic chemistry in RNA [54]. Such ribozymes could have been used in the early evolution of life on the planet. Their role may then have been totally subsumed by protein enzymes, or some active remnants may still remain. Finding these would be far from easy, but would be a very exciting development were it to come about.

Acknowledgments

I wish to acknowledge the critical contributions of Drs. Tim Wilson and Yijin Liu to the work discussed here, and Cancer Research UK for financial support.

References

[1] H. F. Noller, V. Hoffarth, L. Zimniak, *Science* **256**, 1416 (1992).
[2] P. Nissen, J. Hansen, N. Ban, P. B. Moore, T. A. Steitz, *Science* **289**, 920 (2000).
[3] J. S. Weinger, K. M. Parnell, S. Dorner, R. Green, S. A. Strobel, *Nat. Struct. Mol. Biol.* **11**, 1101 (2004).

[4] D. A. Kingery, E. Pfund, R. M. Voorhees, K. Okuda, I. Wohlgemuth *et al.*, *Chem. Biol.* **15**, 493 (2008).
[5] S. M. Fica, N. Tuttle, T. Novak, N. S. Li, J. Lu *et al.*, *Nature* **503**, 229 (2013).
[6] C. Guerrier-Takada, K. Gardiner, T. Marsh, N. Pace, S. Altman, *Cell* **35**, 849 (1983).
[7] E. Kikovska, S. G. Svard, L. A. Kirsebom, *Proc. Natl. Acad. Sci. USA* **104**, 2062 (2007).
[8] R. Przybilski, S. Graf, A. Lescoute, W. Nellen, E. Westhof *et al.*, *Plant Cell* **17**, 1877 (2005).
[9] C. Seehafer, A. Kalweit, G. Steger, S. Graf, C. Hammann, *RNA* **17**, 21 (2011).
[10] K. Salehi-Ashtiani, A. Luptak, A. Litovchick, J. W. Szostak, *Science* **313**, 1788 (2006).
[11] M. Martick, L. H. Horan, H. F. Noller, W. G. Scott, *Nature* **454**, 899 (2008).
[12] C. H. Webb, N. J. Riccitelli, D. J. Ruminski, A. Luptak, *Science* **326**, 953 (2009).
[13] A. Roth, Z. Weinberg, A. G. Chen, P. B. Kim, T. D. Ames *et al.*, *Nat. Chem. Biol.* **10**, 56 (2014).
[14] Z. Weinberg, P. B. Kim, T. H. Chen, S. Li, K. A. Harris *et al.*, *Nat. Chem. Biol.* **11**, 606 (2015).
[15] F. H. C. Crick, *J. Mol. Biol.* **38**, 367 (1968).
[16] L. E. Orgel, *J. Theor. Biol.* **123**, 127 (1986).
[17] T. J. Wilson, D. M. J. Lilley, *Science* **323**, 1436 (2009).
[18] S. Shan, A. V. Kravchuk, J. A. Piccirilli, D. Herschlag, *Biochemistry* **40**, 5161 (2001).
[19] J. L. Hougland, A. V. Kravchuk, D. Herschlag, J. A. Piccirilli, *PLoS Biology* **3**, e277 (2005).
[20] M. R. Stahley, S. A. Strobel, *Science* **309**, 1587 (2005).
[21] M. Marcia, A. M. Pyle, *Cell* **151**, 497 (2012).
[22] J. E. Thompson, R. T. Raines, *J. Am. Chem. Soc.* **116**, 5467 (1994).
[23] R. T. Raines, *Chem. Rev.* **98**, 1045 (1998).
[24] J. Han, J. M. Burke, *Biochemistry* **44**, 7864 (2005).
[25] M. Martick, W. G. Scott, *Cell* **126**, 309 (2006).
[26] D. J. Klein, M. D. Been, A. R. Ferré-D'Amaré, *J. Am. Chem. Soc.* **129**, 14858 (2007).
[27] T. J. Wilson, A. C. McLeod, D. M. J. Lilley, *EMBO J.* **26**, 2489 (2007).
[28] J. C. Cochrane, S. V. Lipchock, K. D. Smith, S. A. Strobel, *Biochemistry* **48**, 3239 (2009).
[29] S. Kath-Schorr, T. J. Wilson, N. S. Li, J. Lu, J. A. Piccirilli *et al.*, *J. Am. Chem. Soc.* **134**, 16717 (2012).
[30] S. Nakano, D. M. Chadalavada, P. C. Bevilacqua, *Science* **287**, 1493 (2000).
[31] A. Ke, K. Zhou, F. Ding, J. H. Cate, J. A. Doudna, *Nature* **429**, 201 (2004).
[32] S. R. Das, J. A. Piccirilli, *Nat. Chem. Biol.* **1**, 45 (2005).
[33] J. H. Chen, R. Yajima, D. M. Chadalavada, E. Chase, P. C. Bevilacqua *et al.*, *Biochemistry* **49**, 6508 (2010).
[34] T. S. Lee, C. Silva Lopez, G. M. Giambasu, M. Martick, W. G. Scott *et al.*, *J. Am. Chem. Soc.* **130**, 3053 (2008).
[35] J. M. Thomas, D. M. Perrin, *J. Am. Chem. Soc.* **131**, 1135 (2009).
[36] D. J. Klein, A. R. Ferré-D'Amaré, *Science* **313**, 1752 (2006).
[37] J. C. Cochrane, S. V. Lipchock, S. A. Strobel, *Chem. Biol.* **14**, 97 (2007).
[38] P. B. Rupert, A. P. Massey, S. T. Sigurdsson, A. R. Ferré-D'Amaré, *Science* **298**, 1421 (2002).
[39] T. J. Wilson, N.-S. Li, J. Lu, J. K. Frederiksen, J. A. Piccirilli *et al.*, *Proc. Natl. Acad. Sci. USA* **107**, 11751 (2010).
[40] T. J. Wilson, D. M. J. Lilley, *RNA* **17**, 213 (2011).

[41] P. B. Rupert, A. R. Ferré-D'Amaré, *Nature* **410**, 780 (2001).
[42] N. B. Suslov, S. DasGupta, H. Huang, J. R. Fuller, D. M. J. Lilley *et al.*, *Nat. Chem. Biol.* **11**, 840 (2015).
[43] Y. Liu, T. J. Wilson, S. A. McPhee, D. M. J. Lilley, *Nat. Chem. Biol.* **10**, 739 (2014).
[44] D. Eiler, J. Wang, T. A. Steitz, *Proc. Natl. Acad. Sci. USA* **111**, 13028 (2014).
[45] A. Ren, M. Kosutic, K. R. Rajashankar, M. Frener, T. Santner *et al.*, *Nature Comm.* **5**, 5534 (2014).
[46] C. S. Gaines, D. M. York, *J. Am. Chem. Soc.* **138**, 3058 (2016).
[47] T. J. Wilson, Y. Liu, C. Domnick, S. Kath-Schorr, D. M. J. Lilley, *J. Am. Chem. Soc.* **138**, 6151 (2016).
[48] Z. Weinberg, P. B. Kim, T. H. Chen, S. Li, K. A. Harris *et al.*, *Nat. Chem. Biol.* **11**, 606 (2015).
[49] G. Sengle, A. Eisenführ, P. S. Arora, J. S. Nowick, M. Famulok, *Chem. Biol.* **8**, 459 (2001).
[50] S. Fusz, A. Eisenführ, S. G. Srivatsan, A. Heckel, M. Famulok, *Chem. Biol.* **12**, 941 (2005).
[51] M. Oberhuber, G. F. Joyce, *Angew. Chemie Int. Ed.* **44**, 7580 (2005).
[52] R. R. Breaker, *Mol. cell* **43**, 867 (2011).
[53] W. Winkler, A. Nahvi, R. R. Breaker, *Nature* **419**, 952 (2002).
[54] T. J. Wilson, D. M. J. Lilley, *RNA* **21**, 534 (2015).
[55] Y. Liu, T. J. Wilson, D. M. J. Lilley, *Nat. Chem. Biol.* **13**, 508 (2017).

CATALYTIC STRATEGIES OF NUCLEOLYTIC RIBOZYMES

DARRIN M. YORK

Center for Integrative Proteomics Research, Laboratory for Biomolecular Simulation Research and Department of Chemistry and Chemical Biology, Rutgers University, Piscataway, NJ 08854, USA

My view of the present state of research on catalysis by ribozymes in molecular machines

The ability of RNA molecules to selectively and efficiently catalyze complex chemical transformation has vast implications. Our understanding of the mechanisms of RNA catalysis have been greatly advanced by the study of small nucleolytic RNA enzymes, or ribozymes, that have evolved naturally in viruses and living organisms, or artificially through high-throughput *in vitro* selection techniques [1]. Experimental structural and mechanistic work, along with computational simulations have provided deep insight into the mechanism of these model ribozyme systems. Very recently, there has been a surge in progress in the determination of crystallographic structures of ribozymes [2–6] that have provided a departure point for theoretical investigations that aim to bridge the gap between structural and mechanistic measurements, and provide a detailed dynamical picture of mechanism at atomic-level resolution. In this way, molecular simulations have the potential to unify the interpretations of a broad range of experimental data and establish a consensus view of mechanism [7]. Ultimately, multiscale simulations, together with experiments, afford the tools needed to gain predictive insight into catalysis, including control factors that regulate selectivity and reactivity, that may guide rational design efforts.

In the present work, results from multiscale molecular simulations are presented for a series of ribozymes for which crystallographic data has recently become available, including the twister [2], Varkud satallite virus (VS) [3] and pistol [4] ribozymes. These results uncover recurring themes, as well as new twists, in the catalytic strategies taken by ribozymes that are apparent only when broadly analyzing their structure, biochemical characterization and detailed mechanisms predicted by molecular simulations. The interpretations of experimental data afforded by new multiscale molecular simulation results uncover general principles and provide predictive insight into the catalytic mechanisms of nucleolytic ribozymes that may guide rational design efforts.

My recent research contributions to catalysis by ribozymes in molecular machines

Recently we have made several major advances in the development of multiscale quantum models for biocatalysis simulations that allow new insight to be obtained into the mechanisms of RNA catalysis. Briefly, these advances include the development of: (1) *ab initio* combined quantum mechanical/molecular mechanical (QM/MM) simulation methods [9] capable of treating rigorous long-range electrostatic interactions under periodic boundary conditions most commonly used to mimic aqueous solution; (2) robust methods for sampling [10] and free energy analysis [11] that allow multidimensional free energy landscapes to be calculated reliably and efficiently; (3) new models for divalent metal ions that accurately describe interactions with nucleic acids [12] in long-time molecular dynamics simulations of ribozyme systems [8, 13]. Together, these advances have made possible the mechanistic study of newly discovered ribozymes that have allowed recurring themes and new twists in the catalytic strategies of nucleolytic ribozymes to emerge [7].

As an example, recent study of the twister ribozyme [8] that has no specific divalent metal ion requirement in catalysis, and combined QM/MM simulations suggest that the active site employs a novel mode of general acid catalysis involving protonation at the N3 position of adenine. Despite this new twist, the twister ribozyme active site architecture determined by molecular simulations bears remarkable resemblance to that of the hairpin [6] and VS [3] ribozymes; two other well-studied nucleolytic ribozymes that have no explicit divalent metal ion requirements for catalysis [14]. Specifically, the active site architecture of these ribozymes has an L-platform scaffold [14] that positions a guanine residue implicated as a general base, and is locked into place by hydrogen bonding with an L-anchor (Fig. 1).

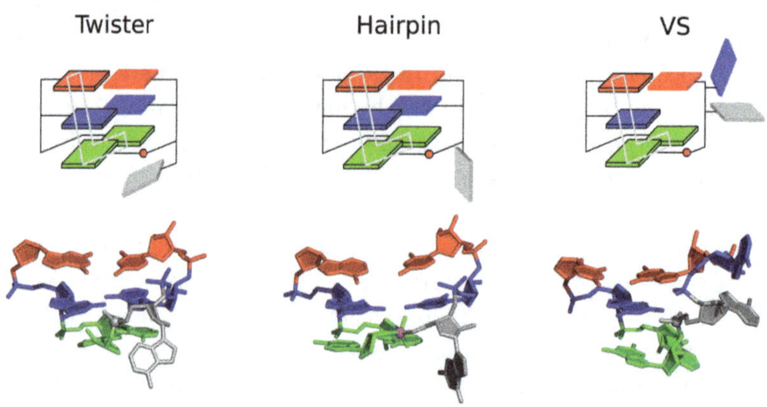

Fig. 1. Active site architecture of the twister, hairpin and VS ribozymes. (Top) Schematic showing the L-platform (indicated by an outlined letter "L") and L-anchor motifs. The scissile phosphate where cleavage occurs in shown as a red dot. (Bottom) Atomistic representation of the active site predicted from molecular simulation [8] (twister) or determined from crystallographic data (hairpin [6] and VS [3] ribozymes).

The pistol ribozyme [4], on the other hand, is a newly discovered ribozyme that has explicit divalent metal ion requirements for catalysis under near-physiological salt conditions, and has a mechanism that has stark similarities to the hammerhead ribozyme [5]. The hammerhead and pistol ribozymes share a similar L-platform scaffold as observed in the twister, hairpin and VS ribozymes, but instead of the L-anchor region that hydrogen bonds with the L-platform, these ribozymes contain a divalent metal ion binding pocket (the L-pocket) that positions a metal ion that assists in catalysis. Taken together, these results support the notion that the catalytic strategies of nucleolytic ribozymes may follow a few simple guiding principles, a detailed understanding of which may ultimately guide rational design efforts.

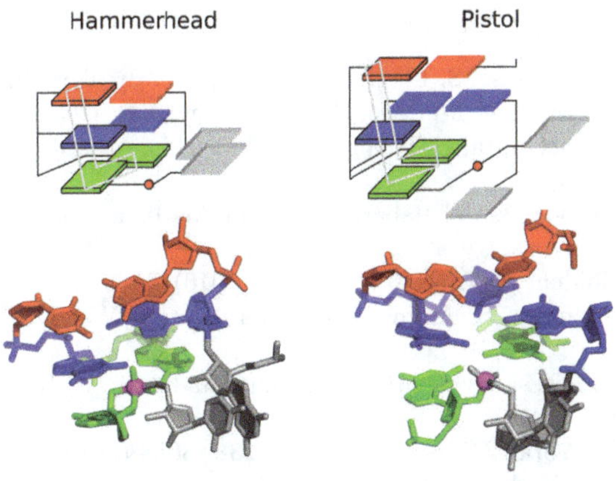

Fig. 2. Active site architecture of the hammerhead and pistol ribozymes. (Top) Schematic showing the L-platform (indicated by an outlined letter "L") and L-pocket motifs. The scissile phosphate where cleavage occurs in shown as a red dot. (Bottom) Atomistic representation of the active site predicted from molecular simulation.

Outlook to future developments of research on catalysis by ribozymes in molecular machines

The future of mechanistic research into catalytic mechanisms of ribozymes is bright indeed. The recent success of theoretical models build upon advances in experimental high-throughput methods, and structural and biochemical characterization of ribozymes. Key to pushing the boundaries of our understanding of these catalytic molecular machines is the close interplay of theory an experiment. Theoretical studies must be pressed to go beyond validation of existing experimental data to make experimentally testable predictions. At the same time, the experimental community should begin to more broadly embrace serious theoretical work and make effort to carry out new experiments motivated by results from realistic physical and chemical models. Finally, the scientific community should work to develop new ways of integrating the information gained by experiment and theory into the rational design pipeline to facilitate discovery.

Acknowledgments

The author thanks Colin Gaines and Ken Kostenbader for valuable contributions to the manuscript and is grateful for financial support provided by the National Institutes of Health (Grants GM62248 and GM107485). This work used the Extreme Science and Engineering Discovery Environment (XSEDE), which is supported by National Science Foundation (NSF) Grant OCI-1053575 (Project TG-MCB110101) and also in part by the Blue Waters sustained petascale computing project (NSF Grant OCI 07-25070 and Petascale Computing Resource Allocations Grant OCI-1036208).

References

[1] K. A. Harris, C. E. Lünse, S. Li, K. I. Brewer, R. R. Breaker, *RNA* **21**, 1852 (2015).
[2] Y. Liu, T. J. Wilson, S. A. McPhee, D. M. J. Lilley, *Nat. Chem. Biol.* **10**, 739 (2014).
[3] N. B. Suslov, S. DasGupta, H. Huang, J. R. Fuller, D. M. J. Lilley *et al.*, *Nat. Chem. Biol.* **11**, 840 (2015).
[4] A. Ren, N. Vušurović, J. Gebetsberger, P. Gao, M. Juen *et al.*, *Nat. Chem. Biol.* **12**, 702 (2016).
[5] A. Mir, B. L. Golden, *Biochemistry* **55**, 633 (2016).
[6] J. Salter, J. Krucinska, S. Alam, V. Grum-Tokars, J. Wedekind, *Biochemistry* **45**, 686 (2006).
[7] M. T. Panteva, T. Dissanayake, H. Chen, B. K. Radak, E. R. Kuechler *et al.*, *Methods Enzymol.* **14**, 335 (2015).
[8] C. Gaines, D. M. York, *J. Am. Chem. Soc.* **138**, 3058 (2016).
[9] T. J. Giese, D. M. York, *J. Chem. Theory Comput.* **12**, 2611 (2016).
[10] B. K. Radak, M. Romanus, T.-S. Lee, H. Chen, M. Huang *et al.*, *J. Chem. Theory Comput.* **11**, 373 (2015).
[11] T.-S. Lee, B. K. Radak, M. Huang, K.-Y. Wong, D. M. York, *J. Chem. Theory Comput.* **10**, 24 (2014).
[12] M. T. Panteva, G. M. Giambasu, D. M. York, *J. Phys. Chem. B.* **119**, 15460 (2015).
[13] T.-S. Lee, B. K. Radak, M. E. Harris, D. M. York, *ACS Catal.* **6**, 1853 (2016).
[14] T. J. Wilson, D. M. J. Lilley, *Prog. Mol. Biol. & Trans. Sci.* **120**, 93 (2013).

KEY CATALYTIC STRATEGIES OF RIBOZYMES

PHILIP C. BEVILACQUA, JAMIE L. BINGAMAN, ERICA A. FRANKEL, KYLE J. MESSINA and DANIEL D. SEITH

Department of Chemistry, Pennsylvania State University, University Park, PA 16801, USA

My view of the present state of research on catalysis by ribozymes

Following is my view of the field, as I write in the Fall of 2016. Ribonucleic acids (RNA) are single-stranded biopolymers that can play both informational and catalytic roles in Nature. This has led to the notion that RNA served a critical role in the emergence of life in the so-called RNA world. The field of RNA catalysis has made enormous strides since the first RNA enzymes, or ribozymes, were identified 35 years ago, discoveries that led to the Nobel Prize in Chemistry in 1989 to Cech and Altman. We now know the structures of many different ribozymes at the atomic level, and the roles of metal ions and nucleobases in catalyzing phosphoryl transfer reactions are becoming clear. Technique improvements have played critical roles in these advances and include the preparation of large quantities of well-behaved RNAs that can be characterized by small angle X-ray scattering (SAXS), rapid kinetics, spectroscopic studies, and atomic substitutions. There have been conceptual advances as well. Data are being understood in terms of different reaction channels and by computational approaches as diverse as bioinformatics, molecular dynamics (MD), and quantum mechanics (QM).

My recent research contributions to catalysis by ribozymes

Studies in the 1980's and 1990's firmly established that metal ions play important roles in ribozyme catalysis [1]. In the early 2000's, our lab helped introduce the notion that small ribozymes can use the nucleobase side chains to catalyze self-cleavage reactions [2], much like the amino acid side chains do in protein enzymes (Fig. 1). Kinetic studies demonstrated that the bases have shifted pK_a's changed from ~ 4 up to neutrality, optimally poised for proton transfer and serving a histidine-like function [2–4]. I presented a statistical thermodynamic formalism that facilitated facile graphical interpretation of rate-pH profiles that has been used widely in the field [5].

In collaboration with the Carey and Golden labs, we established Raman crystallography of ribozymes as a way to directly measure the pK_a's of catalytic bases and showed that the pK_a of C75 of the HDV ribozyme was indeed shifted to neutrality [6]. Raman crystallography was subsequently used on several other small

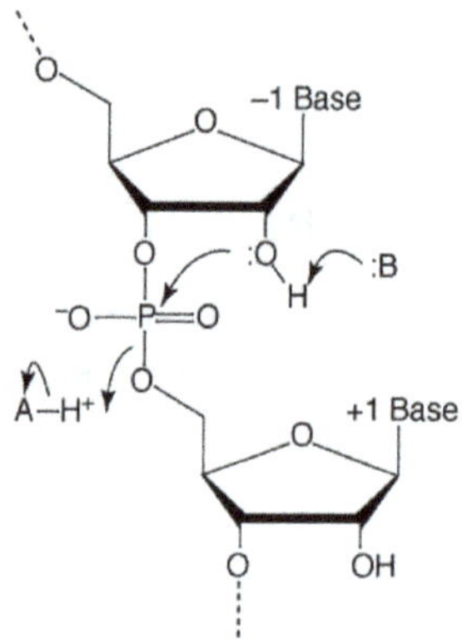

Fig. 1. General mechanism for self-cleavage of small RNAs. The general base ":B" can be a hydrated metal ion or a nucleobase. The general acid "AH^+" is $C75H^+$ in the case of the HDV ribozyme.

ribozymes [7, 8], and similar shifts towards 7 were identified, showing that this catalytic strategy of ribozymes is general. Both Raman crystallography and transient kinetics showed that the pK_a of C75 couples to the presence of Mg^{2+} in an anticooperative fashion [2, 6]. Using electrostatic and later QM methods in collaboration with the Hammes-Schiffer lab, we found that a highly negative electrostatic potential drives pK_a shifting to neutrality [9] and that the pK_a coupling to Mg^{2+} helps drive proton transfer from C75 to the leaving group [10] (Fig. 2).

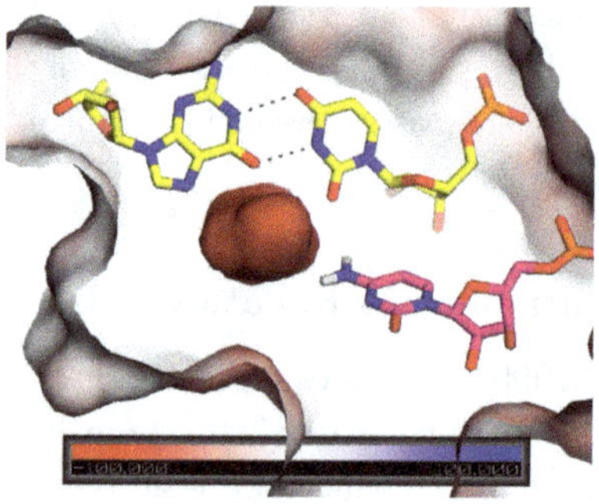

Fig. 2. Non-linear Poisson Boltzmann calculation of the active site of the HDV ribozyme showing the extreme negative potential (red patch) near C75 (magenta sticks) that helps drive its pK_a from 4.2 to $\sim$ 7. Adapted from Ref. [9].

We also went on to show that ribozymes catalyze reactions by several different reaction channels, sometimes using monovalent and divalent ions in different ways [3, 11]. This led to the notion of multichannel mechanisms for ribozymes. A combination of proton inventory and inverse thio effects showed that the major reaction channel varies with ionic conditions, wherein Mg^{2+} ions drive a concerted reaction, but diffuse monovalent ions support a stepwise reaction with a phosphorane intermediate [12, 13]. In other words, the solution conditions can change the catalytic pathway.

Other contributions from my lab include RNA catalysis relevant to early Earth chemistry. In collaboration with the Keating lab, we established that simple compartmentalization of RNA in aqueous two-phase systems (ATPS) can drive RNA catalysis by nearly 100-fold [14]. More recently, we discovered nucleotide-polyamine coacervates that concentrate nucleotides and metal ions to molar or greater amounts; moreover, these coacervates compartmentalize large RNAs by $\sim$ 10,000-fold [15]. Such systems have the potential to favor the polymerization of nucleotides into RNA templates, as well as favor non-enzymatic template-mediated polymerization of RNA. We also used conformationally restricted nucleotides of 8BrG to enforce the *syn* conformation of the bases and judged the catalytic relevance of high-resolution structures of a catalytic RNA [16]. In so doing, we established that computationally derived structures can be more catalytically relevant than high resolution experimental structures derived from NMR and X-ray crystallography. Subsequent studies surveying structures of functional RNAs strongly supported the key role that *syn* bases play in the active sites of ribozymes and riboswitches [17].

Outlook to future developments of research on catalysis by ribozymes

There are a number of new areas of growth for ribozyme catalysis. These include establishing the roles cofactors play in augmenting the catalytic function of naturally occurring RNA. Given that RNA is a functionally poor biopolymer, one can imagine that RNA would benefit as much or more than proteins do from cofactors. There have been astonishing advances in the identification of catalytic RNAs [18], however the relationships among the active sites of these RNAs are not clear. Computational approaches are needed to classify ribozyme active sites and understand their similarities and differences. Lastly, the extent to which charged bases function in RNA catalysis is unclear. Cationic cytosine is established in proton transfer in the HDV ribozyme, but what about cationic versions of A and anionic versions of G and U? Do these charge states of the bases exist and are they catalytic?

Another outstanding question is the role of RNA in the emergence of life. How did chemical catalysis give rise to the first polymers of RNA? And how did these give rise to accurate and processive RNA polymerization without enzymes? Lastly, advances in RNA structural biology are needed. Nearly all of the ribozyme structures that have been solved of the pre-cleaved state of RNA have the 2′OH nucleophile inhibited in some fashion. Thus the first step of catalysis, involving activation of the nucleophile, is opaque. New methods are needed to visualize the nucleophile in an intact ribozyme.

The field of RNA catalysis has made great advances in the last 35 years. It is astonishing to see the versatility of RNA and the similarity of its chemical catalysis to that of proteins. Many questions remain unanswered in the field of ribozyme catalysis and it will be exciting to see those answers emerge in the coming years.

Acknowledgments

We thank the NIH, NSF, and NASA for supporting RNA catalysis research in the Bevilacqua lab over the past 20 years. We are also indebted to our many excellent collaborators, especially the labs of Paul Carey, Barbara Golden, and Sharon Hammes-Schiffer.

References

[1] V. J. DeRose, *Curr. Opin. Struct. Biol.* **13**, 317 (2003).
[2] S. Nakano, D. M. Chadalavada, P. C. Bevilacqua, *Science* **287**, 1493 (2000).
[3] S. Nakano, D. J. Proctor, P. C. Bevilacqua, *Biochemistry* **40**, 12022 (2001).
[4] S. Nakano, P. C. Bevilacqua, *J. Am. Chem. Soc.* **123**, 11333 (2001).
[5] P. C. Bevilacqua, *Biochemistry* **42**, 2259 (2003).
[6] B. Gong, J. H. Chen, E. Chase, D. M. Chadalavada, R. Yajima *et al.*, *J. Am. Chem. Soc.* **129**, 13335 (2007).
[7] M. Guo, R. C. Spitale, R. Volpini, J. Krucinska, G. Cristalli *et al.*, *J. Am. Chem. Soc.* **131**, 12908 (2009).
[8] B. Gong, D. J. Klein, A. R. Ferré-D'Amaré, P. R. Carey, *J. Am. Chem. Soc.* **133**, 14188 (2011).
[9] N. Veeraraghavan, A. Ganguly, J. H. Chen, P. C. Bevilacqua, S. Hammes-Schiffer *et al.*, *Biochemistry* **50**, 2672 (2011).
[10] A. Ganguly, P. C. Bevilacqua, S. Hammes-Schiffer, *J. Phys. Chem. Lett.* **2**, 2906 (2011).
[11] S. Nakano, A. L. Cerrone, P. C. Bevilacqua, *Biochemistry* **42**, 2982 (2003).
[12] A. Ganguly, P. Thaplyal, E. Rosta, P. C. Bevilacqua, S. Hammes-Schiffer, *J. Am. Chem. Soc.* **136**, 1483 (2014).
[13] P. Thaplyal, A. Ganguly, S. Hammes-Schiffer, P. C. Bevilacqua, *Biochemistry* **54**, 2160 (2015).
[14] C. A. Strulson, R. C. Molden, C. D. Keating, P. C. Bevilacqua, *Nat. Chem.* **4**, 941 (2012).
[15] E. A. Frankel, P. C. Bevilacqua, C. D. Keating, *Langmuir* **32**, 2041 (2016).
[16] R. Yajima, D. J. Proctor, R. Kierzek, E. Kierzek, P. C. Bevilacqua, *Chem. Biol.* **14**, 23 (2007).
[17] J. E. Sokoloski, S. A. Godfrey, S. E. Dombrowski, P. C. Bevilacqua, *RNA* **17**, 1775 (2011).
[18] Z. Weinberg, P. B. Kim, T. H. Chen, S. Li, K. A. Harris *et al.*, *Nat. Chem. Biol.* (2015).

PROSPECTS FOR RIBOZYME DISCOVERY AND ANALYSIS

RONALD R. BREAKER

Department of Molecular, Cellular and Developmental Biology;
Department of Molecular Biophysics and Biochemistry, Howard Hughes Medical Institute,
Yale University, New Haven, CT 06520, USA

Abstract

After the fast-paced discoveries of natural ribozymes and their mechanisms in the 1980s, the publication of new classes of RNA enzymes had slowed to a rate of about one per decade. In following text I discuss the reasons for the scarcity of novel class validations, and note some technical advances that have greatly accelerated ribozyme discovery and analysis very recently. These advances create the right conditions for researchers to maintain an accelerated rate of novel ribozyme discovery in the coming years. Such a renaissance in ribozyme exploration creates tantalizing opportunities to discover more natural ribozymes that promote chemical transformations beyond phosphate ester chemistry.

Introduction

Ribozymes are structured nonconding RNAs (ncRNAs) that enhance the rate constants for chemical reactions many orders of magnitude higher than background. Since the first reports of such catalytic RNAs [1, 2], the list of distinct and well-validated classes uncovered from natural sources has grown to 14 (Fig. 1A). Known ribozyme classes primarily catalyze chemical transformations involving phosphoester bonds (Fig. 1B), namely phosphoester transfer and phosphoester hydrolysis. The one prominent exception is the reaction catalyzed by the RNA component of the large subunit of ribosomes, which accelerates the formation of amide bonds (peptide bonds) at the expense of carbon ester bonds from aminoacyl-tRNAs [3, 4].

Each of these natural ribozyme classes, as well as all the diverse examples of engineered ribozymes [5, 6], serve as proof that RNA can promote chemical reactions with high speed and high selectivity, much like their protein enzyme counterparts. Indeed, the fact that ribosomal RNAs catalyze the production of all genetically-encoded proteins in all organisms strongly supports the RNA World theory [7, 8]. This theory holds that modern cells emerged from primitive organisms that ultimately ran complex metabolic states based on RNA catalysts and other functional ncRNAs, long before the emergence of modern genetically encoded proteins.

The complexity of the most advanced versions of these ancient RNA World organisms is a major factor when considering how many ribozyme classes might

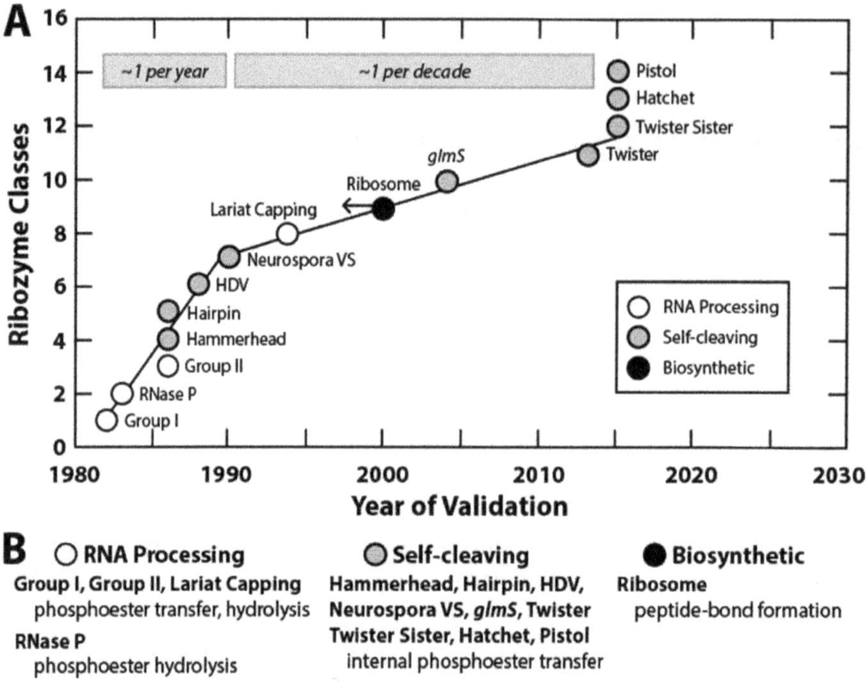

Fig. 1. The discovery of natural ribozyme classes and their biochemical functions. (A) The timeline for ribozyme class discovery. Each point reflects the first publication of experimental data validating biochemical function. The arrow associated with the ribosome point indicates that ribozyme activity was suspected many years earlier (see the main text). (B) A summary of the biochemical functions of various ribozyme classes.

remain to be discovered in modern cells. If our RNA World ancestors indeed were catalyzing and controlling complex metabolic pathways, then sophisticated ncRNAs that exploit the ability of RNA to form diverse structures to catalyze reactions [9] or serve as molecular receptors [10] would have been abundant. Some direct descendants of these ncRNAs might remain to be discovered, or perhaps at least traces of these primordial RNA activities will remain in modern cells as represented by the emergence of additional functional ncRNA much later in evolution.

Another major factor to consider when contemplating how many ribozyme classes exist in modern cells is the stiff completion functional RNAs face in the form of very proficient protein enzymes. If evolution has been particularly selective against functional RNAs since the emergence of proteins, then future ribozyme discovery opportunities might be very poor. However, if a variety of functional RNAs have survived the evolutionary extinction game with proteins, then where are all the long-lost ribozymes and other functional RNAs? Have the ancient ribozymes indeed been purged from modern cells by the stiff competition from proteins? Or, are there deep, dark regions of the genomes of many organisms that still hide novel ribozymes that remain relevant to modern organisms? In the following text, I briefly discuss some of the challenges faced by those who seek to find additional classes of natural ribozymes and comment on the future potential for ribozyme discovery.

The ribozyme discovery crisis

Starting with the first ribozyme class discovery in 1982 [1], new classes were being described at a rate of almost one per year (Fig. 1A). Note that although ribosomal RNAs were discovered decades earlier than plotted [11], by the late-1980s many (but certainly not all) RNA researchers anticipated that someday the RNA components of this giant ribonucleoprotein complex would be proven to catalyze peptide bond formation. This rapid pace of discovery caused Leslie Orgel and Francis Crick, two early proponents of the RNA World theory, to declare [12] that they were too conservative with their original speculation [13, 14] that RNA might only be marginally capable of catalytic function.

Unfortunately, this golden age of ribozyme discovery was followed by a ribozyme discovery desert, where new classes were being reported at the meager rate of about one per decade. For biologists and biochemists who study ribozymes, this trickle of new classes starved the field of objects for in-depth analysis, which otherwise might reveal more secrets about how RNA bridges us from life's beginning to the modern DNA-RNA-Protein World that we live in today. Also, it calls into question the functional capability of RNA. If only a dozen or so ribozyme classes are all that remain from the RNA World, then perhaps our earliest pre-protein ancestors were quite pathetic biochemists. Moreover, we cannot even trust that all extant ribozymes have descended from RNA World organisms because some might have evolved long after ribosomes began to make proteins [9].

There are several possible reasons for why so few ribozymes have been discovered in recent years: (i) No additional classes exist; (ii) We lack the skills or the motivation to discover the remaining classes; (iii) New classes will be difficult to find (rare) and/or difficult to validate (unknown function). Below I will address each of these possibilities, and make a case for the third reason.

The known limits of RNA catalytic diversity

One of the major problems for natural ribozymes is the striking lack of chemical diversity for unmodified RNA. Most known ribozyme classes use only their own functional groups, along with some help from divalent metal ions, to catalyze their chemical reactions. Among the 14 validated ribozyme classes, 12 promote simple RNA phosphoester transfer reactions (Fig. 1B). Two of these are self-splicing (group I and group II) ribozymes, one is a self-branching ribozyme (lariat capping), and nine are small self-cleaving ribozymes (hammerhead, hairpin, *Neurospora* VS, HDV, *glmS*, twister, twister sister, pistol, and hatchet). One of the remaining two classes is RNase P, whose biochemical role is to directly hydrolyze RNA, although some group I and group II ribozymes also can efficiently promote RNA hydrolysis and

DNA phosphoester transfer. Unfortunately only one ribozyme class, ribosomes, promote a reaction that does not involve phosphoester chemistry. This ribozyme promotes the rather facile reaction of amide bond formation at the expense of a carbon ester bond.

It is apparent from the collection of natural ribozymes that RNA is very capable of processing RNA transcripts by manipulating phosphoester bonds. Given recent advances in ribozyme engineering, it is clear that RNA also can be made to function as a reasonably efficient RNA polymerase enzyme [15]. The key question therefore remains unanswered: Are polymers made of RNA well suited to promote many of the chemical transformations needed to biosynthesize most metabolites? If this limitation is rooted in the nature of catalysis by RNA, then the diversity of metabolism in a purely RNA World would have been greatly curtailed, perhaps so much that metabolism might have been far less diverse than that found in modern cells.

Even given only the limited set of known natural ribozyme classes, we can be confident that RNA has the catalytic potential to promote a far greater diversity of reactions. Because catalysis can be achieved by precisely positioning functional groups in three-dimensional space, and because RNAs can for a great diversity of structures, it follows that there must be many additional ribozymes that can exist in the vastness of RNA shape-space. Indeed, even if biology is not exploiting a diverse collection of ribozyme in modern cells, ribozyme engineers have proven over the years that additional reactions can be catalyzed by RNA [16]. Such advances strongly indicate that the catalytic potential of RNAs, sometimes carrying modified nucleotides or specialized cofactors other than just divalent metal ions, is both large and largely unexplored. Are there other clues from nature that also support this conclusion?

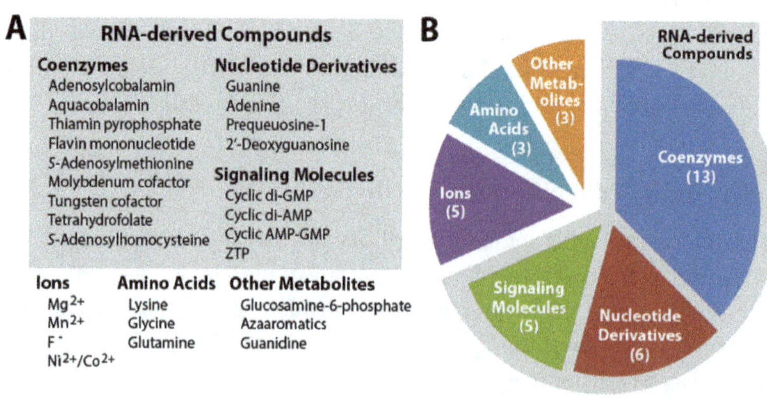

Fig. 2. The diversity of riboswitch ligands and classes. (A) List of known riboswitch ligands. (B) The number of distinct riboswitch classes arranged by ligand group reveals that the largest single group responds to coenzymes or their derivatives. Moreover, the majority of classes sense derivatives of RNA compounds (gray shading). Note that some ligands (*S*-adenosylmethionine, prequeuosine-1, cyclic di-GMP, and Mg^{2+}) are each sensed by multiple distinct riboswitch classes.

More natural ribozyme classes must have existed

There are several powerful arguments supporting the notion that many different ribozyme classes once existed before the emergence of proteins. For example, because RNA forms the active site of ribosomes, it seems reasonable to speculate that ribozymes predate the emergence of genetically encoded proteins. If true, then the early metabolic processes needed to give rise to genetically encoded proteins likely would have been facilitated by enzymes made of RNA. Thus, numerous ribozyme classes might have had to maintain a metabolic state of sufficient complexity to produce amino acids, chemically activate them, and eventually link them together via peptide bonds to create proteins with specific amino acid sequence. This would require diverse types of ribozymes with catalytic functions that are not found among the known natural classes or among those created by directed evolution methods.

The putative RNA World origin of modern coenzymes also is consistent with a once-robust ribozyme-mediated metabolism. Carl Woese [17] and Harold White III [18] independently proposed that the striking similarity between the chemical structures of coenzymes and RNA nucleotides might be due to their origin in an RNA World. This general hypothesis has many variations, but a common conclusion drawn from speculating on the origin of coenzymes is that primitive ribozymes might have extensively exploited these compounds to carry out biochemical transformations much like modern protein enzymes do today.

Intriguingly, there is abundant evidence demonstrating the rich potential for diverse interactions between natural RNA structures and numerous coenzymes (Fig. 2). Riboswitches represent an abundant form of gene regulation in modern cells, wherein RNA directly forms a receptor for various ligands [10, 19]. Strikingly, the largest single category of ligands sensed by the ~ 35 validated riboswitch classes respond to various coenzymes that presumably originated in the RNA World. Thus, there are abundant RNA structures that can selectively bind to coenzymes, although none of these examples have yet been shown to exploit the bound coenzyme to catalyze a chemical transformation. Regardless, since many of these riboswitch classes are among the most widespread, and therefore likely the most ancient, it appears that living systems have long exploited RNA-coenzyme interactions.

With so many natural RNA-coenzyme complexes likely present in ancient cells, it is hard to imagine that these would not have been exploited to form ribozymes that catalyze more complex reactions. Similarly, coenzyme-responsive ribozymes exist in abundance in modern cells [10], and more classes are certainly going to be discovered [19]. Given the rapid pace of genomic DNA sequencing, it is very likely that additional natural classes of ribozymes also will be discovered. Will such new discoveries reveal the existence of coenzyme dependent ribozymes that can catalyze multiple turn-over metabolic reactions? At this time, the only natural ribozymes that come close to these characteristics are ribosomes (multiple turn-over amide bond formation) and *glmS* self-cleaving ribozymes (use glucosamine-6-phosphate as a cofactor) [20].

How metabolic ribozymes might appear

Usually, the discovery of a new ribozyme class occurs by chance. The earliest ribozymes were discovered by researchers who were studying some biological process, and then they determined that the process intimately involved a structured ncRNA. More recently, ribozyme classes have been discovered by first encountering a conserved RNA motif via bioinformatics searches. Subsequent experiments then lead to validation of the ribozyme's biochemical function.

To date, only one type of ribozyme class has proven to be amenable to focused discovery approaches. The discovery of new self-cleaving ribozymes can be purposefully carried out, either by using selection methods to identify natural self-cleaving RNAs [21], or by using bioinformatics to search for ncRNA motifs that reside near genes that commonly associate with ribozymes [22]. Are there other signatures available that would permit the purposeful discovery of novel ribozyme classes?

One obvious search method is to pursue the discovery of strange sequence or structural variants of known riboswitch classes — preferentially those that bind coenzymes. Normally, ligand-binding riboswitches in bacteria reside immediately upstream of the genes they regulate. However, variants of a coenzyme-binding riboswitch class that oddly do not associate with specific genes might be transcribed separately and function as an independent ribozyme that exploits its ligand as a coenzyme. This search approach could also extend beyond just coenzyme-sensing riboswitches. For example, fluoride-binding riboswitches [23] form a ligand-binding pocket that is very similar to the active site of phosphatase enzymes [24]. Perhaps members of this and other riboswitch classes are only a few mutations away from forming active sites that promote biochemical transformations.

Another approach that could be used to reveal the existence of novel ribozyme classes is the search for large ncRNAs that reside near genes that code for metabolic enzymes. Many bacterial genomes extensively exploit operon arrangements wherein genes associated with the same biological process, such as a specific metabolic pathway, are clustered together and are co-expressed. Imagine that a ribozyme catalyzed one part of a multi-step pathway. It seems likely that this ribozyme might be generated from the same operon that gives rise to the mRNAs for the protein enzymes in this same pathway. If large ncRNAs are found in such operons, then researchers will have an excellent clue regarding the function of the RNA when beginning their experimental validation efforts.

Intriguingly, several large ncRNAs already have been discovered in recent years [25, 26]. The most conservative hypothesis for the functions of these RNAs is that they operate as complex ribozymes. Simply put, all well-conserved ncRNAs that are of the size (greater than $\sim$ 400 nt) and structural complexity as these mysterious ncRNAs are ribozymes. However, these RNAs do not reside in operons whose biochemical roles have been clearly defined. Perhaps other large ncRNAs will be discovered in the future that do reside in operons linked to metabolic pathways, which would facilitate the design of experiments that more quickly lead to assigning function.

Why metabolic ribozymes might be largely extinct

It is widely accepted that ribozymes would have experienced strong competition from protein enzymes. Even if there once was a robust RNA World filled with coenzyme-utilizing ribozymes and ncRNAs with many other functions, these might have been swept to extinction once proteins entered the biosphere. It is usually assumed that proteins will win these evolutionary battles because of the catalytic power and diversity of polypeptides compared to nucleic acids. Of course, there is much merit to this view because, although the chemical diversity of amino acids is also quite limited, it is greater than that of the four unmodified natural ribonucleotides. However, there are additional reasons why RNA is a less attractive biocatalyst.

Others have noted that the RNA fitness landscape is likely to be far rougher that the evolutionary landscape for proteins. It is easy to see that mutating a leucine to isoleucine could give a gentle tweak to a protein structure, where as a guanosine to adenosine change (or any such mutation) will alter multiple functional groups that could easily destroy any local secondary and tertiary interactions. There are even more fundamental advantages that proteins have over RNA. Nucleotides, on average, are about one-third the size of an amino acid, and so functional RNAs and proteins of comparable chain length will be dramatically different in size. Ribozymes will demand much space per unit than their protein counterparts, but would also demand more atoms of carbon, nitrogen, and phosphorus. This means that ribozymes would be far more costly in terms of energy and mass compared to proteins, even if they were catalytically equivalent polymers.

Another disadvantage for ribozymes is the fact that, with only four standard nucleotides, RNA frequently comes in contact with stretches of nucleotide sequence that are complementary. This is advantageous when forming necessary substructures such as hairpins and pseudoknots, but could be disastrous for an RNA World organism that is working to maintain a high concentration of ribozymes in an enclosed cellular environment. It might be much easier for proteins to evolve enzyme surfaces that avoid unwanted interactions with other proteins, even at extremely high concentrations.

Concluding remarks

Despite some of the disadvantages of RNA as a biological catalyst, there are many reasons to believe that they were once common, and that nature might exploit many ribozymes even in modern cells. With the growth of genomic sequencing data from diverse species, there has been an increase in the discovery rate for novel riboswitch classes. This pace should continue as more genomic data becomes available, as these sequencing efforts examine more branches of the tree of life, and as new biological niches are sampled.

Bioinformatics search approaches, which seem most useful to find diverse types of ncRNAs, do not always immediately inform researchers of the RNA's biochemical function. New ribozyme classes that are important to extant organisms might already be in hand [25, 26], but their functions have yet to be established. Thus, a combination of powerful discovery methods coupled with determined experimental validation methods will be needed to further explore the true scope of catalytic function by natural RNAs.

Acknowledgments

I thank many members of the Breaker Laboratory for numerous discussions on the theoretical and validated functions of ncRNAs, and for their contributions of the discovery of new classes of ribozymes and riboswitches. RNA research in the Breaker laboratory is supported by grants from the National Institutes of Health, the Howard Hughes Medical Institute, and Yale University.

References

[1] K. Kruger, P. J. Grabowski, A. J. Zaug, J. Sands, D. E. Gottschling *et al.*, *Cell* **31**, 147 (1982).
[2] C. Guerrier-Takada, K. Gardiner, T. Marsh, N. Pace, S. Altman, *Cell* **35**, 849 (1983).
[3] P. Nissen, J. Hansen, N. Ban, P. B. Moore, T. A. Steitz, *Science* **289**, 920 (2000).
[4] T. A. Steitz, P. B. Moore, *Trends Biochem. Sci.* **28**, 411 (2003).
[5] D. S. Wilson, J. W. Szostak, *Annu. Rev. Biochem.* **68**, 611 (1999).
[6] G. F. Joyce, *Annu. Rev. Biochem.* **73**, 791 (2004).
[7] W. Gilbert, *Nature* **319**, 618 (1986).
[8] G. F. Joyce, *Nature* **338**, 217 (1989).
[9] S. A. Benner, A. D. Ellington, A. Tauer, *Proc. Natl. Acad. Sci. USA* **86**, 7054 (1989).
[10] R. R. Breaker, *Cold Spring Harb. Perspect. Biol.* **4**, a003566 (2012).
[11] G. E. Palade, *J. Biophys. Biochem. Cytol.* **1**, 59 (1955).
[12] L. E. Orgel, F. H. Crick, *FASEB J.* **7**, 238 (1993).
[13] F. H. C. Crick, *J. Mol. Biol.* **38**, 367 (1968).
[14] L. E. Orgel, *J. Mol. Biol.* **38**, 381 (1968).
[15] D. P. Horning, G. F. Joyce, *Proc. Natl. Acad. Sci. USA* **113**, 9786 (2016).
[16] G. F. Joyce, *Annu. Rev. Biochem.* **73**, 791 (2004).
[17] C. R. Woese, *The genetic code. The molecular basis for genetic expression* (Harper & Row, 1967).
[18] H. B. White III, *J. Mol. Evol.* **7**, 101 (1976).
[19] R. R. Breaker, *Mol. Cell* **43**, 867 (2011).
[20] W. C. Winkler, A. Nahvi, A. Roth, J. A. Collins, R. R. Breaker, *Nature* **428**, 281 (2004).
[21] K. Salehi-Ashtiani, A. Lupták, A. Litovchick, J. W. Szostak, *Science* **313**, 1788 (2006).
[22] Z. Weinberg, P. B. Kim, T. H. Chen, S. Li, K. A. Harris *et al.*, *Nat. Chem. Biol.* **11**, 606 (2015).
[23] J. L. Baker, N. Sudarsan, Z. Weinberg, A. Roth, R. B. Stockbridge *et al.*, *Science* **335**, 233 (2012).
[24] A. Ren, K. R. Rajashankar, D. J. Patel, *Nature* **486**, 85 (2012).
[25] E. Puerta-Fernandez, J. E. Barrick, A. Roth, R. R. Breaker, *Proc. Natl. Acad. Sci. USA* **103**, 19490 (2006).
[26] Z. Weinberg, J. Perreault, M. M. Meyer, R. R. Breaker, *Nature* **462**, 656 (2009).

THE RIBOSOME AS A CATALYST OF GTP HYDROLYSIS BY TRANSLATIONAL GTPASES

MARINA V. RODNINA

Department of Physical Biochemistry, Max Planck Institute of Biophysical Chemistry, 37077 Goettingen, Germany

My view of the present state of research on catalysis on the ribosome

The ribosome, a molecular machine that synthesizes proteins in all living cells, catalyzes three chemical reactions: peptide bond formation, peptidyl-tRNA hydrolysis, and GTP hydrolysis [1]. At its evolutionary conserved ancient core, the ribosome is a template-dependent amino-acid polymerase which catalyzes peptide bond formation. Each time a codon of the mRNA template is read by the adaptor molecule aminoacyl-tRNA, its aminoacyl group enters the peptidyl transferase center of the ribosome, where it reacts with the peptidyl-tRNA, thereby adding the next amino acid to the growing polypeptide chain. When the mRNA is translated to the end of the coding region, the ribosome — assisted by translation termination factors — catalyzes the hydrolysis of peptidyl-tRNA, thereby releasing the completed protein from the ribosome. Peptide bond formation and peptide release are catalyzed by the same active center of the ribosome, albeit by somewhat different mechanisms. In addition, another catalytic site of the ribosome acts as an activator for rapid GTP hydrolysis in translational GTPases, which include translation factors that promote all steps of protein synthesis. The active site for GTP hydrolysis is located on the translation factors; the role of the ribosome is to donate groups that activate GTPase cleavage [2].

Although the ribosome consists of both RNA and proteins, its catalytic sites are composed exclusively of rRNA residues; thus, the ribosome is a ribozyme [3]. The ribosome is not a very efficient catalyst: it can accelerate peptide bond formation, peptidyl-tRNA hydrolysis or GTP hydrolysis by only up to seven orders of magnitude compared to the respective spontaneous reactions [2, 4], much less than a protein-based catalyst can achieve [5]. This is because RNA has a limited repertoire of chemical groups that can act as proficient catalysts at physiological conditions. Furthermore, the activation of some translational GTPases is tightly controlled: for those GTPases that deliver aminoacyl-tRNAs to the ribosome, GTP hydrolysis serves as a control checkpoint for selecting the aminoacyl-tRNA that matches the mRNA triplet presented for decoding [6]. This group of translation factors includes the bacterial elongation factors Tu (EF-Tu) and SelB, a factor specialized in the delivery of selenocysteine to the ribosome, and their eukaryotic homologs eEF1A and EFsec, which bind to the ribosome in the respective ternary complexes with aminoacyl-tRNA and GTP. The GTPase activation depends on the correct pairing between the mRNA codon and the anticodon of the aminoacyl-tRNA, which is sensed in the decoding center of the ribosome. However, the decoding center is located about 70 Å away from the active site of the GTPases. Thus, key questions are not only how these GTPases work and how the ribosome promotes the reaction, but also how correct codon-anticodon interaction allosterically regulates GTPase activation.

My recent research contributions to catalysis on the ribosome

EF-Tu is the bacterial GTPase that delivers all elongator aminoacyl-tRNAs to the ribosome. The mechanism of GTP hydrolysis by EF-Tu has been addressed by structural, theoretical and biochemical studies [7–17]. In general, the reaction proceeds through a nucleophilic attack of a water molecule on the γ-phosphate of GTP. Similarly to many other GTPases, EF-Tu possesses a very low intrinsic GTPase activity on its own, i.e. not bound to the ribosome, with a rate in the order of 10^{-5} s^{-1} [18]. We have shown that this reaction most likely proceeds through a dissociative mechanism [18]. The negative charge developing on the β-γ bridging oxygen is probably stabilized by the side chain of a universal Lys24 and multiple main-chain interactions with the P-loop residues of EF-Tu, as shown for other GTPases [19]. A monovalent ion coordinated by the side chain of Asp21, another P-loop residue, has a mild stimulatory effect [18]. This mechanism remains largely unchanged on the ribosome, provided there is no cognate codon-anticodon interaction, except that the effect of the monovalent ion is no longer observed. However, codon recognition by aminoacyl-tRNA accelerates the reaction by six orders of magnitude and dramatically alters the mechanism of GTP hydrolysis [18]. The reaction on the ribosome depends on the presence of His84 in the switch II region and the side chain of Asp21 of EF-Tu, as well as the phosphate group of A2662 in the so-called sarcin-ricin loop (SRL) of 23S rRNA [18, 20, 21]. A2662 is responsible for the stabilization of a conformation where His84 is rotated towards the nucleotide that is observed in the activated state [9]. Asp21 further stabilizes the transition state by coordinating a Mg^{2+} ion close to the crucial A2662 [22–24]. Asp21 favors the movement of the negatively charged phosphate group of A2662 toward His84 [13]. Thus, the ribosome appears to accelerate GTP hydrolysis by EF-Tu by rearranging the catalytic site into a conformation that provides the optimal electrostatic stabilization. This is consistent with the exceptionally high entropic contribution to catalysis [25], the lack of a pH-dependence, and with the kinetic solvent isotope effect of the reaction [18].

Further evidence for the mechanism of codon-activated GTP hydrolysis in translational GTPases comes from the recent high-resolution cryo-EM structure of a complex of the specialized factor SelB with its ligand Sec-tRNA^{Sec} captured on the ribosome in the pre-hydrolysis state [24]. In that complex, the universally conserved residue G2661 of the SRL stabilizes the active-site His61 of SelB (homologous to His84 in EF-Tu) in the flipped-in conformation, pointing towards the water molecule that is aligned for the attack on the γ-phosphate. A hydrophobic residue, Val9, stacks onto the His61 imidazole ring, providing additional stabilization of the active conformation. The phosphate groups of A2662 and Asp10 in SelB (homologous to Asp21 in EF-Tu) coordinate a Mg^{2+} ion. Several charged residues of SelB further stabilize the interactions with the SRL, albeit none of them takes part in the chemical steps directly. Given the evolutionary conservation of the residues constituting the GTPase-activating center, GTP hydrolysis is likely to follow the same universal pathway for all translational GTPases [24].

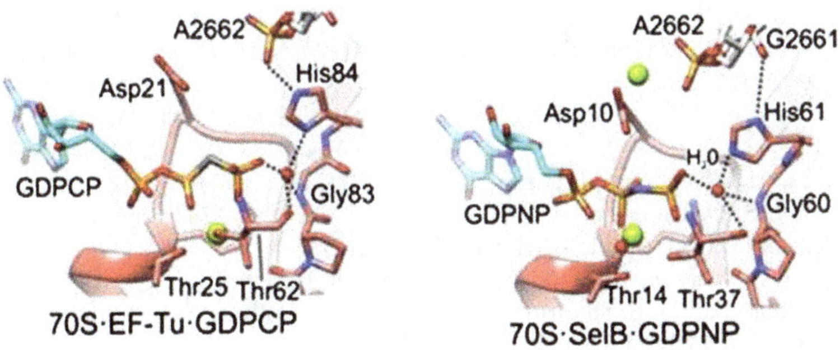

Fig. 1. Catalytic sites of the GTPases EF-Tu (left panel) and SelB (right panel) bound at the GTPase-activating site of the ribosome (residues 2662 and 2661 in the SRL of 23S rRNA). The hydrolytic water molecule (red sphere) and Mg^{2+} ions (green spheres) are indicated. The EF-Tu structure is from PDB ID 4V5L[35]; the SelB structure from [24]. Figure adapted from Ref. [24].

The cryo-EM structures of the ribosome with the SelB–Sec-tRNA^{Sec} complex bound to it also reveal how the GTPase is activated [24]. Correct codon-anticodon interaction triggers local closure of the decoding site and a global domain closure of the small ribosomal subunit. The closed conformation facilitates movement of the tRNA and docking of the GTP binding pocket of SelB on the SRL, thereby aligning key residues of SelB for GTPase hydrolysis. The docking of GTPases onto the SRL as a result of correct codon-anticodon complex formation and small ribosomal subunit domain closure may represent a common mechanism by which the ribosome ensures the pre-hydrolysis selection of the cognate aminoacyl-tRNA delivered by SelB, EF-Tu, or their eukaryotic homologs. These recent findings suggest how the ribosome solves the problem of low efficiency of rRNA-based catalysis [26]. The million-fold acceleration of GTP hydrolysis brought about by the ribosome is achieved solely by stabilizing and shielding effects [2, 18]. This would be similar to the catalysis at the peptidyl transferase center of the ribosome [4, 27], which may provide yet another example of how an ancient RNA-based catalyst solves the problem posed by the limited catalytic power of RNA.

Acknowledgments

I thank Cristina Maracci, Niels Fischer and Holger Stark for their contributions to the work presented here. Research in my laboratory is supported by the Max Planck Society and grants of the Deutsche Forschungsgemeinschaft (DFG).

References

[1] M. V. Rodnina, W. Wintermeyer, *J. Mol. Biol.* **428**, 2165 (2016).
[2] C. Maracci, M. V. Rodnina, *Biopolymers* **105**, 463 (2016).
[3] P. B. Moore, T. A. Steitz, *Cold Spring Harb. Perspect. Biol.* **3**, a003780 (2011).
[4] M. V. Rodnina, *Curr. Opin. Struct. Biol.* **23**, 595 (2013).
[5] A. Radzicka, R. Wolfenden, *Science* **267**, 90 (1995).
[6] M. V. Rodnina, W. Wintermeyer, *Annu. Rev. Biochem.* **70**, 415 (2001).
[7] A. J. Adamczyk, A. Warshel, *Proc. Natl. Acad. Sci. USA* **108**, 9827 (2011).
[8] B. R. Prasad, N. V. Plotnikov, J. Lameira, A. Warshel, *Proc. Natl. Acad. Sci. USA* **110**, 20509 (2013).
[9] R. M. Voorhees, T. M. Schmeing, A. C. Kelley, V. Ramakrishnan, *Science* **330**, 835 (2010).
[10] A. Liljas, M. Ehrenberg, J. Aqvist, *Science* **333**, 37 (2011).
[11] G. Wallin, S. C. Kamerlin, J. Aqvist, *Nature Comm.* **4**, 1733 (2013).
[12] A. Aleksandrov, M. Field, *RNA* **19**, 1218 (2013).
[13] J. Aqvist, S. C. Kamerlin, *Biochemistry* **54**, 546 (2015).
[14] B. L. Grigorenko, M. S. Shadrina, I. A. Topol, J. R. Collins, A. V. Nemukhin, *Biochim. Biophys. Acta* **1784**, 1908 (2008).
[15] T. Schweins, M. Geyer, K. Scheffzek, A. Warshel, H. R. Kalbitzer *et al.*, *Nat. Struct. Mol. Biol.* **2**, 36 (1995).
[16] K. A. Maegley, S. J. Admiraal, D. Herschlag, *Proc. Natl. Acad. Sci. USA* **93**, 8160 (1996).
[17] G. Li, X. C. Zhang, *J. Mol. Biol.* **340**, 921 (2004).
[18] C. Maracci, F. Peske, E. Dannies, C. Pohl, M. V. Rodnina, *Proc. Natl. Acad. Sci. USA* **111**, 14418 (2014).
[19] C. Allin, K. Gerwert, *Biochemistry* **40**, 3037 (2001).
[20] T. Daviter, H. J. Wieden, M. V. Rodnina, *J. Mol. Biol.* **332**, 689 (2003).
[21] M. Koch, S. Flur, C. Kreutz, E. Ennifar, R. Micura *et al.*, *Proc. Natl. Acad. Sci. USA* **112**, E2561 (2015).
[22] D. S. Tourigny, I. S. Fernandez, A. C. Kelley, V. Ramakrishnan, *Science* **340**, 1235490 (2013).
[23] Y. Chen, S. Feng, V. Kumar, R. Ero, Y. G. Gao, *Nat. Struct. Mol. Biol.* **20**, 1077 (2013).
[24] N. Fischer, P. Neumann, L. V. Bock, C. Maracci, Z. Wang *et al.*, *Nature* **540**, 80 (2016).
[25] J. Aqvist, S. C. Kamerlin, *Scientific Reports* **5**, 15817 (2015).
[26] J. A. Doudna, J. R. Lorsch, *Nat. Struct. Mol. Biol.* **12**, 395 (2005).
[27] L. K. Doerfel, I. Wohlgemuth, V. Kubyshkin, A. L. Starosta, D. N. Wilson *et al.*, *J. Am. Chem. Soc.* **137**, 12997 (2015).

LESSONS FROM CATALYSIS BY RNA ENZYMES

DANIEL HERSCHLAG[1] and RAGHUVIR SENGUPTA[1,2]

[1] *Departments of Biochemistry and Chemical Engineering and Chemistry, Stanford ChEM-H (Chemistry, Engineering, and Medicine for Human Health), Stanford University, Stanford, CA 94305, USA*

[2] *Beckman Center, 279 Campus Drive B400, Stanford, CA 94305, USA*

Chemical, historical, and evolutionary perspectives on research of RNA catalysis

We can set the stage for discussion of RNA catalysis by considering the inception of this field less than 25 years ago and asking the following question: Why was catalysis by RNA a surprise to the scientific community? At the time, 1982, there was strong opposition to Cech's conclusion that an RNA sequence, and not a protein, was responsible for cleaving and joining RNA regions in a process he referred to as 'self-splicing' (Fig. 1A) [1]. The objections ranged from 'there must be a protein contaminant' to 'it's not a real enzyme because it doesn't carry out a multiple turnover reaction [2].' But the data were strong, the original group I self-splicing intron carried out self-splicing, the intron was readily converted into an RNA enzyme (or "ribozyme;" Fig. 1B) [3], and Pace and Altman demonstrated the following year that an RNA is the catalytic component of RNase P, a multi-turnover enzyme that processes tRNAs to allow them to function in protein synthesis [4].

While biological RNA catalysis was a surprising result, why was it so difficult to accept? From a chemical perspective it would seem reasonable that RNA could be a catalyst [5]. The field of biomimetic chemistry was in full swing, and there were examples of catalytic lipids (in the form of micelles) [6], and there were even examples of template-directed oligonucleotide synthesis [7, 8]. While "real" or biological enzymes were much better catalysts, these results demonstrated that you didn't have to be a protein to perform catalysis, a straightforward point from a chemical perspective.

It seems that the reticence to view Cech's results on the face of the data arose from dogma, and in this case not just any dogma, but the "Central Dogma of Biology:"

$$\text{DNA (information)} \rightarrow \text{RNA} \rightarrow \text{protein (function)}$$

Crick's model, especially when presented in 1958 [9], was enormously useful for describing and teaching biology, organizing knowledge, and framing much needed research for the coming decades. In it, RNA is an intermediary, carrying information

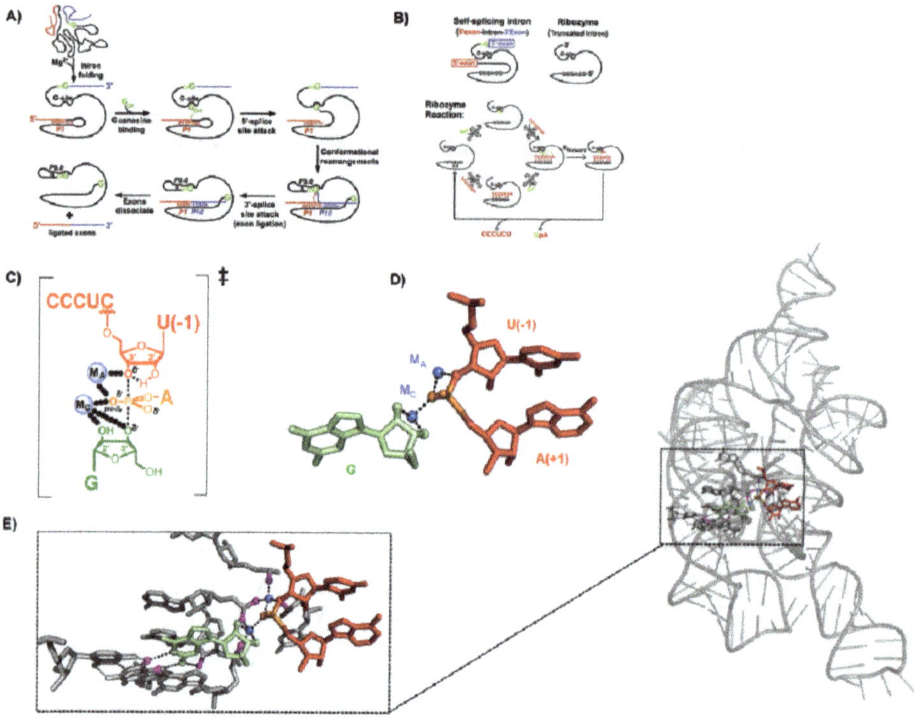

Fig. 1. Group I intron catalysis. (A) Cartoon representation of self-splicing. (B) Comparison between the self-splicing group I intron and the truncated form ("ribozyme"). Images from (A) and (B) were taken with permission from [34]. C–E) Model of the group I ribozyme active site. (C) Transition state model derived from biochemical and structural data (see Ref. [36] and references therein). Closed circles and hatched lines represent metal ion interactions and hydrogen bonds, respectively. Partial-negative charges in are represented by "δ^-." (D) Model of the ground state E•S•G complex [36]. (E) Model of the network of interactions within the active site of the E•S•G complex of the group I ribozyme [36]. Atoms highlighted in magenta contact M_A, M_C, and G. For The location of this network within the overall structure of the *Azoarcus* group I ribozyme is shown on the right.

from the code, DNA, to the protein synthesis machinery that assembles proteins to perform the encoded functions. However hard we try, our thoughts, opinions, and analyses are affected by what we already know — and what we already *think* we know, and so it is likely that scientists were influenced by the dogma of the day, as well as RNA's known, seemingly uninteresting role. Indeed, "dogma" can take on lives of its own, and Crick later regretted using this appellation, noting that he did not at the time know the word's definition [10].

It is informative to consider, in this environment, how three scientists, including Crick, independently proposed RNA as a biological catalyst, more than a decade prior to Cech's discovery, in three single-author papers published in the late 1960s in the Journal of Molecular Biology [11–13]. First, rather than accepting a dogma as fact, Crick, Orgel, and Woese were thinking about biology — how biological systems might operate and how they might have evolved. Second, whereas mRNA in the central dogma was (and is) often drawn as a squiggly line, tRNA was known to be structured and these authors recognized "where there's structure, function can follow." Their central idea, later dubbed the "RNA World" by Wally Gilbert [14], stemmed from the need for information to be stored and that information to be copied, or replicated, for life to occur and the corresponding "chicken and egg" problem of which came first, molecules for information storage or functional molecules to copy that information; RNA could solve this problem by serving both functions.

As implied above, this idea was not widely discussed until after Cech's 1982 discovery, though now the RNA World might be considered a dogma of its own. Nevertheless, learning from the insights of Crick, Orgel, and Woese, we were inspired to think further about the molecular features of RNA — in particular the ability to form highly stable local structure with a small number of monomeric units — how these features may have been particularly amenable to early life forms, and how these properties may have also set the stage for the later takeover by proteins as the predominant bio-catalysts in modern-day life on Earth [15, 16].

Progress in RNA catalysis research and contemporary questions

Given the above, it is not surprising that it took the field some time to demystify the behavior, properties, and catalytic mechanisms of RNA enzymes, but nevertheless, overall progress has been rather rapid. Key contributions came from developing pre-steady state kinetic approaches for RNA to build kinetic and thermodynamic frameworks that allowed individual reaction steps to be investigated [17, 18]; the extension of metal ion rescue that Cohn and Eckstein had used with protein enzymes to quantitative "thermodynamic fingerprint analysis" that allowed individual metal ions could be functionally assessed among the sea of metal ions bound in the RNA ion atmosphere [19–25]; the identification of ribozymes that acted without a requirement for metal ions and thus the necessity to consider additional catalytic strategies [26–28]; an illuminating exposition on the ability of nucleic acid side chains to act in general acid-base catalysis despite their non-optimal pK_a values relative to protein side chains from Bevilacqua [29]; and, also building on work from protein enzymology, a clear exposition of the factors that can contribute to RNA cleavage from Breaker [30]. With these foundational tools, a combination of structural studies and very clever chemical biology approaches to manipulate properties of groups potentially involved in catalysis has led to reasonable models for catalytic interactions for nearly all known ribozymes [31–34].

Consider the catalytic interactions and active site architecture depicted in Figs. 1C–1E for the group I ribozyme, which are supported by a substantial interplay of functional and structural results [34–36]. One Mg^{2+} ion activates the guanosine (G) nucleophile and the other Mg^{2+} ion stabilizes charge development on the leaving group oxygen, and both Mg^{2+} ions interact with a nonbridging phosphoryl oxygen atom, likely making favorable electrostatic interactions and providing a template for the transition state. The nucleophile and leaving groups are in binding sites, held in position by multiple interactions that position the substrates and greatly increase their reaction probability and the reaction's specificity. The metal ions are themselves held in place by RNA as well as substrate ligands, and networks of interactions to accomplish their positioning have been demarcated. Overall the reaction is catalyzed by $\sim 10^{11}$–10^{14} (depending on the chosen comparison state), well within the range of catalysis by protein enzymes [17, 37].

The above and related research has placed RNA on roughly similar footing to protein enzymes. And these studies have also taught us much about RNA folding, dynamics, and structure, by using catalysis as a convenient and powerful readout. Here I focus on unique opportunity for deep insights provided by the discovery of catalytic RNA. Specifically, comparing and contrasting two distinct macromolecules that have been used by Nature to carry out catalysis with high rate enhancements and high specificity can help reveal what is common — and fundamental. Conversely, the differences help us better understand each macromolecule, its behaviors, capacities, and limitations. Indeed, our research has been greatly enriched by these comparisons and will continue to be, and I briefly highlight important examples and current challenges below.

Most broadly, RNA and protein enzymes share the ability to fold into distinct, globular structures with indentations or pockets that serve as binding sites for substrates. Groups involved in carrying out chemical catalysis, such as general acids and bases, are positioned near to where they function, greatly increasing the reaction probability, as are groups with charge complementarity to the reactions' transition states. These features fortify foundational principles of enzymology, and the examples below extend the value of these comparisons to deepen our conceptual understanding of enzyme energetics.

A major conundrum facing early investigators of enzymatic catalysis and energetics was the special ability of enzymes to accelerate reaction of just the "right" substrate. It was particularly baffling that *smaller* substrates would be excluded — *e.g.*, how is hydrolysis of activated compounds like ATP prevented and group transfer allowed? Jencks provided strong support for usage of binding interactions for catalysis and a coupling of their energetics — i.e., reaction of the correct substrate is favored because it's binding is used for catalysis [38]. In many ways this concept is now self-evident — enzymes have specific binding sites and form Michaelis complexes with bound substrates. But the linkage to rate enhancement (i.e., catalysis) and energetic consequences have been less clear to researchers and virtually absent from textbooks. Indeed, it has been common over the years to assign functions to enzyme residues in binding *or* catalysis rather than to seek to understand their interplay.

We have used the known energetics of base pair formation and the structural rigidity of RNA duplexes to probe the use of binding energy for catalysis with two RNA enzymes (Fig. 2). For the group I ribozyme (Fig. 2A), we were able to

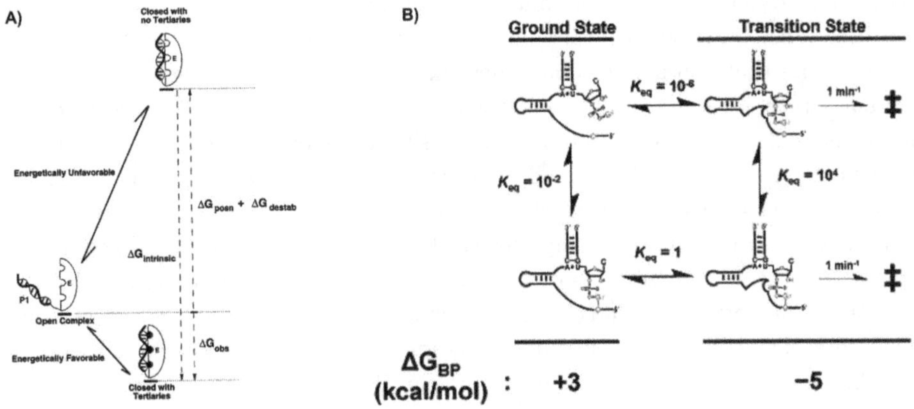

Fig. 2. Use of intrinsic binding energy by the group I (A) and hammerhead (B) ribozymes. Figures taken from or adapted with permission from Ref. [5].

show that several substrate functional groups remote from the site of chemistry could contribute to catalysis within the Michaelis complex, thereby demonstrating a linkage of binding and catalysis. More directly, we could also show that those same functional groups, in the presence of a different constellation of functional groups elsewhere on the substrate helix, could instead contribute to binding [5, 39]. Thus, the same functional group can contribute to binding or catalysis, a dichotomy predicted by Jencks [38] and accounted for by a simple energetic model based on known properties of this ribozyme.

In a second example, we were able to show that a single canonical Watson–Crick base pair could contribute > 5 kcal/mol to catalysis, whereas base pairs provide only 2–3 kcal/mol of binding stabilization [5, 40]. This difference again corresponds to "intrinsic binding energy" as described by Jencks, where binding energy from base pair formation is used to align groups for reaction so that this binding energy does not show up (is not "expressed") in a ground state complex but is expressed in the reaction's transition state. While Jencks presented compelling evidence for this concept for protein enzymes — *e.g.*, the ability of remote side chain binding sites in elastase to aid catalysis and binding of transition state analogs but not affect substrate binding [38] — RNA's properties have allowed more direct demonstration of this energetic interplay.

Outlook to future developments of research on RNA catalysis

There are multiple exciting and important challenges in current ribozyme research. These are linked to evolutionary questions, including why RNA has remained active in certain roles in modern biology, and whether RNA is more or less adept at performing certain reactions and have these capabilities affected the course of evolution? On the functional side, can we use our growing knowledge of RNA folding and dynamics to understand and even predict RNA conformational states and functions? Further, are there additional comparisons between RNA and protein enzymes that can help evaluate mechanistic proposals? The fields of RNA and protein catalysis have, rightly so, focused on active sites; we are now ready for the next-generation question of how is an active site, RNA or protein, constructed, from its surrounding scaffold? We already have results that indicate that remote tertiary interactions have effects on distinct steps in the group I reaction [41], and new tools will allow us to deeply and comprehensively map and test these interactions and their origins. Thus, we are poised to enter a new structure-function era to elucidate the structural, dynamic, and energetic properties of RNA and protein enzymes, knowledge that will deepen our understanding of biology and enzyme function and our ability to manipulate and engineer biological molecules and systems.

References

[1] K. Kruger, P. J. Grabowski, A. J. Zaug, J. Sands, D. E. Gottschling *et al.*, *Cell* **31**, 147 (1982).
[2] T. R. Cech, *Angew. Chem. Int. Ed. Engl.* **29**, 759 (1990).
[3] A. J. Zaug, T. R. Cech, *Science* **231**, 470 (1986).
[4] C. Guerrier-Takada, K. Gardiner, T. Marsh, N. Pace, S. Altman, *Cell* **35**, 849 (1983).
[5] G. J. Narlikar, D. Herschlag, *Annu. Rev. Biochem.* **66**, 19 (1997).
[6] J. H. Fendler, *Acc. Chem. Res.* **9**, 153 (1976).
[7] R. Lohrmann, P. K. Bridson, L. E. Orgel, *Science* **208**, 1464 (1980).
[8] T. Inoue, L. E. Orgel, *J. Am. Chem. Soc.* **103**, 7666 (1981).
[9] F. H. Crick, *Symposia of the Society for Experimental Biology* **12**, 138 (1958).
[10] H. F. Judson, *The Eighth Day of Creation: Makers of the Revolution in Biology*, Expanded Edition (Cold Harbor Spring Laboratory Press, 1996).
[11] C. Woese, *The Genetic Code: The Molecular Basis for Genetic Expression*, (Harper & Row, 1967).
[12] F. H. C. Crick, *J. Mol. Biol.* **38**, 367 (1968).
[13] L. E. Orgel, *J. Mol. Biol.* **38**, 381 (1968).
[14] W. Gilbert, *Nature* **319**, 618 (1986).
[15] D. Herschlag, *J. Biol. Chem.* **270**, 20871 (1995).
[16] Z. Tsuchihashi, M. Khosla, D. Herschlag, *Science* **262**, 99 (1993).
[17] D. Herschlag, T. R. Cech, *Biochemistry* **29**, 10159 (1990).
[18] K. J. Hertel, D. Herschlag, O. C. Uhlenbeck, *Biochemistry* **33**, 3374 (1994).
[19] S. L. Wang, K. Karbstein, A. Peracchi, L. Beigelman, D. Herschlag, *Biochemistry* **38**, 14363 (1999).
[20] S. Shan, A. Yoshida, S. Sun, J. A. Piccirilli, D. Herschlag, *Proc. Natl. Acad. Sci. USA* **96**, 12299 (1999).
[21] S. Shan, A. V. Kravchuk, J. A. Piccirilli, D. Herschlag, *Biochemistry* **40**, 5161 (2001).
[22] J. L. Hougland, A. V. Kravchuk, D. Herschlag, J. A. Piccirilli, *PLoS Biology***3**, e277 (2005).
[23] M. Forconi, J. Lee, J. K. Lee, J. A. Piccirilli, D. Herschlag, *Biochemistry* **47**, 6883 (2008).
[24] J. K. Fredericksen, N. S. Li, R. Das, D. Herschlag, J. A. Piccirilli, *RNA* **18**, 1123 (2012).

[25] E. L. Christian, *Handbook of RNA Biochemistry*, R. K. Hartmann, A. Bindereif, A. Schön, E. Westhof (eds.), 319 (Wiley-VCH, 2005).
[26] J. B. Murray, A. A. Seyhan, N. G. Walter, J. M. Burke, W. G. Scott, *Chem. Biol.* **5**, 587 (1998).
[27] S. Nakano, D. M. Chadalavada, P. C. Bevilacqua, *Science* **287**, 1493 (2000).
[28] J. A. Doudna, T. R. Cech, *Nature* **418**, 222 (2002).
[29] P. C. Bevilacqua, *Biochemistry* **42**, 2259 (2003).
[30] Y. F. Li, R. R. Breaker, *J. Am. Chem. Soc.* **121**, 5364 (1999).
[31] M. J. Fedor, *Annu. Rev. Biophys.* **38**, 271 (2009).
[32] D. M. Lilley, *Biochem. Soc. Trans.* **39**, 641 (2011).
[33] M. Marcia, A. M. Pyle, *Cell* **151**, 497 (2012).
[34] J. L. Hougland, J. A. Piccirilli, M. Forconi, J. Lee, D. Herschlag, *RNA World, Third Ed.* **43**, 133 (2006).
[35] S. V. Lipchock, S. A. Strobel, *Proc. Natl. Acad. Sci. USA* **105**, 5699 (2008).
[36] R. N. Sengupta, S. N. Van Schie, G. Giambasu, Q. Dai, J. D. Yesselman *et al.*, *RNA* **22**, 32 (2016).
[37] R. Wolfenden, M. J. Snider, *Acc. Chem. Res.* **34**, 938 (2001).
[38] W. P. Jencks, *Adv. Enzymol. Relat. Areas Mol. Biol.* **43**, 219 (1975).
[39] G. J. Narlikar, D. Herschlag, *Biochemistry* **37**, 9902 (1998).
[40] K. J. Hertel, A. Peracchi, O. C. Uhlenbeck, D. Herschlag, *Proc. Natl. Acad. Sci. USA* **94**, 8497 (1997).
[41] T. L. Benz-Moy, D. Herschlag, *Biochemistry* **50**, 8733 (2011).

STRUCTURAL BIOLOGY OF PROTEIN FOLDING ON THE RIBOSOME

JOHN CHRISTODOULOU

Institute of Structural and Molecular Biology, University College London and Birkbeck College, London, WC1E 6BT United Kingdom

Nascent polypeptide chain folding during biosynthesis

The ability for proteins to successfully acquire their 3D structures is an essential requisite for biological activity within all cells. An ultimate devastating consequence of the failure of proteins to fold correctly is the manifestation of a class of diseases, more broadly referred to as "conformational diseases", that includes several associated with neurodegeneration [1].

Our rich current knowledge of the fundamental principles underlying the chemical kinetics of protein folding has been acquired almost exclusively from the studies of the folding of *isolated* proteins using experimental and theoretical approaches. It is well understood, however, that *in vivo*, protein folding can begin at the earliest stages of biosynthesis, which is carried out by the ribosome [2] a macromolecular machine.

A typical *E. coli* cell contains *ca.* 50,000 ribosomes, compromising 25% of the cell's mass, engage in protein translation at any one time [3]; the nascent polypeptide chain (NC) is assembled at the ribosome's catalytic centre at a rate of 5–20 amino acids per second and progressively emerges from the ribosomal exit tunnel. It is upon its exit that that the highly dynamic NC has its first opportunity to acquire its three-dimensional structure. Folding could therefore be initiated at the N-terminus and proceed in a vectorial manner before the C-terminus has fully emerged from the ribosome.

A structural and mechanistic understanding of the ribosome and of the mechanism of peptide bond formation is also both rich and comprehensive [4], although our knowledge of the nascent chain is far less well described. The foundations of co-translational folding phenomena were established in array of seminal studies applying elegant biochemical experiments [5] and recent interest in this area has flourished with the use of elegant biophysical approaches such as single molecule fluorescence to identify putative folding intermediates [6].

High-resolution structural studies of protein biosynthesis captured as a series of snapshots: the role of the ribosome in co-translational protein folding

Experimental observations of stalled ribosome-nascent chain complexes (RNCs) are used to explore the chain length-dependence of the free energy landscape. Structural approaches using cryoEM [7] have revealed the propensity for the nascent chain to form simple structural motifs even within the confines of the 10–20 Å narrow tunnel, however the majority of complex tertiary structure forms *beyond* the ribosome. The highly dynamic nascent chain however, precludes its study by either crystallography or cryoEM. As a result, the molecular details of how a nascent polypeptide chain acquires its native structure and avoids misfolding during translation, remain sparse. The dynamic entity has been explored by NMR spectroscopy [8, 10], an ideal technique to explore has been the folding processes that occur. To explore the structure of emerging NCs during protein translation, we developed an *in vivo* strategy to explore the nascent polypeptide using *E. coli* and in particular studied the biosynthesis of a tandem pair of immunoglobulin domains (FLN5 and 6). Using a translational-arrest motif from the *E. coli* protein *SecM*, we generated stalled ribosomes to capture the process of biosynthesis of FLN5 nascent chains as tethered to the ribosome via FLN6, in a series of discrete steps (Fig. 1). We used these biosynthetic "*snapshots*" as a means of studying the FLN5 nascent polypeptide chain as it progressively elongates and exits the ribosome. In addition, by manipulating *E. coli* growth conditions, we have selectively, isotopically-labelled the nascent chain within the 2.4 MDa ribosomal complex, using labelling strategies that specifically target either the peptide backbone (using ^{15}N labelling) or individual side chains of methyl-containing amino acids (*via* perdeuteration and selective ^{13}C labelling). This enabled us to apply rapid acquisition NMR spectroscopy based upon *SOFAST* methodology as well as implementing *methyl TROSY* experiments to probe exclusively both the structural and dynamic features of the ribosome-bound nascent chain at a residue-specific level [8]. Each step of biosynthesis therefore, reported upon a structural snapshot of the elongating nascent chain (Fig. 1). Moreover, by coupling this strategy to translational diffusion experiments and rigorous biochemistry, we were able to ascertain that our observations arose exclusively from intact ribosomal complexes [9].

Using each of our structural snapshots, we use resonance intensity measurements within our ^{15}N correlation spectra to quantify the *loss* of nascent chain disorder, and use complementary *direct* observations of native structure formation in ^{13}C correlation spectra. This approach has enabled us to structural define during biosynthesis, at which the nascent chain can form native structure. Surprisingly, we have found that the nascent chain does not fold immediately once the amino acid sequence is available, despite the isolated FLN5 protein being capable of forming native-like structure, when 90% of its sequence is present [8]. Instead, we observe a "*delay*" (or *offset*) for folding of *ca.* 43 amino acids and by also taking into account that the

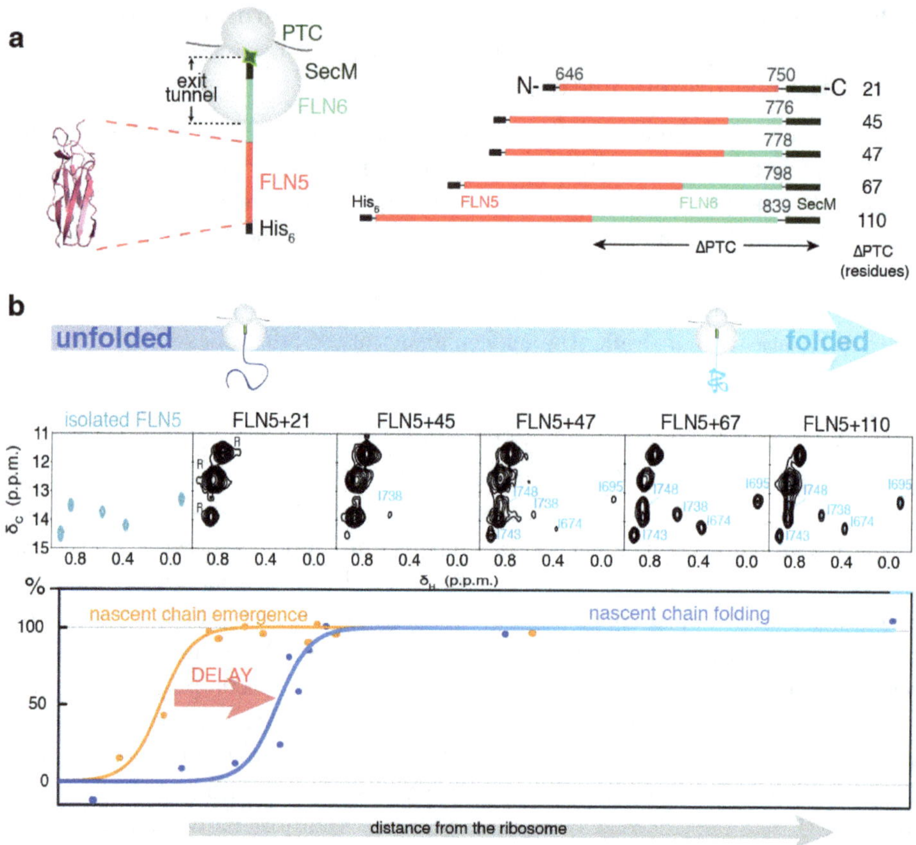

Fig. 1. (a) A panel of DNA constructs used to capture the biosynthesis of FLN5 in discrete steps, as generated *via* SecM-based translationally-arrested ribosome-nascent chain complexes. (b) Structural snapshots of FLN5 as monitored by ^{13}C methyl NMR spectroscopy (blue). Nascent chain folding is "delayed" relative to the full emergence of the amino acid sequence from the ribosomal exit tunnel, as determined using site-specific PEGylation experiments (orange). Adapted from [NSMB].

ribosomal exit tunnel can hold *ca.* 30–33 amino acids as determined by our biochemical assays using it appears that the ribosome actively *prevents* folding from taking place (Fig. 1). This is likely to be a mechanism by which aberrant folding processes are mitigated, to encourage efficient folding [8], in large and/or multi-domain protein systems and particularly in cases of the latter in which a high degree of similarity (> 40%) between domains can have a deleterious effect on folding.

In addition, we used NMR data of the dynamic nascent chain as structural restraints within molecular dynamics simulations (Fig. 2) to generate a 3D structural ensemble of a globular nascent chain tethered to its parent ribosome; the structures show a number of contacts made with several ribosomal proteins near the exit tunnel which is perhaps providing some early insights of the mode by which folding is being modulated by the surface of the ribosome.

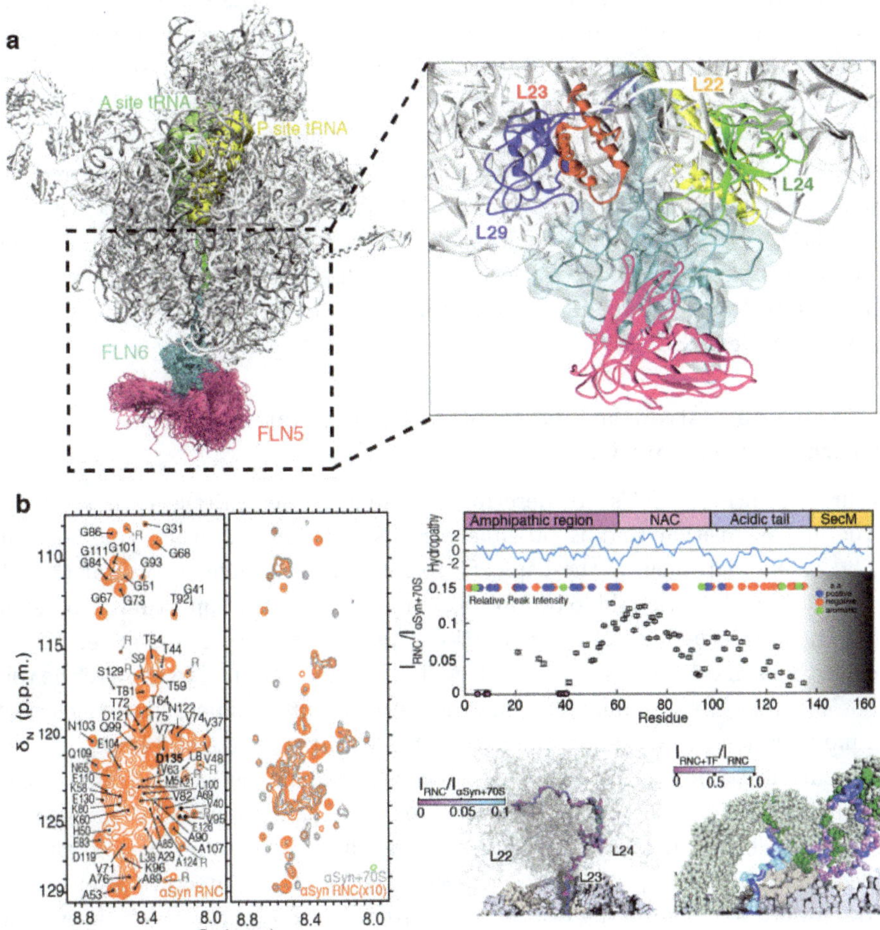

Fig. 2. (a) An NMR and cryo-EM restrained structural ensemble of folded FLN5 tethered to its parent ribosome. A number of key interactions between the nascent chain and the ribosome at the ribosomal exit are seen (adapted from NSMB). (b) (Left) ^{15}N NMR spectroscopy of α-synuclein ribosome-nascent chain complexes. (Right, upper) NMR intensity measurements show amino acid specific interactions exist between the α-synuclein nascent chain and the ribosomal surface. (Right, lower) Molecular dynamics simulations of α-synuclein ribosome-nascent chain complexes both in the presence and absence of the ribosome-associated molecular chaperone, trigger factor (adapted from PNAS).

In our parallel studies, we have also examined by NMR, ribosome-nascent chain complexes a *folding incompetent* protein, α-synuclein, an intrinsically disordered protein. These show that even without the complication of folding needing to take place, that the nascent chain inherently makes transient interactions the ribosomal surface in a sequence-dependent manner, as revealed through resonance intensity measurements; this implicates specific residue types [10] related to both charge and aromaticity (Fig. 2). The generality of these findings in the context of assisting and/or modulating nascent chain dynamics and folding are currently the subject of ongoing studies within the laboratory using advanced relaxation-based as well as 3D strategies.

Moreover, to consider broader facets of *in vivo* protein folding, we have also examined the impact of molecular chaperones. Specifically, we have considered the ribosome-associated protein, trigger factor (TF), which is first to interact with the nascent chain. We observe the capacity of TF to shape nascent chain dynamics and folding processes further, and we observe using both NMR and molecular dynamics simulations, that such modulations are possible, even in the case of a non-substrate such as α-synuclein (Fig. 2).

Perspective on co-translational protein folding: integrated structural biology approaches

Structural studies of RNCs are describing at high-resolution, the earliest events of structure acquisition that is possible for nascent chains as they emerge from the ribosome. Increasingly, the capacity to integrate multiple structural (NMR, cryoEM) and computational biology approaches will offer, the opportunity to define the fundamental principles of co-translational protein folding of the nascent chain and of the ribosome in exquisite detail. It will enable for example, to explore the role of ribosome-associated auxiliary factors such as molecular chaperones and nascent chain modification processes, all which compete with the effects of folding.

Our current studies for example, are revealing that the distinct co-translational folding behaviour identified for an immunoglobulin domain implicates the ribosomal surface itself. The functionality of the ribosome therefore extends beyond that of a molecular machine of protein biosynthesis and its 3D architecture is emerging as a framework for modulating downstream processes of structure acquisition, and increasing a means of quality control. These studies have important implications for understanding the events of nascent chain *misfolding* events, the earliest of which can occur during biosynthesis. Such studies will provide high-resolution insight into the cellular homeostasis which is integral to the biological activity of all cells.

Acknowledgments

This work is supported by a Wellcome Trust Investigator Award (097806/Z/11/Z) to JC.

References

[1] F. Chiti and C. M. Dobson, *Annu. Rev. Biochem.*, **75**, 333 (2006).
[2] L. D. Cabrita, C. M. Dobson and J. Christodoulou, *Curr. Opin. Struc. 2. Biol.*, **20**, 33 (2010).
[3] S. Bakshi, A. Siryaporn, M. Goulian and J. C. Weisshaar, *Mol. Microbiol.*, **85**, 21 (2012).

[4] T. M. Schmeing and V. Ramakrishnan, *Nature*, **461**, 1234 (2009).
[5] J. Frydman, H. Erdjument-Bromage, P. Tempst and F. U. Hartl, *Nat. Struc. Biol.*, **6**, 697 (1999).
[6] W. Holtkamp, G. Kokic, M. Jäger, J. Mittelstaet, A. A. Komar and M. V. Rodnina, *Science*, **350**, 1104 (2015).
[7] O. B. Nilsson, R. Hedman, J. Marino, S. Wickles, L. Bischoff, M. Johansson, A. Müller-Lucks, F. Trovato, J. D. Puglisi, E. P. O'Brien, R. Beckmann and G. von Heijne, *Cell Rep.* **12**, 1533 (2015).
[8] L. D. Cabrita, A. M. Cassaignau, H. M. Launay, C. A. Waudby, T. Wlodarski, C. Camilloni, M. E. Karyadi, A. L. Robertson, X. Wang, A. S. Wentink, L. S. Goodsell, C. A. Woolhead, M. Vendruscolo, C. M. Dobson and J. Christodoulou, *Nat. Struc. Mol. Biol.*, **23**, 278 (2016).
[9] A. M. Cassaignau, H. M. M. Launay, M. E. Karyadi, X. Wang, C. A. Waudby, A. Deckert, A. L. Robertson, J. Christodoulou and L. Cabrita, *Nat. Protoc.*, **11**, 1492 (2016).
[10] A. Deckert, C. A. Waudby, T. Wlodarski, A. S. Wentink, X. Wang, J. P. Kirkpatrick, J. F. Paton, C. Camilloni, P. Kukic, C. M. Dobson, M. Vendruscolo, L. D. Cabrita and J. Christodoulou, *Proc. Natl. Acad. Sci. USA*, **113**, 5012 (2016).

SESSION 6: CATALYSIS BY RIBOZYMES IN MOLECULAR MACHINES

CHAIR: DAVID LILLEY
AUDITORS: A. GARCIA PINO[1], R. LORIS[2]

[1] *Université libre de Bruxelles, Campus de Gosselies (Biopark), ULB CP 300, rue des Professeurs Jeener et Brachet 12, 6041 Charleroi, Belgium*
[2] *VIB and VUB Center for Structural Biology, Pleinlaan 2, 1050 Brussels, Belgium*

Discussion among the panel members

David Lilley: Does anyone have a question to one of our speakers, please?

Ron Breaker: This is a question for John or Marina. In ribosomes of course you have got the chemical process of linking together amino acids to make polypeptides. But an important part of the enzyme is the ability to release substrate and ribosomes are unique in that they have this long peptide exit channel where there is potential for interaction between the peptide product and the ribosome channel, which is I think largely made of RNA. Is there some special surface chemistry that is formed by RNA that allows that peptide to slip out without getting hung up in the exit channel?

Marina Rodnina: So the nice thing about the tunnel is that it has lots of different groups, which may or may not interact with a nascent peptide. So, in most cases, there are no interactions, all these interactions are very transient. This allows any peptide — and you have to keep in mind that on every ribosome potentially there will be a different peptide synthesized — to slip through. However, there are some cases where peptides have evolved to interact with the tunnel wall and this then has very interesting consequences for the ribosome. This may lead to stalling and it can help some remarkable gymnastics of ribosome, for example hopping over the large pieces of mRNA and continue translation somewhere else. And at this time, the peptide is held in the tunnel and this is the way that the ribosome is held and you can continue this way then. So, in other words, the tunnel is usually neutral but in some cases it's not and if it's not this is for a reason.

John Christodoulou: In the work that I have presented, we showed that we particularly used a stalling sequence to allow us to have the nascent polypeptide on the ribosome and that makes particularly strong interactions. There are some particular residues that make some strong interactions with the upper part of the exit tunnel. In most cases, you also have the competing process of protein synthesis that is going on that is actually driving the nascent polypeptide down the tunnel. And so the interactions with the tunnel are competing against this and it is only in a few cases in particular that the interactions are strong enough to hold synthesis.

Daniel Herschlag: So, for those of us who were in the RNA field early on, I know it doesn't look that old, but the advances and the mechanisms have been astounding. You know to the level of looking at protons being shifted around in the end and the set up of active sites. Obviously there are extraordinary advances at the level of the ribosome and understanding its work. There seems to be clearly laid out questions and next steps, and of course that is extraordinarily important for biology. We are very far behind in understanding another extraordinarily important biological ribozyme, which is the spliceosome, which is responsible for most of the genetic diversity in humans. So given that backdrop, I am curious for the entire panel: what are the next challenges in this field? And possibly particularly with respect to what this audience is interested in, in terms of catalysis. Or are there new and other directions? Because it seems like there has been remarkable work and we might want to be thinking about what is the next chapter or chapters.

Philip Bevilacqua: One of the challenges is to try to obtain an idea of what the active form of a ribozyme or any enzyme or any catalyst for that matter looks like, and the question is how do you do that. Darrin is right in that this is where we really need collaboration with computational chemists to use our imagination and our experiments to move from experimental structures closer to what those transition states look like. We can do a better job on the experimental structures though. I think every structure that has ever been solved of a ribozyme is of an inactive one by definition. These are not true enzymes. Almost every single one is a single turnover reaction, it is changed in the reaction. So they have to be inhibited during the reaction in order to freeze out, in order to have the scissile phosphate be present. So the nucleophile is gone and if it could be present, if we can trap these structures with all the native nucleotides, we can at least get closer and work toward that goal. So there is a need for advances experimentally and corresponding advances computationally.

Kurt Wüthrich: I think in the interest of our colleagues who are not into biological systems so much, it is important to stress that catalytic actions are not finished when the chemical structure of the polypeptide chain is made and we cannot discuss this extensively here. But I think it may be helpful for you to understand what is going on, that we really need an additional step of catalysis and that is the chaperoning of proper folding of the result of the ribozyme catalysis. This has a big impact today in pharmaceuticals because many pharmaceutical products nowadays are biologicals, are proteins. And it is a very big problem to make biosimilars because of this final step in the folding of the protein. The FDA is lost at the moment, for example, in defining criteria by which biosimilars should be allowed on the market in order to replace very expensive biologicals that we are in the process of using.

Marina Rodnina: I have two comments on that. First of all, protein folding is definitely extremely important. Also to the colleagues who are coming from chemistry: the proteins are synthesized as linear polymers but then they have to fold. This happens partially on the ribosome and partially off the ribosome and there are many factors that are involved in that. If this process does not occur correctly, you have disease, aging and these are diseases from Alzheimer, Parkinson down toward cancer. So this is extremely important. However, is that catalysis? And this is the question that I would like to discuss with Dan; can you define that as catalysis? I am not sure. I would like to raise that as a discussion point. For the second comment, I would like to return to Dan's point about what we need and I think what we badly need are better computer models and molecular dynamics simulations, better understanding of what happens in the active site of the enzymes. And this is in its infancy even in relatively small systems. But if you take a system like a ribosome, this very rapidly becomes a disaster. And I have heard from people who are very competent on that that, if you take just a part of the ribosome, what you get out of that is potentially just crap. So you really have to take all of the ribosome with this more than two million kD to really model that. And this is a really formidable challenge.

John Christodoulou: I completely agree with all of the recent comments. I think that I mentioned the idea of the ribosome's surface as a chaperone for instance. And as I mentioned, there is a significant amount of misfolding that occurs at the point of synthesis, with a range of devastating disease consequences in many cases. I think that the structural analyses that are emerging, in particular of combining structure with a really detailed atomistic understanding of the dynamic motions of for example the nascent polypeptide and the downstream chaperone, is actually going to be critical to tackling many of the future problems including potentially the capacity to have antibodies engineered to fold better on the ribosome and avoid some of the problems associated with the preparation.

Daniel Herschlag: In response to Kurt's comments, there is a really interesting emerging area in biology that I think this group of chemical community chemists will be interested in, especially those in heterogeneous chemistry. There is now compelling evidence for phase transitions within cells and they involve proteins that exist often as natively unfolded proteins, so those are the exception that proves the rule. And these particles are transition phases. The biology is not clear, some of them tend to form under stress but they are clearly there and they follow the physics of what is expected for phase transitions. They often involve RNA and RNA binding proteins and I think one of the challenges is understanding the physical and chemical features and properties of these. And of course another challenge is understanding whether they have a role in biology or if it is simply something that occurs. It seems that there is some evidence for roles and then they also raise the interesting question, as you know Marina mentioned folding diseases, so why are they so common? Well, if some of your proteins have to aggregate in a phase transition in order to function, then you are at this precipice. If they don't do it enough, you don't have function and if they do it too much, you forget what you were doing and you have that disease. So those might be interesting areas to explore in terms of these different phases for where one can put enzymes and attach things to enzymes and potentially other catalysts.

Ron Breaker: I wanted to mention a couple of things. Dan asked about what might be needed next and what might come next in the field of ribozymes. I'll mention two things quickly. One is I believe we are at a stage now where no structured noncoding RNA will remain undiscovered in the genomes that have been sequenced. I think we can find everything that Nature has put out there. But if we want to have an understanding of a new class of ribozymes quicker than one per decade, we have to get much better at understanding what these mystery RNAs do. I think the only way we are going to get there is through very challenging biochemical, genetic and bioinformatic approaches to figure out what each one of these does. And then the second thing is, I think everybody should watch closely the ribozyme engineering space. I know there are individuals out there who would like to make ribozymes that might be useful for therapeutic applications. But I think a very interesting step forward, and when this work is done I am sure everybody would read the paper, is the creation of an RNA that can copy RNA or copy itself. And so I think researchers are getting very close to making an RNA that catalyzes its own reproduction, which then, RNAs can evolve their own functions in the test tube if given the right conditions in the right supply of activated monomers.

David Lilley: I think what Phil Holliger has done with improving the Bartel ribozyme now replicating up to hundreds of nucleotides is similar to this, and really shows the way forward with that sort of thing. But I was coming back to what you said about expanding the range of chemistries and using coenzymes. It is by comparison relatively easy to find something that cuts itself. But once you are looking for something that is acting in trans on a substrate, for example a TPP catalyzed reaction, how on earth do we figure out what the substrate is, to go looking?

Ron Breaker: Certainly, we have thought very carefully about how do you identify the function of a ribozyme or a candidate ribozyme that you might find by bioinformatics and the challenge is that you can find the RNAs nowadays without knowing what biology they are involved in. Whereas some pretty special ribozymes have been identified and validated in the past because people were working on how translation occurs or people were working on how certain RNAs are processed. And so ribosomes and RNase P, both those classes work on substrates in trans and they do multiple turnover but their functions were established very quickly because people were studying those processes and determined that an RNA was involved. And so I don't know if I have any easy answers for how do you do that, when you identify the RNA and you know it is special but you don't know what biology it is involved in. I think the only way forward is good biochemistry and good genetics and an open mind.

Philip Bevilacqua: Going back a couple comments to self-replicating RNAs: you know with this community here, I am thinking about catalysis. One of the most important and perhaps one of the most difficult questions is how did life begin and so you can keep going backwards in that thought. If you start to think about an RNA world or whatever, you think about how life began from inorganic material. It really is the ultimate question of catalysis and so it is before biocatalysis. So not only how did RNA replicate itself but how did molecules replicate themselves in the presence of templates and then where did the templates come from and where did the monomers come from? Then you are back to small molecule catalysis and organization. And this is something we don't understand. It is hard even to ask those questions properly. Can you ask them in a way where even if you answer them, they are satisfying? It is a huge challenge for all types of catalysis in chemistry and biology, just more of a comment than anything.

David Lilley: That is a vast problem it seems to me because once you have got some sort of primitive however crude replicating system, a means of encoding information into the future, you have got evolution and you can move forward. Until you have got to that point you could make a bag of chemistry that is doing OK and then it goes "bleh" and it is like it never existed because there is no way of encoding that into the future. We clearly did get over that problem but I am having trouble imagining how that occurred.

Darrin York: Kind of at the other end of the spectrum, one of the things that were mentioned as an overarching goal was design. And I think one of the things that are important for being able to guide design is what Dan was mentioning as trying to come up with predictive insight. So, as experiments become more and more interesting and clever and aggressive, so is the need to be able to use some type of modeling to be able to interpret those experiments. I think inherently theory kind of takes a bottom up approach by looking very small at quantum mechanical interactions or short range interactions. And then we hope that when we put all these models together, we can predict near macroscopic phenomena whereas experimental approaches often take a look at very complex phenomena and try to interpret down to the atomistic details. What we really need is more of a meet in the middle approach where these two things are more intimately stitched together so that theory can be leveraged to gain the type of predictive insight that would be really useful to design new types of technologies.

Daniel Herschlag: So follow up on that, I really agree and there is some progress that I just wanted to mention from Rhiju Das, colleague and former student at Stanford, who has been able to design very simple RNA structures from a predictive standpoint in an automated computer way and now validate those structures, show that they fold and show that they form. Those are much simpler than catalysts. So I think bringing together the beautiful mechanistic insights that have been described here along with making those algorithms more sophisticated and bringing in more knowledge from experiments will be critical to bring along and combine with types of models that Darrin is bringing up.

David Lilley: If there are no more immediate questions, perhaps this is a good time to take a break.

General discussion

David Lilley: I think we should restart. We are into the final straight. The last hour of discussion we will throw open the discussion now. Certainly I will be interested to hear what the real enzymologists think of RNA enzymes, these pathetic little enzymes that can only do a million fold rate enhancement and I am interested to hear from the general catalysis community, what questions they have for us, so please, questions?

Kurt Wüthrich: To my surprise, the principal of allostery was completely eliminated from the discussion on protein enzymes yesterday. Is there any indication or clear cut evidence for allosteric effects in ribozyme biochemistry? I mean I think that allostery gives us answers or at least indications for answers why we need minimal sizes for these machines. And here we are dealing with some very big machines. Is there an indication that allosteric effects play a role?

Ron Breaker: There are very very few examples of ribozymes in nature that are controlled allosterically and I very briefly showed one in my presentation. This is an RNA complex that uses cyclic Di-GMP as the allosteric effector and the allosteric domain controls the function of a ribozyme that self splices. We demonstrated that biochemically and genetically. There are several additional examples similar to this that we can find by bioinformatics but they haven't been demonstrated experimentally. However, years before this, my lab and others demonstrated I think, I hope, convincingly that you can engineer allosteric ribozymes quite easily to respond to all sorts of small molecules. This is a technology that works really well from an engineering standpoint. It is a technology that we know biology exploits to a certain degree in modern cells and it is certainly a technology that would have been important if indeed there was an RNA world.

David Lilley: Many of the ribozymes that we study are single turnover. For example, the hammerhead and hairpin, all they have to do, as the RNA is synthesized by a rolling circle mechanism, is to cut themselves. A ribozyme is only going to cleave (or ligate) once, so it really needs no control. So that is perhaps why it doesn't need to evolve allosteric control.

Marina Rodnina: Of course the ribosome is a typical case of something that is regulated from all sides. And for the peptidal transferase there is one thing that is particularly spectacular and this is binding of a signal recognition particle to the outside of the exit tunnel which then may regulate this synthesis. This can happen, but what I have shown was mostly about the GTPase so this is the allosteric relationship between the codon-anticodon interaction and GTPase activation. There are many many things of that. But this is related to the flow of information. So you have to regulate the flow of information. Therefore you have to couple different parts of the ribosome so that they work all together.

Philip Bevilacqua: I have a comment mostly to follow up on David's comment and that is that these are single turnover molecules. And in the GlmS ribozyme which Ron discovered, we have done some recent studies on that. The activity I showed briefly seems to be controlled mostly by a single hydrogen bond competition of the nucleophile, the $2'$ OH, for a non-bridging oxygen and the nucleophile is tied up until the co-factor comes in and competes for that oxygen and frees it up and then it attacks. And that regulation is a million fold. That is big. It is a big effect that seems to mostly come from a hydrogen bond. And it doesn't seem to be allostery at all. The structures with and without the cofactor are nearly identical. And that regulation is much bigger than you find in allosteric protein enzymes like the ones involved in tryptophan synthase where often the regulation is 10 or 20 fold and I think the biological reason is the one that David gave because you really want to keep it off. It is sort of a one trick pony. It does the job and then it is over.

Daniel Herschlag: I think we can think about allostery in RNA in a broader sense and say that it is quite common. A place to start thinking about that is that there is a simple answer and a more complex answer to why there was or might have been an RNA world. The simple answer is RNA can be both a catalyst and code genetic information but that is not going to be enough. And so if you are thinking about trying to get started in a primitive world you need to have something that, with a small number of monomeric units, can form a stable structure and that is what RNA can do. And so the probability of just mixing together a small number of things and getting some activity is considerably higher for RNA than it is for proteins. Everything that has a positive also has a negative, except for you David. It is only positive, sorry to clarify that. And so it also has a tendency to get stuck in structures. We can ask then, and that is something that we have hypothesized was involved in the transition from an RNA to a protein world, that there were RNA chaperones made of proteins first that helped the RNA work and then ultimately took over. But now you can ask the question of why have RNA machines continued like the ribosome and the spliceosome? Perhaps it is because RNA is very good at getting stuck in individual conformations that you can then use to regulate and do something complicated like the steps in the ribosome process for the regulation that you need for splicing and for choosing alternative splice sites. So, like in the beautiful pictures that Rudolph Marcus showed us of the F1-ATPase working by going through several steps, that is something that RNA seems to be very good at and maybe it's why it has been kept around for some of these functions. That is from a mechanistic point of view the same as allostery because you have one factor affecting the conformation at a distance.

Donald Hilvert: Co-factors, transition metals greatly extend kinds of reactions that polypeptides catalyze. Ron mentioned that there are vestiges of the RNA world perhaps in the RNA-based co-factors that are found at riboswitches. To what extent has the chemistry of those co-factors been explored? You suggested that there was a pocket that might be able to accommodate substrates. But does the thiamine-dependent riboswitch do benzoin condensations and the like?

Ron Breaker: As you noted, a tremendous number of the known riboswitch classes do bind to coenzymes and these are the same compounds that are being used by many many protein enzymes to do modern chemistry. We don't see any indication that the number of riboswitches is going to be small. There are going to be probably thousands of classes of additional riboswitches to be discovered and many of them might bind some of the additional coenzymes that you didn't see on the list I showed. Right now there is no evidence that any of those bind the cofactor and do chemistry with it and one could spot them bioinformatically by finding a weird looking riboswitch that is not in the genome in a position to control gene expression. It could be a free floating, riboswitch-like molecule in that it binds a coenzyme and it does chemistry in some metabolic process. There are probably informatic ways of identifying exactly I think what you are looking for. The interesting arrangements of some of the structures that bind coenzymes, those interesting arrangements could allow for a substrate to be positioned and do carboxylation or decarboxylation reactions or methylation reactions. It is not clear whether these RNAs then are distant relatives of former ribozymes or whether they are simply trying to avoid the hot spot in the co-factor that otherwise might inadvertently do chemistry on the RNA itself.

Donald Hilvert: But you suggested that at least in the case of the thiamine ribozyme, that the thiazolium is actually exposed. If so, it might be able to interact with a substrate molecule.

Ron Breaker: That is exactly right. Although we don't know if biology is exploiting that pocket in some distant relative of the known riboswitches. A better way to put this would be that if I, as a scientist, were forced to create a thiamine-utilizing ribozyme, I would start with the TPP riboswitch and would try to evolve another domain on that RNA that would bring in a substrate and exploit that hole and hope to get catalysis in that way. It might be even easier to use something like the SAM class one riboswitch. You have got a hot methyl group that is ready to be donated. You just got to get the RNA to evolve a domain that will position substrate there and that methyl group transfer should happen.

Judith Klinman: In terms of the methyl transfer from the structure, is the activated methyl group pointing into an interior region of the riboswitch or out toward the solvent? And in principle even if it is slow, you could just add in a methyl group acceptor to see if you get any chemistry at all.

Ron Breaker: It is quite extraordinary that in that riboswitch class, the methyl group is not pointing at all towards the RNA. The RNA looks like it is working very hard to stay far away from that methyl group and again it might be that that class is so popular in biology — it is the most common riboswitch class of among five different riboswitches that bind S-adenosyl methionine — it might be the most popular because it is trying to avoid modifying an RNA base so that the cell would have to clean that modification up when it is trying to recycle the monomers. But that means then that the space is there for the RNA to evolve to position a substrate. Others and we have talked about that kind of an experiment where you would evolve an artificial ribozyme that would do methylation. I think with the people that I have talked to about this we all believe it is so likely to happen and easy that we decided to do other things and not do that experiment. But it should work.

Judith Klinman: Well I was thinking of a different experiment: you just add in an acceptor and see the chemical reactivity of the co-factor in the presence or absence of the riboswitch, just to see if there is anything going on there that could lead forward.

Ron Breaker: And I think it is something that could be done. I don't think anybody has done it yet.

Christophe Copéret: Some of you actually mentioned metal ion catalysis, so I wanted to clarify what we mean by metal ion catalysis. So first of all what is the definition of metal, is it always you know barium, calcium, magnesium and sodium or can it be open to other ions? This is my first question. And my second question is that these cations are only here to glue things in the right position, so can we call that a catalyst or not? Or are they involved basically in making a proton a bit more acidic by modulating the proton acidity by having a Lewis acid? So I ask a question to all of you basically.

Daniel Herschlag: The answer is yes. So our metal ions are really boring. There is no redox chemistry. And it is very likely biologically all magnesium because the sites aren't that highly chelated. And for magnesium, you are able to take advantage of nitrogen specificity for zinc or think that way. There are few examples of bound potassium ions because potassium is also there as well. And then what they are doing is just what he said: they are acting electrostatically. Sometimes gluing together a structure, sometimes they have been shown interacting with nucleophilic leaving groups to stabilize charge development or a charge that develops and sometimes to assist, as Phil had shown, in deprotonation events.

Ron Breaker: Just to add to this, if you turn again to riboswitches, it is clear that other types of metal ions are highly recognized by RNA. There are very selective nickel/cobalt binding riboswitches. There are riboswitches for molybdenum and tungsten cofactors and there are riboswitches that control all of the cobalt homeostasis by binding to coenzyme B-12 which of course has a cobalt chelated in the Corrin ring. So there are other tricks that RNA can use to expand the number of metal ions that it can interact with. Then the question is whether they use those metals for chemistry and that is for future research to reveal.

Robert Grubbs: Yes, and so most of the transition states you showed were all linear transition states but as Westheimer showed back when he studied phosphate substitution long ago, there were examples that were very important involving pseudo rotation of the five cornered intermediates. Are any of those things involved in any of these reactions?

Daniel Herschlag: Very likely not for two reasons: one is that the pseudo-rotation occurs for phosphate tri-esters when you have three substitutions and that gives enough stability to the phosphine species, the P5 species. And the other reason is likely that on protein, and there is no evidence for that on any protein enzyme either, if you have things positioned already, and a lot of where your catalysis is coming from, you can call it a template effect, then you are not necessarily going to spin things around unless you really need to.

Philip Bevilacqua: We also agree with what Dan said and we actually went and looked for such things and some of the evidence for that is if you have that sort of rotation. Then you can actually go backwards and make a 2'–5' bond with an isomerization in the HDV ribozyme which seems to go through under certain monovalent ion conditions through some sort of intermediate. We tried to look for such a species. Now we didn't find it, but that doesn't mean it doesn't happen.

Bert Weckhuysen: I have two questions. Let's first start on this so many fold rate enhancement. For heterogeneous catalysis people we have now learned something new at this thing and you made a nice comparison between the 10^{27} versus the 10^{10}.

When you look into the splitting of these parameters, then you see that the leaving group stabilization and a number of hydrogen bonds, that is actually the difference and that links to Christophe Copéret on the leaving group stabilization. There you see Zn^{2+} versus Mg^{2+}. So that too is right away for me something like Lewis acidity aspect maybe. So that would mean that if you would like to upgrade the qualities of your ribosome, of this catalyst, that you should try to intervene on that level and the second thing is on the position of the hydrogen atom. That means that for your genetic engineering (where I don't know anything about, but I can imagine that you can start to design in a computer somehow how it should be and hopefully you can fold it right away), then you have the perfect pocket with hopefully then the perfect metal ion and then you can hopefully also beat the proteins. Is that science fiction or possible?

David Lilley: In the nucleolytic ribozymes, we only get a million fold enhancement rate but the rates, the k_{obs}, that we observe for these ribozymes are typically maybe ten per minute. I mean they are not going fast. However, if you look at the pK_a of the groups that are involved, G+A, only about one molecule in a thousand is actually active at any given moment. So if you correct for that, the intrinsic rate of an active molecule cleaving is going say 10^4 per minute. Which is within a log or so of RNase A. So actually they are not doing that badly. And, as I say, they only have to cut once so there is no evolutionary pressure to make a superfast ribozyme.

Bert Weckhuysen: That is true. But let's now assume that some of these could be used in pharmaceutical or in a chemical industry. I mean I am more from a practical side. If you would be an industrial scientist, you would at some point not care if it is a heterogeneous, homogeneous or a biocatalyst. You just think: let's make the product and then it becomes important if you can tune. You say: I want to have this Lewis acid in it, metal ion, I have position control over my hydrogen so I can make a certain pharmaceutical intermediate, no?

Ron Breaker: Many of us in the field tend to focus on ribozymes that have only the four common natural nucleotides and magnesium as a typical divalent metal ion. If you go into applications that do not necessarily have to go into human cells, for example, for therapeutics, then the gloves are off. You can use any kind of metal ions. I would use things like manganese or cobalt or zinc etc. at higher concentrations. And then, you have the whole world of modified bases, artificial modified bases, not just the 150 or so that are known in biology. You can use then any kinds of chemical modification to really change RNA as a polymer into something more akin to a plastic polymer that has all the wonderful functional groups you might want to have for industrial enzymes. I think that's where some of the people who are working on RNA engineering would like to turn, to use more modified nucleic acids and more extreme conditions including metal ions.

Kyoko Nozaki: During the coffee break, I heard a little bit of discussion. In a sense of application oriented, do you think the ribozyme is a good starting point for industrial applications to transform petrochemicals to some products, or is this a bit of a crazy idea?

David Lilley: One obvious point is that RNAs actually are rather intrinsically unstable molecules because the 2' hydroxyl is like a dagger poised to its heart at every position. So it wouldn't be the molecule I would choose to make an industrial catalyst.

Philip Bevilacqua: With that said, there are analogues of RNA, and the simplest one is DNA. Single stranded DNA, not double stranded, can do an amazing array of chemistries. Ron has done some of that, and Scott Silverman, I think, leads the charge on that right now. DNA can cleave DNA and there are lots of advantages to evolving and modifying with the nucleic acid base system. So I don't know the answer but I don't think it's a crazy idea. I mean maybe it's crazy, but it's not an impossible idea to me, to think about doing that. Some of those DNAzymes — now that the first structures came out in Nature this year or last year — use zinc ions and so on.

Graham Hutchings: If we're talking about industrial applications, coming from the heterogeneous side there are three things we look for: activity, selectivity (which obviously enzymes and ribozymes have got) and also lifetime. We have catalysts that last for a few seconds or last for years, but the engineering then comes into play. Throughout all this discussion, yesterday about enzymes as well, there's not been any mention of how long these things go on for. Whereas we, I think, from the chemical catalysis side, we did approach it a bit because we say these things change. They deactivate. There are induction periods and all this sort of thing. So, are there any comments on this? Do they keep going on forever?

Philip Bevilacqua: Just a quick comment. If you turn them into multiple turnover by engineering them from single to multiple, which is pretty trivial to do, they can do tens of thousands of turnovers before they die, if they die at all. They can last a long time. It's possible to take time points out to many days, even a week under the right conditions. If you were to not use RNA, it could last a very long time. DNA is very stable, even to boil it, unless you put it in acid conditions.

Avelino Corma: I appreciate very much to see the specificity that you get and also to see those molecular machines in the way they are working. It is just beautiful. I see that you can also modify them. If you can modify them, then how far can you go in widening the scope of the reactants that you are using? Have there been successes? Are there possibilities for that? Or are they specific for one molecule and for the next one you have to go for another, for the next one for yet another? How does it work?

Marina Rodnina: I think there's a difference between the ribosome and the ribozymes. The ribosome will take almost everything which you can charge on a tRNA. So as soon as you can bring your substrate into the ribosome it will most probably take it. This is one of the reasons why you can introduce modified amino acids or even not amino acids but something else into the peptides and you can make, (also answering another application question), you can design proteins which have elements which are entirely unnatural. This, of course, also has industrial applications, because by that you can develop proteins for diagnostics, for labelling, for treatments. So this is very nice. There you can do almost everything, until it fits to the peptidyl transferase pocket. But for the ribozymes I think it's different.

Dan Herschlag: People have tried that in a number of ways: adding in an imidazole off of a base. There have been modest successes. My sense is it's really an area that is virtually infinite untapped potential and it's really limited by technology. Each time you do a selection and try to read out what these bases are and which ones work from a selection experiment or a pool, this is quite difficult. I don't think there's much work in it, until someone comes along with a way to rapidly screen through. The synthesis is slow of the basis you want to put in, and then the selections are cumbersome to get winners. It's a phenomenal idea and lots of people have had that idea and tried that. But right now it's slow. So we need new technology.

Steven Boxer: Just a quick follow up question. Are there catalytic PNAs?

Dan Herschlag: Nobody knows of any, but a trivial example would be what Leslie Orgel did with RNA very early on, which is if you hybridize two oligos together you get rate enhancements. So presumably that would happen in a templated way and you can get great enhancements of a hundred to a thousand by doing that, or even larger if you're looking at very diluted solutions. But no one has really looked for it. Or if people have looked they haven't published it or it hasn't worked.

Steven Boxer: So just to follow up, I mean it would seem that from the perspective of understanding, a mechanistic understanding, especially the role of the backbone, that would be a useful thing or is that just there? They tend to be much more hydrophobic and harder to work with. Interesting.

Dan Herschlag: I think it's because they tend to aggregate and they tend to be extraordinarily expensive to get a hold of. And then you can't do the genetic manipulations, what's headed down but it would be phenomenal because you could manipulate where you have charge and also have to fight less entropy.

Ron Breaker: Of course, any polymer that can form diverse structures should be able to do a catalysis. And so, many modifications you could imagine you could make to the parent RNA polymer might be wonderful for forming alternatives to ribozymes. The problem is in replicating the information that you might be creating in that polymer. So we often work with RNA and DNA, and of course proteins as well, because there are ways of copying the successful catalyst. Now, if one had the ability to copy PNA through some sort of enzymatic fashion, then that gives you the ability to technically carry out the engineering or evolution that you want. And this is why I like the idea of a self-replicating RNA because if you could evolve that enzyme to take alternative chemistries, then the RNA or the modified RNA that can replicate, can of course replicate other polymers. And so you could expand the tools needed to evolve alternative polymers, which would be fantastic.

JoAnne Stubbe: I don't follow this area so closely, so this is probably a naive question. But in the old days when I did follow it, the key issue with small pieces of RNA was getting the right folding. In fact most of it didn't fold correctly. Then the second key issue is magnesium, which is probably one of the most interesting metals but you can't study it because it's in rapid exchange and you have no spectroscopic handle. And then that also leads to the third question about your pH rate profiles, which people in enzymology have used for a long time but are extremely challenging to interpret because they're related to conformational changes and not related to catalysis. So I'd like to hear some information about where we are.

Philip Bevilacqua: In our most recent work we actually spent a long period of time, I would say almost a year, getting RNA, (there were several questions in there but the first was about folding), getting the RNA to be well-behaved and that's not uncommon in our lab. So too, in the HDV ribozyme, getting it to form crystals and being able to have those crystals be catalytically relevant. We look for characteristics because it's a single turnover where the rates are fast, they are single exponential and they go to completion. So that's really important to avoid some of these, to allow us to be able to think clearly about what's going on. In terms of interpreting rate pH profiles, it's a complex problem for sure to think about those. There are many interpretations for why, many different reasons why the profile can level off other than titrating the protonation of an atom, a change in a rate limiting step, or kinetic pKa and things like that. One looks at the plateau region and whether chemistry is still rate limiting, doing solvent isotope effects and the like, but then I think in the end it's always helpful to have some sort of spectroscopic method to determine that pKa and that's were we turned towards Raman spectroscopy on crystals to be able to do that just because no matter how many controls you do, you're never sure.

Darrin York: Just to also follow up on that question. I think it's the interpretation of things like activity pH profiles that are very important because they are complicated. They are a valuable tool in trying to say something about mechanism and they're coupled with many things. In the case of RNA, there is a high degree of conformational heterogeneity that is controlled by metal ion binding and other factors. This is a perfect example of one of the things where, in principle, theory can step in and help out with. The methods that we used to be able to sample alternate conformations and protonation states at the same time allow us to ascertain directly what are the conditional probabilities of a system being in an active state. And then we can dissect that and say which conformational basins are of catalytic relevance and which ones are not. This brings in the whole aspect of even allosteric effects. I mean, one of the things that are interesting about looking at RNAs is that they can fold up in different ways. Oftentimes there's a handful of stable conformations that are all thermally accessible. And if you can figure out ways of preferentially stabilizing one of those forms relative to another you've designed a switch, as Ron was mentioning. And so, this is something that, in my opinion, hasn't really been fully exploited.

David Lilley: I don't think there is a simple answer to JoAnne's question about interpreting pH. You've got to take it on a case by case basis. You do have to be incredibly careful and there have been horrible mistakes made that have confused the field. But we can dissect this to a degree by things like atomic mutagenesis and seeing how this changes the profiles. And by dissecting individual steps by, for example, phosphorothiolate substitution, which means you no longer have the acid working, just the base, and so on. You know, we are as careful as we possibly can be. I think in a number of cases with a ribozyme it's clear that we are looking at the chemical step of these things.

Donald Hilvert: In that context, is there anything like co-transcriptional folding of the RNA molecules?

Dan Herschlag: Yes there is, and actually Olke Uhlenbeck showed two and a half decades ago that if you try to refold RNA, you get a different answer, you end up in different conformations and conformational traps. It's sort of funny because in the RNA field, we purify our RNAs denatured generally on a gel, while for proteins very few people purify their proteins in a denatured state and re-nature them. It's something you avoid like the plague. So you're starting from this other state and certainly at the sort of classic case of coacher inscriptional folding or the ribo-switches which are making their decision while they're being made or during translation. So there is also a biological role for that and those can differ. I also wanted to go back to Joanne's question and maybe sort of rephrase a little bit what David said. I'd like to just say that, while agreeing that the chemical tools have been extraordinarily valuable and, whereas RNA has the disadvantage that every residue is an acid or base, albeit with pKa 4.9, you have this horrible mess. On the other hand, you can site-specifically substitute those with groups that perturb that pKa, and so really know what's involved and that's easier to do with proteins. Similarly, for the magnesium, what Joe Petrelli and I were able to do, was extend methods for proteins that were pioneered by Mildred Cohen and Fritz Eckstein for phosphorothioates using basically a background of magnesium and then with the phosphorothioate, you would be able to then quantitatively look at rescue by manganese or other thiophilic metal ions. Those chemical tricks have really been at the core of our ability to move forward with RNA mechanism.

Judith Klinman to Darrin York: I have a couple of questions, the first is the conformational dynamics in RNA. Could you give us some idea what kind of time scales you're talking about for the remodeling of the different structures because that's going to be relevant in the context of the actual catalytic rates or turnovers when they turn over. Could you say anything about that in terms of time scales?

Darrin York: The time scales for local types of re-arrangement like we see going from states that have been trapped in the crystal to how things might relax in solution occur on the tens to hundreds of nanoseconds scale. Other more substantial conformational rearrangements like the equilibrium in the minimal sequence hammerhead ribozyme occur on longer time scales that are more on the order of microseconds.

Judith Klinman: Ok, but they're all fairly fast really. My second question had to do with this statement that six to thirty percent of nascent chains are hydrolysed as they're being synthesized on the ribosome. It just seems surprising that there's so much hydrolysis going on. You mention ubiquitin: is there actually a role for ubiquitination in the nascent chain? Just some clarification on that point.

John Christodoulou: This goes also back to the question of translational speed. We can improve the efficiency of the ribosome but at the end of it, the nascent polypeptide still has to fold, so the downstream elements really all very much depend on it. If there's too much of the nascent polypeptide emerging, it increases chances of misfolding. With the linear synthesis, I guess you are going to have many exposed hydrophobic residues and there's a high chance of misfolding events. A significant number of nascent chains is found to be ubiquitinated. There was a recent paper, a study by Bernd Bukau, where they identified a quality control complex that first selectively scans for misfolding polypeptides. And then another part of the same polypeptide actually participates in the ubiquitination to the UPS. This is a significant activity within itself.

Marina Rodnina: I have to say that this number, how many peptides are misfolded, this is one of the most debated numbers in the literature. There's Bukau, there is Hartl, there are other people. I think they will get different numbers depending on which techniques are used. What you have to keep in mind is that when you have a synthesis of the polypeptide, it comes out and this is then checked by the oldest quality control machineries of the cell: these are chaperones, these are of course proteases, there is the proteasome. This is also influenced by where these proteins are. Are they targeted to the membrane? This will be one type of quality control machinery. Are they in the cytosol? This will be a different type of quality control machinery. All these things are quite diverse. They are different in different types of cells. And this is one of the major determinants of proteostasis in the cell. I think that there is no single answer to that.

David Lilley: Can I just comment on the first part of your question about the rates of folding? This was an eye opener for me. This was with the hairpin ribozyme. The hairpin ribozyme is based on a four-way junction in RNA and there are two loops and this has to sort of swivel like a pair of scissors, bringing the loops together, that dock and create the active site. We can follow in the millisecond region the rates of this docking and undocking process. I mentioned this briefly yesterday, but to my immense surprise molecules could differ, the same molecules, well, molecules relative to their neighbors, by two logs in those rates. What the structural basis of that is I have no idea, but it just amazes me how heterogeneous apparently simple RNA molecules can be. Whereas, if you take away the loops and you just look at the plain junction dynamics, that is homogeneous.

Martina Havenith: Maybe to the time scales: we discussed yesterday that we have a range from picosecond to the millisecond time scale. Now you mentioned nanosecond and microsecond. Is there a fundamental difference or are the faster ones of special importance?

Darrin York: Both timescales are important and some of the equilibrium does range into the millisecond and even greater time domain depending on the specific sequences of RNA that are used to create the constructs that are used for ribozyme catalysis. The stuff that occurs on the very fast time scale motion is often times critical for tuning the residues in the active site to be able to promote catalysis. So, having the right hydrogen bonding network and the right fluctuations that help to push you over a barrier are very critical. And yet you also need to have the overall architecture of the active site in place. Like in the example that Dan was illustrating early on, where biochemical experiments identified regions that crystallographically were very distant but biochemically it was inferred that these things needed to come together in some way. In that particular case, it was the sequence that was used, which was in a more stable but catalytically irrelevant conformation that was what got crystallized.

So, both time scales are very important. One of the things that distinguishes RNA a little bit from a lot of protein systems is that there can be this equilibrium between oftentimes quite different states. Another example is the L1 Ligase, which has actually been crystallized in two different conformations where a stem loop is positioned 80 Å away and only one of those conformations is catalytically relevant. I don't know the time scales of that equilibrium but certainly it's probably millisecond or beyond.

David Lilley: Can I just point out that we are in the last five minutes of this discussion. We might focus our answers a little.

Bert Weckhuysen to Dan Herschlag: A question to Dan. Can you explain again: we all hear homogeneous, heterogeneous biocatalysis and you made at the end a comment towards heterogeneous catalysis, something with the precipitation. Can you repeat what you exactly meant because it was too fast, for me at least? And what is then the question to us? How could we then help you as to heterogeneous catalysis?

Dan Herschlag: In cells, there are regions that appear, from the standard physics phase transition, that appear to be driven by natively unfolded regions of natively unfolded proteins. These often have RNAs in them and they appear to have biological function. What we are used to thinking about in a cell: you have a membrane and that separates things, but you have another way of separating things in cells without membranes. That's really interesting. Obviously, there are phase transitions just as an aqueous or organic boundary. There may be other properties of these phase transitions to bring in, with the different reactivity. I don't have anything specific in mind other than here is a formulation that we're not used to thinking about that has properties that people are probably more used to thinking about. Micelles and other motions that happen in biological systems and that we might be able to bring in the functionality of proteins and RNAs.

Donald Hilvert: The ribosome is clearly this absolutely amazing machine, but one of the most extraordinary properties is the accuracy with which it translates the message. Marina talked about this 10^6 fold increase in activity upon formation of the correct codon:anticodon pair. Is that the entire basis of the accuracy or are there additional proofreading mechanisms?

Marina Rodnina: Certainly there are additional proofreading mechanisms. The source of the fidelity is definitely the quality of the codon:anticodon complex. But then this is sensed by the decoding site of the ribosome and this originates these series of these allosteric contractions which lead to the activation of the GTPase. But of course the difference between the stability of cognate and near-cognate codon:anticodon complex remains to be there. These are already two sources of fidelity. Then this can be repeated twice. First on the level of initial selection where the ternary complex brings the tRNA and at that point for the first time the ribosome is looking at that. Then, in the second time, when EF-Tu already dissociates after GTP hydrolysis. But then the tRNA has to move from its position outside the ribosome into the peptidyl transferase centre. This is another step where exactly the same kinetic partitioning applies. There is faster incorporation in the peptidyl transferase centre for the cognate tRNA and faster dissociation of tRNA for near-cognate tRNA. If you repeat this twice, then you can reach relatively high fidelity.

There is also some other potential source of fidelity which is relatively new. Rachel Green worked on that, but we also have something in this direction. If the ribosome makes a mistake and even moves this wrong peptidyl tRNA from the A site to the P site, and it makes it twice, so it tends to accumulate mistakes at this point; and at a certain point the release factors start to work on the place where they should not work on the sense codon. This then just breaks translation, so then you have a near-cognate peptidyl tRNA dissociated from the ribosome and this is removed.

David Lilley: It will be the last question I'm afraid, sorry.

Christophe Copéret: A philosophical question maybe. Yesterday and today we said that we have many proteins and RNAs that we don't know the function of. My question is: do we need a function to exist? This may be just created and bizarre: we see them, but do they need a function in the RNA or protein world?

Ron Breaker: I think I can try to answer that. My answer for RNA would be the same as for protein. If biology carries a conserved coding element either for protein or for RNA it almost certainly has a function that is essential for the long term survival of that species. And so, if we see an RNA that is present in the genomes of very disparate organisms, that element is going to be very important sometime in the life cycle of that species. I think the answer is yes. Certainly in bacteria, since they turn over quite fast and they tend to eliminate components that are not essential. If you see them in bacteria, they are almost certainly going to be important for biology.

David Lilley: There is the famous quotation that nothing in biology makes sense except in the light of evolution.

Dan Herschlag: There is one really interesting potential exception to that, that has been discussed, which is that some of the RNAs that don't fall into the category so clearly described by Ron, that aren't conserved, could be just for that organism or could be products that are being made in order to control the polymerase or even open up the DNA. So they can be involved in the function of transcribing other things, where the RNA product isn't what's functional, but the remodeling and physical function within that region of DNA is what's important. That's one idea that's out there, that's really interesting.

David Lilley: Ok that's it. I think Mark wants to make some general points to finish off. But before they do that, I would like to, on behalf of those of us who are the participants at this meeting, thank a number of people because of whom we have enjoyed a spectacular meeting. So obviously, Kurt Wüthrich and Robert Grubbs. We thank you deeply for putting this meeting together. It's been wonderful. But also clearly it wouldn't have happened without a lot of hard work that we've all benefited from and seen, so I also want to thank Anne De Wit, Marc Henneaux, Dominique Bogaerts and Isabelle Van Geet. So thank you.

Mark Davis: So thanks. I'd also like to express my thanks especially for the group that's helped us put this together. Many of us are not very good at answering emails etc., so I appreciate the patience that everyone has shown as we've moved forward. For us to come from the area of chemical catalysis it's been really exciting to watch, to see the movies, to see how complicated things really are. And also to observe the details which, in the area of biology, you're worrying about, solvent effects etc. We tend to worry about sort of bigger things, about changing mechanisms. And also how we think different about catalysts and that we never think about things like 10^{26} because most of our reactions don't happen at all, so the numbers are sort of infinity so it's a very different way of thinking about things. But I think it's really important for us in the chemical area to start thinking a lot more about the details and to get much more into how solvents and environmental effects control our reactions and how we can use those to get very good effects. So again, thanks everyone. Thanks for all your help in coming and putting this all together. Thank you for getting me involved in this operation.

Kurt Wüthrich: There is now a big danger that I will still repeat what others have said but maybe there are a couple of points that seem relevant. Jean-Marie Solvay indicated to me that at the last physics meeting they had an additional one hour around where each participant would briefly formulate what his impressions were of the meeting. Now I will not go into this exercise here for time reasons but I would like to encourage you to provide us feedback by email, one or two sentences, especially if you see possibilities to improve things. This is an always evolving enterprise. We will have another meeting three years from now and we will certainly consider all the input that you may have. The same applies of course, for your collaboration in preparing the proceedings. We will depend very much on your response to what Anne De Wit and her collaborators will send to you during the upcoming months.

I would also like to add some thanks, particularly to the session chairs. Without the help of the session chairs we would not have been able to generate a program of the breadth that we had here and I really appreciate the great help that we got from the six session chairs. I would like to thank the auditors, under the guidance of Professor Anne De Wit, who have been in the background here and who will be responsible for the successful editing of the proceedings. And then I really want to say that working with Isabelle Van Geet and Dominique Bogaerts has been a pleasure all along. It was very easy to organize the meetings. They did everything to perfection. They are great, many thanks to you. And then, I owe heartfelt thanks to the Solvay family for the great support that they have given to this meeting.

Marc Henneaux: So I'm afraid I'm going to be a bit repetitive in the thanks. Everyone tells me that this has been a great conference and I'm very pleased to hear that. In a sense, for us it's the best feedback that we can get. The program was wonderful. I understand it was a challenge because people are coming from different communities. But the challenge was successfully met. We are extremely pleased with the international Solvay Institutes, because this is our goal, to organize conferences which are peculiar and where this chemistry works.

The success would not have been possible without the careful work that went into the preparation of the conference and this preparation started three years ago right after the previous Solvay conference on chemistry and I would like to thank all the scientists that have been involved in its preparation.

Our scientific committee in particular Kurt Wüthrich, the conference chair and Robert Grubbs the conference co-chair, and all the session chairs, as it had been said, who meticulously prepared the discussions and the presentations. I would also like to thank the speakers and all the participants who made the scientific discussions proceed vividly in the spirit of the Solvay conferences. We attach a great importance to the discussions and while I could not attend the conference unfortunately, I was here this morning and I could see that the discussions proceeded vividly, and that you had to cut or otherwise it would still be going on which is a very good sign.

Now as you know we will publish proceedings. I understand that the proceedings committee already did a remarkable work because all the presentations are already here in print. So, this has to be praised and I would like to thank Anne De Wit and Thierry Visart de Bocarmé for this splendid job that they have done. Also since discussions are central they will be reproduced in the proceedings. Everything that you said will be put in print, with minor editing. Transcribing discussions that are sometimes passionate to some good written text requires a lot of editorial work and so I'm grateful to the auditors, whose work is only starting now, who will prepare the discussions; and who will be in touch with you for reading the discussions. At some point, when things are ready we would make them available to you, wait for your feedback and then send things to the publishers. If there is no feedback that will mean that you agree. Well, one important thing is that you cannot change what you said here. We will be very careful about that.

I would also add to the thanks, I already thanked them at the banquet but I will repeat that indeed it's a pleasure to work with Dominique and Isabelle. They don't count their hours and moreover, as I also said at the banquet, they always smile, so it is a great pleasure to work with them and they are extremely efficient.

The success of the 24th Solvay Conference makes us look forward with confidence to the next Solvay Conference on Chemistry in three years. I can tell you that the committee already started thinking about the themes for the conference in 2019 and so I'm very optimistic that it will again be a success. We will try to improve what can be improved. But while the level, I understand, is already extremely high. With these optimistic closing words I would like to thank you all again and I wish you a very nice trip back home.

Index

www.ingramcontent.com/pod-product-compliance
Lightning Source LLC
LaVergne TN
LVHW081931200926
848484LV00009B/2403
* 9 7 8 9 8 1 3 2 3 7 1 6 2 *